A PERSPECTIVE ON HOLISTIC ENGINEERING MANAGEMENT

Learning, Adapting and Creating Value

World Scientific Series in R&D Management

Print ISSN: 2591-7498
Online ISSN: 2591-7501

Series Editor: Tugrul U Daim *(Portland State University, USA)*

Published

Vol. 7 *A Perspective on Holistic Engineering Management:*
Learning, Adapting and Creating Value
by Robert J Aslett, John M Acken and Siva K Yerramilli

Vol. 6 *Digital Transformation: Evaluating Emerging Technologies*
edited by Tugrul U Daim

Vol. 5 *Managing Medical Technological Innovations: Exploring Multiple*
Perspectives
edited by Tugrul U Daim and Alexander Brem

Vol. 4 *Managing Mobile Technologies: An Analysis from Multiple*
Perspectives
edited by Tugrul U Daim and Alexander Brem

Vol. 3 *Cooperative Innovation: Science and Technology Policy*
by Frederick Betz

Vol. 2 *Technology Roadmapping*
edited by Tugrul U Daim, Terry Oliver and Rob Phaal

Vol. 1 *Managing Technological Innovation: Tools and Methods*
edited by Tugrul U Daim

World Scientific Series in R&D Management – Vol. 7

A PERSPECTIVE ON HOLISTIC ENGINEERING MANAGEMENT

Learning, Adapting and Creating Value

Robert J Aslett

3e8 Consulting, USA

John M Acken

Portland State University, USA

Siva K Yerramilli

Synopsys Inc., USA

W **P** **World Scientific**

NEW JERSEY · LONDON · SINGAPORE · BEIJING · SHANGHAI · HONG KONG · TAIPEI · CHENNAI · TOKYO

Published by

World Scientific Publishing Co. Pte. Ltd.

5 Toh Tuck Link, Singapore 596224

USA office: 27 Warren Street, Suite 401-402, Hackensack, NJ 07601

UK office: 57 Shelton Street, Covent Garden, London WC2H 9HE

Library of Congress Cataloging-in-Publication Data
Names: Aslett, Robert J., author. | Acken, John M., author. | Yerramilli, Siva K., author.
Title: A perspective on holistic engineering management : learning, adapting and creating value /
    Robert J Aslett, 3e8 Consulting, USA, John M Acken, Portland State University, USA,
    Siva K Yerramilli, Synopsys Inc., USA.
Description: Singapore ; Hackensack, NJ ; London : World Scientific, [2021] |
    Series: World Scientific series in R&D management, 2591-7498 ; vol. 7 |
    Includes bibliographical references.
Identifiers: LCCN 2020046878 | ISBN 9789811228322 (hardcover) |
    ISBN 9789811228339 (ebook for institutions) | ISBN 9789811228346 (ebook for individuals)
Subjects: LCSH: Research, Industrial--Management. | Industrial management.
Classification: LCC T175.5 .A83 2021 | DDC 620.0068--dc23
LC record available at https://lccn.loc.gov/2020046878

British Library Cataloguing-in-Publication Data
A catalogue record for this book is available from the British Library.

For any available supplementary material, please visit
https://www.worldscientific.com/worldscibooks/10.1142/12043#t=suppl

Desk Editor: Amanda Yun

Typeset by Stallion Press
Email: enquiries@stallionpress.com

Printed in Singapore

"My big regret: not having this book as a young engineering manager, as a newly appointed R&D director, as a rapidly learning technology VP, and yes, as a global high-tech CEO! I love how the authors interweave leadership and management and show that culture, strategy, and execution are not merely additive ingredients but multipliers on the path to great engineering management!"

— Aart de Geus,
Chairman and co-CEO, Synopsys, Inc.

"Technology is such an opportunity to innovate and transform all aspects of life for the better. The outcome is technology; the actors are the people who shape and deliver the results. Earning the collective and woven discretionary energy of an R&D organization is paramount. With *A Perspective on Holistic Engineering Management*", Siva Yerramilli, Robert Aslett, and Prof. John Acken provide a wonderful and comprehensive blueprint for how to lead and manage disruptive and impactful R&D teams."

— Aicha Evans,
CEO, Zoox, Inc.

"The increasing pace of technology advancement is relentless with growing interdependencies between different engineering organizations and companies with a significant increase in design complexity. To excel, engineering managers must understand and harness a highly dynamic environment that stretches beyond their organization on a global basis. This book is an essential reference for aspiring engineering managers who want to know how to lay the proper foundation, conduct core engineering functions, connect an entire R&D system and develop their own leadership skills to create a high velocity engineering organization that not only survives but excels in today's competitive world."

— Amir Faintuch,
Senior VP, GM, GlobalFoundries, Inc.

"For those who want to lead an engineering organization that thrives, *A Perspective on Holistic Engineering Management* will be an indispensable addition to their library. This book not only explains each piece of the puzzle but describes how they connect. Most importantly, it describes how the proper cultural foundation, tightly connected core engineering functions and enlightened leadership combine to extract the most value from an R&D system that extends well beyond the manager's organization. I wish this reference had been available when I learned many of these important concepts through a much more painful trial and error method."

— William M. Holt,
former Executive VP,
GM, Intel, Corporation

Dedication

Rob dedicates this book to his wife Anita and
sons Kevin and Colin.

John dedicates this book to Charlotte Acken.

Siva dedicates this book to his loving wife,
Sandhya Yerramilli, his mother,
Yerramilli Kusumakumari, and his sister, Nagalatha Ryali.

Foreword

Collaborative innovation is the lifeblood of business and the engine of great engineering. We like to romanticize the lone genius inventing a breakthrough on their own, but in truth, virtually every successful engineering project is driven by a collective effort. More often than not, success hinges on multiple teams coming together to achieve a shared goal.

If collaboration fuels good engineering, the next question becomes: what's the key to effective collaboration? In my experience, it comes down to leadership. I often think of leadership as the "operating system" that makes collaboration possible. Like a computer OS that orchestrates disparate resources into connected and interoperable parts, leadership empowers teams to execute at scale. And just as successful digital products must run on great code, leaders must skillfully manage the "software" and "hardware" of corporate culture and strategy. That includes orchestrating multiple initiatives that need to function and flourish at the same time.

The best leaders are not experts in everything; they often do not have fancy titles, and many are not even managers yet. What they have in common is an ability to identify a good idea and then shape it into a cohesive vision and a concrete action plan. Effective leaders connect the dots across diverse teams to build trust and instill a shared sense of purpose.

Unfortunately, if you look at who is making key decisions in business today, it remains a very homogeneous group of leaders. Especially in tech, we are not nearly as diverse as we should be. We can and we must do better. As engineers, we need to double-down on our commitment to diversity and inclusivity. That means getting comfortable with the uncomfortable. In my own organization, I recognize that begins with me. I am deeply committed to building diverse teams in which every team member is empowered to show up and contribute.

At VMware, we describe our Leadership Code as the operating system of our culture and our business. It's a holistic approach that defines what we do (our skills), how we think (our mindsets) and who we are (our values). I refer to these pillars as the Body, Mind and Soul of our organization. Each one is critical on its own, yet together they shape how we execute upon every task of each day.

I had the pleasure of working with John M. Acken, Robert Aslett and Siva Yerramilli on a variety of design technology development projects during my tenure at Intel. As it happens, the day I began my quest for a master's degree at Stanford marked the very same day that John began his Ph.D. program at the University. Over the years, I personally witnessed these three leaders exemplify the core principles of holistic engineering laid out so persuasively in this book: The ability to build cohesive teams, serve as role models and drive collaboration across a diverse ecosystem of partners.

For too long, businesses tended to think of innovation as happening "in pockets" — by default it was often the job of the R&D team. Today's reality is that every function and every business unit across your organization must innovate across organizational boundaries. Holistic engineering reminds me of my favorite African proverb: "If you want to go fast, go alone. If you want to go far, go together."

Pat Gelsinger
CEO, VMware
Pat Gelsinger became the Intel CEO in 2021

Preface

This book is the result of a desire to share our perspective on managing an engineering R&D organization. From our positions in the semiconductor industry and academia, we have been fortunate to have played a part in the rapid advance of technology required for the development of integrated circuits over the last 35 years. During that time, we have managed engineering organizations that ranged from 6 to over 1,000 engineers where we experimented with different management approaches. We also had the tremendous fortune to work with and learn from some of the best and brightest managers in our industry. We have seen that successful managers create a strong foundation of a common culture that enables learning, value creation, diversity and inclusion. They create organizations that tightly connect the core engineering functions of strategic planning, research and development and are able to comprehend and direct a broader R&D system that stretches well beyond their own organization's boundary. Doing all of this to extract the greatest value in the least amount of time is what we call *holistic engineering management*. These managers are part leader, coach, engineer, system analyst, behavioral psychologist, economist and business strategist. This belief in the role of the engineering manager is a major theme of the book. Although many excellent references deal with each individual topic in this book, we feel that all of them need to be considered together to form a complete picture of the holistic approach to engineering management.

The genesis of this book is a lecture given to students of Dr. Tugrul Daim's graduate-level engineering and technology management class at Portland State University (PSU). Therefore, the primary audiences for the book are students of engineering management and current engineering managers. However, we also believe individual engineers and managers who work with engineering organizations will find the information useful.

The content for this book is based on over 105 years of combined experience working in a rapidly changing industry long enough to identify important trends that stand the test of time. In most chapters, we provide practical examples of the concepts provided. These mostly refer to engineering R&D managers of reasonably large organizations with multiple levels of organizational hierarchy reporting to them. This allows us to discuss several topics that are unique to that size and scope and not often addressed in detail in other books. Because of our background in the semiconductor industry and academia working on technologies to enable the development of integrated circuits (primarily microprocessors) several of our examples come from that domain. However, we feel the concepts they highlight are relevant to engineering management in general.

We want to acknowledge the many people who helped us with this book. We want to thank Tugrul Daim for his guidance during the writing of this book. We want to thank the people we interviewed or gave us extensive feedback including Marilyn Wolf, Joy Mutare, Lauren Krueger, Carolina Gomez-Montoya, Marina Laurette, Noel Menezes, Deirdre Hanford, Sonali Fernando and Sanjay Panditji. We want to thank Penny Harbin and Nanda Kuruganti for their helpful comments. Finally, we want to thank Colin Aslett for proofreading the content and Sandhya Yerramilli for proofreading the content and contributing two drawings. We hope this book will give food for thought and answer some questions.

Robert J. Aslett
John M. Acken
Siva K. Yerramilli

Contents

Foreword   ix
Preface   xi
List of Tables   xix
List of Figures   xxi

Chapter 1   Introduction   1

1.1 Book Overview   5
    1.1.1 Foundational Concepts   5
    1.1.2 Core Engineering Functions and Their
        Organization   7
    1.1.3 Special Topics   13
1.2 Important Terms   15
1.3 A Simple Example of an R&D System   21
References   26

Chapter 2   Culture   29

2.1 What is Culture?   31
2.2 The Importance of Values   32
2.3 Stated Values vs. the Actual Culture   37
2.4 Setting the Culture   44
2.5 Summary   50
2.6 Key Points   50
2.7 Case Studies   51
    2.7.1 Changing the Values at Intel Corp   51
    2.7.2 Changing the Culture of an R & D Organization 54
    2.7.3 Changing the Culture at Two Companies 58
References   62

Chapter 3   Diversity and Inclusion   65

3.1 Applying Diversity and Inclusion   66
3.2 Diversity, Inclusion and Value   66
3.3 Focus on the Individual   67
3.4 Policies and Culture   72
3.5 Diversity and Inclusion Over Time   76
3.6 Proper Communication   78
3.7 Summary   80
3.8 Key Points   81
References   81

Chapter 4   Value, Waste and Velocity   83

4.1 Value   84
4.2 Waste   94
4.3 Sources of Waste   97
4.4 Shifting Time Spent on Waste to Value   101
4.5 Velocity   103
4.6 Summary   107
4.7 Key Points   108
4.8 Case Studies   108
    4.8.1 Increasing Velocity Using a Value Stream Map   108
References   115

Chapter 5   Learning Cycles   117

5.1 The Importance of Learning Cycles   118
5.2 Learning Cycles in the Engineering R&D System   120
5.3 Enabling Learning Cycles   125
5.4 Executing Learning Cycles   130
5.5 Summary   142
5.6 Key Points   143
5.7 Case Studies   143
    5.7.1 Manufacturing, CAD and Design Learning Cycles   143
References   147

Chapter 6   Strategic Planning   149

6.1 Strategy and Strategic Actions   150
6.2 Strategic Planning Scope   153
6.3 Continuous Strategic Planning   155
   6.3.1 Observation   158
   6.3.2 Orientation   160
   6.3.3 Decision   169
   6.3.4 Action   175
6.4 Summary   178
6.5 Key Points   179
References   180

Chapter 7   Development   181

7.1 Development and Value   182
7.2 Development Scope   183
7.3 The Importance of Standards   185
7.4 Development Velocity   191
7.5 Continuous Improvement   194
7.6 Managing Development Choices   199
7.7 Build vs. Buy   206
7.8 Including Customers in Development   207
7.9 Summary   209
7.10 Key Points   210
7.11 Case Studies   211
   7.11.1 Standardization of an Engineering
       Environment   211
References   219

Chapter 8   Engineering Research   221

8.1 Engineering Research and Value   222
8.2 Research Scope and Types   224
8.3 Research Funding   232
8.4 The Research Pipeline   234

8.5 Learning Cycles and Rapid Prototyping   242
8.6 Pathfinding   244
8.7 Performance Management   248
8.8 Summary   249
8.9 Key Points   250
8.10 Case Studies   251
     8.10.1 Pathfinding   251
References   260

Chapter 9   The R&D Organization   261

9.1 The Role of R&D Organization   262
9.2 Strategy   266
9.3 Structure   268
9.4 Processes   275
9.5 Rewards   278
9.6 People   281
9.7 A Practical Example   283
9.8 Governance   288
9.9 Summary   295
9.10 Key Points   296
9.11 Case Studies   296
     9.11.1 Intel Operating Segments 2005–2018   296
References   299

Chapter 10   The R&D System   301

10.1 The R&D System Scope   302
10.2 The Role of the Manager   306
10.3 The Shared Objective of the R&D System   308
10.4 Collaboration and the R&D System   312
10.5 Summary   327
10.6 Key Points   328
10.7 Case Studies   329
     10.7.1 Background: Three Eras in a CAD R&D System   329
     10.7.2 Era 1: Rapid Growth (1988–1997)   334

10.7.3 Era 2: Customer Autonomy (1998–2004)   340
10.7.4 Era 3: New Devices and Platform Scope
(2005–2012)   346
References   355

Chapter 11   Engagement with Academia   357

11.1 The R&D Organization and University Goals   357
11.2 Conditions for Engagement   360
11.3 Providing Value in Joint Projects   362
11.4 Engagement Examples   364
11.5 When to Engage, or Not to Engage   366
11.6 Information Security and Publication   369
11.7 Summary   369
11.8 Key Points   370
11.9 Case Studies   370
11.9.1 The Evolution of Academic Engagement
at Intel   370

Chapter 12   Information Security   375

12.1 The Importance of Information Security   376
12.2 Privacy and Information Security   378
12.3 Implementing Information Security   380
12.4 Operations, Technology and Culture   383
12.5 Information Security and the Law   385
12.6 Perimeter Defense is Insufficient   388
12.7 Some Information Security Technology Basics   393
12.8 The Importance of a Security Culture   397
12.9 Summary   398
12.10 Key Points   399
12.11 Case Studies   400
12.11.1 Insider Threat   400
12.11.2 Wide Mix of Data Security   401
References   404

Chapter 13   Leadership   **407**

13.1 Articulating Why, What and Enabling How   408
13.2 Articulating the Why   410
13.3 Articulating the What   417
13.4 Enabling the How   422
13.5 Common Topics for Why, What and How   429
       13.5.1 Velocity and Efficiency   429
       13.5.2 Leadership and Learning   431
       13.5.3 Leading People   435
13.6 Summary   443
13.7 Key Points   444
13.8 Case Studies   445
       13.8.1 Challenges in Aligning Multiple Teams   445
References   450

Chapter 14   Closing Thoughts   **451**

14.1 Key Directives for the R&D Manager   451
       14.1.1 Foundational Concepts   451
       14.1.2 Core Engineering Functions and their
              Organization   453
       14.1.3 Special Topics   454
14.2 Life is More Than Work   455
       14.2.1 Investing in Personal Health   456
       14.2.2 Enabling Employees Balance   457
       14.2.3 The COVID-19 Disruptive Event   459
References   462

Glossary   463
Index   479
About the Authors   487

List of Tables

Table 4.1. Example of engineering activity value vs. consumer. 87

Table 4.2. The seven wastes in manufacturing. 97

Table 4.3. The seven wastes of software development. 98

Table 4.4. The eight wastes of management — Aslett, Acken and Yerramilli. 98

Table 5.1. Traditional performance critiques vs. AAR. 136

Table 5.2. Growth in manufacturing masks as transistor dimensions shrink. 144

Table 7.1. Different levels used to determine engagement. 206

Table 8.1. Wright brothers' chronology. 244

Table 9.1. An example of common processes in an R&D organization. 277

Table 9.2. Example governance and process elements in the matrix. 286

Table 10.1. Supplier-partnering hierarchy. 312

Table 10.2. Pisano and Verganti collaboration types. 317

Table 10.3. Collaboration types for the R&D organization. 318

Table 10.4. Cost and expected benefits of each collaboration type. 319

Table 10.5. Strategic planning collaboration types. 321

Table 10.6. Research collaboration types. 322

Table 10.7. Pathfinding collaboration types. 324

Table 10.8. Development and collaboration types. 325

Table 12.1. Security assessment for general schematics and proprietary IP. 383

Table 12.2. Summary of roles and tasks. 385

List of Figures

Figure 1.1.   The foundation of the R&D organization.   10
Figure 1.2.   The R&D system.   11
Figure 1.3.   The R&D ecosystem.   12
Figure 1.4.   The formal reporting hierarchy and the intrinsic network.   16
Figure 1.5.   Entities, groups, working groups, individuals and organization.   17
Figure 1.6.   Tasks, workflows and learning cycles.   18
Figure 1.7.   Lines of communication vs. team size, highlighting a span of 6.   19
Figure 1.8.   The number of levels of hierarchy vs. the number of employees.   20
Figure 1.9.   Interaction of functions across the R&D system.   22
Figure 1.10.   Interaction of the workflows across the R&D system.   23
Figure 1.11.   Information flow for a request.   24
Figure 1.12.   Information flow to satisfy a request.   25
Figure 2.1.   Culture is at the heart for the R&D organization.   32
Figure 2.2.   The Toyota Way as summarized by Liker and Hoseus.   34
Figure 2.3.   The values of anticipation and fearlessness.   36
Figure 2.4.   The correlation of stated and actual cultural values.   37
Figure 2.5.   Enron's stated values and actual culture.   40
Figure 2.6.   Wells Fargo's stated values and actual culture.   43
Figure 2.7.   ABC Inc. and XYZ Tools current state.   60
Figure 2.8.   ABC Inc. and XYZ Tools proposed state.   61

Figure 4.1.   Types of value.   86
Figure 4.2.   Value, necessary waste and unnecessary
                waste.   96
Figure 4.3.   Increasing value over time.   101
Figure 4.4.   Velocity and efficiency trade-offs.   106
Figure 4.5.   Original value stream map.   110
Figure 4.6.   Identified improvements to value
                stream map.   113
Figure 4.7.   The new value stream map.   114
Figure 5.1.   A representative 1-week SCRUM sprint.   119
Figure 5.2.   Software development time, 1970s
                to 2000+.   120
Figure 5.3.   Concurrent learning cycles.   121
Figure 5.4.   A generic workflow with learning cycles.   122
Figure 5.5.   An example of connected circuit design
                workflows.   123
Figure 5.6.   An example of a concurrent engineering
                process.   124
Figure 5.7.   An example of an intrinsic network.   126
Figure 5.8.   Dr. W.E. Deming's Plan-Do-Study-Act
                (PDSA) model.   131
Figure 5.9.   Example A3 document (first half).   133
Figure 5.10.   Example A3 document (2nd half).   134
Figure 5.11.   A depiction of Boyd's OODA loop.   138
Figure 5.12.   Allen Ward's Look-Ask-Model-Discuss-
                Act (LAMDA) model.   140
Figure 5.13.   Tim Brown's Design Thinking model.   140
Figure 5.14.   Typical Si manufacturing learning cycle.   145
Figure 5.15.   Co-optimization on a two-year cadence.   146
Figure 6.1.   Strategic planning and the R&D
                ecosystem.   154
Figure 6.2.   The OODA loop for continuous planning.   156
Figure 6.3.   The observation function in the OODA loop.   159
Figure 6.4.   The process of orientation in the OODA
                loop.   161

Figure 6.5.	Michael E. Porter's five competitive forces.	162
Figure 6.6.	The value net model and the R&D organization.	165
Figure 6.7.	Robert Burgelman's rubber band model.	168
Figure 6.8.	Preparing for the strategy planning decision meeting.	170
Figure 7.1.	A typical engineering cycle during development.	183
Figure 7.2.	Development and the R&D ecosystem.	184
Figure 7.3.	An example of engagements needed for development.	185
Figure 7.4.	A non-standard and standard development environment.	187
Figure 7.5.	Components of a design automation environment.	190
Figure 7.6.	The development flow through the R&D system for a customer.	191
Figure 7.7.	An example development flow with delays.	194
Figure 7.8.	An improved development flow.	196
Figure 7.9.	Interaction of the workflows across the R&D system.	198
Figure 7.10.	A depiction of concurrent engineering.	204
Figure 7.11.	A single place for visualizing requests and commitments.	205
Figure 7.12.	An example of an IC design environment.	211
Figure 7.13.	An example of the contents in a design environment.	212
Figure 7.14.	Uniqueness of each design environment.	213
Figure 7.15.	Problems with Integrating IP across groups.	214
Figure 7.16.	Initial design kit content.	216
Figure 7.17.	Initial transfer and deployment.	216
Figure 7.18.	Typical development and integration cycle.	218

Figure 8.1.   Engineering research scope.   224
Figure 8.2.   Sustaining innovation along a series of
               S curves.   227
Figure 8.3.   Research and pathfinding aligned to
               technology generations.   228
Figure 8.4.   A depiction of sustaining and disruptive
               innovation.   230
Figure 8.5.   Distinct and interdependent
               engineering research paths.   235
Figure 8.6.   An example of a research maturity model.   240
Figure 8.7.   Alan Ward's LAMDA cycle.   243
Figure 8.8.   A pathfinding process for interdependent
               research.   247
Figure 9.1.   The R&D organization.   262
Figure 9.2.   A depiction of the Galbraith 5-star model.   264
Figure 9.3.   The sequence for constructing an
               organization.   265
Figure 9.4.   Strategic planning affects the design of
               the organization.   266
Figure 9.5.   The rubber band model.   267
Figure 9.6.   A basic matrix structure and the levers
               used to balance power.   269
Figure 9.7.   The formal reporting hierarchy and the
               intrinsic network.   271
Figure 9.8.   Example transition to a matrix operation.   274
Figure 9.9.   Key R&D processes already discussed.   278
Figure 9.10.   A depiction of a highly matrixed
               organization.   284
Figure 9.11.   A depiction of a siloed organization.   284
Figure 9.12.   An example of roles in a simple
               organization structure.   287
Figure 9.13.   A depiction of the Cynefin framework.   291
Figure 9.14.   Intel operating segments for 2005 to 2018.   297
Figure 10.1.   The R&D system.   302
Figure 10.2.   The R&D ecosystem.   303

Figure 10.3. The interdependent R&D models, workflows and networks. 304

Figure 10.4. Depiction of value, cost and profit for all parties. 308

Figure 10.5. Four eras of central CAD and microprocessor interaction. 333

Figure 10.6. Central CAD circa 1988. 336

Figure 10.7. Central CAD circa 1997. 337

Figure 10.8. R&D system in the first era. 338

Figure 10.9. Central CAD in the second era. 343

Figure 10.10. The R&D system in the second era. 345

Figure 10.11. Central CAD in the third era. 351

Figure 10.12. The R&D system in era 3. 353

Figure 12.1. The R&D security organization perimeter. 386

Figure 12.2. Perimeters around the R&D organization and the company. 389

Figure 12.3. Access control responses to requests. 394

Figure 12.4. False-positive rate (FPR) and false-negative rate (FNR) tradeoff. 395

Figure 13.1. R&D manager behavior for why, what and enabling how. 408

Figure 13.2. R&D manager behavior and the OODA loop. 409

Figure 13.3. Thinking model to help with the prioritization. 433

Figure 13.4. Elephant and the rider. 437

Figure 13.5. Individuals with their eyes closed and an elephant. 439

CHAPTER 1

Introduction

> Even when I am composing instrumental music, my custom is
> to have the whole in view.
> — **Ludwig van Beethoven [1]**

After the U.S. patent for the telephone was granted in 1876 it took over 60 years for it and the required infrastructure to evolve to the point where it was adopted by 40% of households in the United States. When the first smartphone was introduced, about 120 years later, it only took ten years for it and the required infrastructure to evolve and achieve the same level of adoption [2]. Today, the development of many different technologies that once took years can be done in months or weeks. In parallel, there has been a dramatic increase in the importance of suppliers. Engineers from Apple design each iPhone and work with companies such as Intel, Micron, Broadcom Texas Instruments, NXP, Samsung, TSMC, Amkor, Corning and many others to define and integrate their components. Apple's list of official suppliers is over 30 pages long [3–5]. There are few examples today where the engineering work is completely contained in one company or one engineering organization. Business strategies have also evolved. The analysis of competitive forces in an industry has matured to include the concepts of disruptive innovation and coopetition (that coined the word, complementor). Today, a prosperous technology company can be disrupted and put out of business in a blink of an eye. Similarly, in an ecosystem characterized by rapid changes in technology and how it is developed, an engineering R&D organization will quickly become irrelevant if it fails to enable the pace of innovation needed to win. This book gives our perspective on how a holistic approach to engineering management creates an R&D system that avoids that fate and thrives.

Merriam Webster's dictionary defines technology as, "a capability given by the practical application of knowledge." Technology can come in many forms, such as methods, software, systems, or devices. Technology that has resulted from engineering includes such things as electrification, the automobile, the airplane and computers. Although our examples tend to focus on integrated circuits or software, the principles we discuss in this book apply to all types of technology.

Engineering covers a lot of territory. We use the definition provided by the American Association for the Advancement of Science, "The systematic application of scientific knowledge in developing and applying technology" [6]. This includes activities related to the design, development, construction and operation of technology. However, for this book, the engineering R&D organization we discuss is only responsible for the design and development of the technology and determining how it is used. More specifically, the R&D organization must develop technology that balances goals related to such things as cost, compatibility, reliability, serviceability, safety and performance and must ensure that those are met over a range of operating conditions. As a result, the engineering process involves iterative cycles of planning, modeling, prototyping, analysis, optimization and testing that results in a final specification that achieves the objectives. For software engineering, this also includes coding the software.

To survive and ultimately thrive, an engineering R&D organization needs to create and deploy value. Engineering activities create value if they contribute to the company's vision in a way that the customer will pay. However, the type of customer and how they pay will vary and, as we will see, this can make determining value complicated. The term velocity is used in different ways when referring to organizations, we define it as the rate of deployed value. To win, the engineering R&D organization must have a velocity faster than its competition. Increasing velocity in an environment with many interdependencies means that the engineering R&D manager must be concerned with more than just the operation of their organization.

Successful engineering managers create a strong foundation of a common culture that enables learning, value creation, diversity and inclusion. They create organizations that tightly connect the core engineering functions of strategic planning, research and development and can comprehend and direct a broader R&D system that stretches well beyond their own organization's boundary. These managers are part leader, coach, engineer, system analyst, behavioral psychologist, economist and business strategist.

Holistic engineering management spans the engineering manager's R&D organization and the more expansive engineering R&D system that includes the suppliers, customers, complementors, consortia, institutions and others that they engage with to complete the work. Going forward, for the sake of brevity, we sometimes use terms R&D organization and R&D system to denote those two domains and we sometimes use the term R&D manager to denote the person who leads the R&D organization and directs the larger R&D system.

A product is a term that is usually used to describe the final usable form of the technology that the end customer uses. It often contains more than technology. Sometimes there is a clear line between the technology and other elements in the final product. Sometimes there is not. To avoid flipping between these two terms and causing confusion we use the term technology to describe what the engineer specifies and what is delivered to the paying customer regardless of what might be wrapped around it.

While reading the book, the reader will notice that we are firm believers in lean thinking. This book does not explain the lean philosophy in detail, but there are several places where we share concepts or give examples derived from our experience trying to put the thinking into practice. If the reader has not already done so, we highly encourage them to begin their own journey on that topic. Agile development principles have also influenced us and we refer to them in some examples that we share. Similar to the case with lean thinking, we do not explain agile in detail but highly encourage the reader to explore the many excellent resources available on the topic.

The book is written from the perspective of an R&D manager with two or more levels of organizational hierarchy reporting to them. This allows us to explain some challenges that are unique to large engineering organizations. With that scope in mind, implementing a holistic approach to engineering management is full of difficult challenges. Different engineering organizations that constitute the R&D system have different cultures, processes, engineering methods, etc. Some of these differences don't matter, but some can create a huge drag on velocity. Within the R&D organization itself, there are similar tough challenges that must be overcome.

We have grouped the chapters into three separate parts of the book. Chapters 2–5 address the foundational topics of culture, diversity, inclusion, value, waste, velocity and learning cycles. Chapters 6–10 discuss the core engineering functions of strategic planning, development, research and how they should be organized and directed. Chapters 11–13 focus on special topics important to the application of holistic engineering management that warrant a deeper dive. They include engagement with academia, information security and leadership. The final chapter provides a summary, a discussion on creating the right work-life harmony and some thoughts on how recent events will shape engineering work in the future.

What follows is a brief overview of each chapter, an explanation of some important terms and a short example of what we mean by *holistic engineering management*.

1.1 Book Overview

1.1.1 *Foundational Concepts*

Chapter 2, Culture: Culture is the muscle memory of an organization that guides the behavior of each employee such that when faced with ambiguous or incomplete information they instinctively know what to do. A strong culture means that employees agree on "how we do things here" and "what we value here". It is at the heart of *holistic engineering management* because it enables the autonomy needed for rapid decision making while maintaining continuity. When the culture aligns with the stated values and strategy of the R&D organization, then it is a powerful catalyst for rapid progress, but when there is significant dissonance, the consequences can be catastrophic. As a result, alignment on a common culture within the R&D organization is necessary. Alignment with other entities in the R&D system does not have to be complete but there are cases where better alignment in a couple of areas can make a tremendous impact on the overall performance of the system and benefit everyone. In those cases, the manager of the R&D organization will need to take the lead and work with others to make it so. Finally, the manager of the R&D organization has the biggest impact on the organization's culture and must lead by example. What a leader says does not resonate with employees as much as what they do. Any perceived dissonance between what is said and what is done is evidence to the employee that the stated values are not taken seriously by the leadership and therefore do not have to be taken seriously by them.

Chapter 3, Diversity and Inclusion: Diversity is measured by the variety of the workforce, whereas inclusion is measured by the perceptions of the individuals within the workforce. Diversity of thought, experiences and perspectives lead to creative insights that open up new opportunities. Fully including everyone and acting on those insights is required to translate diversity into greater value to the customer. An important collaboration between the R&D organization and the rest of the R&D system is also enhanced when the diversity of the R&D organization mirrors the rest of the world. As a result, fully embracing diversity and inclusion will increase the velocity of the R&D system. The manager of the R&D organization sets the policies for diversity and inclusion. However, far more importantly, the manager must provide the leadership required to fully enable diversity and inclusion through their actions, active listening, and personal role modeling. Although recruitment and hiring are the means to create a diverse workforce, leading by example is how a manager creates an inclusive culture. Similarly, while there are legal requirements for diversity, meeting the legal requirements and creating policies and procedures will only create an environment of inclusion when the culture supports inclusive values.

Chapter 4, Value, Waste and Velocity: Employees operate more effectively when they understand what constitutes value and how their work contributes to it. This helps them identify the time spent on activities that do not add value and constitute a form of waste. When coupled with a culture that motivates everyone to learn and improve, a clear understanding of value will reduce waste and either increase the time spent on engineering work that creates value or reduces the time it takes to deploy existing value. The rate of deployed value vs. time is what we call *velocity* and continuously increasing that is required to enable the pace of innovation required to win. However, determining the value of an engineering activity is not always straightforward for R&D organizations that deliver to internal customers. Also, all value is not equal and all waste elimination is not equal. Decisions need to be made considering the entire R&D system and what provides the greatest value from the perspective of the customer. At the end of the day, these are the type of trade-off discussions that need to be directed by the manager of the R&D organization and it all starts with a clear agreement in what is value.

Chapter 5, Learning Cycles: Whether it is jet fight combat, software development, or strategic planning, the reason for short learning cycles is aptly summed up by Colonel John Boyd's comment, "He who can handle the quickest rate of change survives." [7]. Translated into our frame of reference, the engineering organization with the shortest learning cycles explores more ideas, makes course corrections sooner, improves faster and delivers more value. To work, this requires that all employees view themselves as scientists and continuously run experiments to improve the workflows and engineering results. The completion of each task in the workflow generates information on the engineering decisions and its operation. When this knowledge is used to improve subsequent engineering decisions or incrementally improve the workflows, then a learning cycle has been completed. Learning cycles are not a new concept and they are prevalent in lean thinking and Agile principles. They are the application of the scientific method of *observation — hypothesis — test* to engineering work. It also means that the role of the engineer is much more than designing new or improved technology; it also includes continually improving the workflows and processes they use. Learning cycles offer every engineer a means to continuously challenge and improve standards with the involvement of others who are also responsible for their execution. More than anything else, implementing learning cycles requires a culture where everyone is involved and believe that it is a critical part of their job. Enlightened managers know that shorter learning cycles lead to better technology and enabling everyone to continuously improve their workflows provides a competitive advantage for the company.

1.1.2 *Core Engineering Functions and Their Organization*

Chapter 6, Strategic Planning: Strategic planning focuses on actions that materially affect the organization's direction. It determines the best mix of value for the engineering organization and the means to achieve it. Whether the R&D organization delivers its output directly to a paying customer or other internal groups, it must understand and take advantage of the entire ecosystem to achieve greater velocity than the competition. Strategic planning forces actions that create differentiable and sustainable value within the context of the capacity, skills and culture of the R&D system. Strategic actions differ from tactical actions in that they commit the organization down a path that will be difficult to reverse. They may be induced top-down by management or they may result from autonomous actions of individuals or groups that do not initially align with the existing strategy. Staying on top of this requires a continuous strategic planning loop. Therefore, the primary product of strategic planning is not the plan but a continuous process that synthesizes new information, creates useful knowledge and translates that into tangible action faster than the competition. The strategic planning process addresses the external and internal forces that affect its destiny. These include the competitive forces that fight over their share of the same pie and the cooperative forces that have the potential to make the pie bigger. Enabling this process and reinforcing a culture that enables autonomous actions and then balancing the internal selection process when it comes to funding decisions is an important part of the R&D manager's job.

Chapter 7, Development: Development includes the execution of the engineering work required to optimize and deliver the final specifications needed to build the technology. The design must balance goals related to such things as cost, compatibility, reliability, serviceability, safety and performance. Development must also ensure that those results are achieved over a range of operating conditions. As a result, development processes and practices involve iterative cycles of planning, modeling, prototyping, analysis, optimization and testing to converge on a final specification. Continuous improvement of the development workflows to increase velocity is the responsibility of everyone in the development function, and velocity is measured from the perspective of the customer. Due to finite resources, saying yes to one request often requires saying no to something else and may also require a decision on whether to build, buy, or collaborate on what is needed. Each decision has a measure of uncertainty and a framework needs to account for that when making initial decisions on what to add or subtract. To prevent paralysis and reduce risk, it is often best to use concurrent engineering to incrementally refine the options and gain consensus as more knowledge is gained and more resources need to be committed. Finally, delivering in batches on an established cadence and utilizing standards are critical to enabling predictability. When utilized correctly, standards are dynamic and focus everyone in the R&D system on improving the same thing. The resulting concentration of effort leads to faster learning and greater innovation.

Chapter 8, Engineering Research: Engineering research increases the power of engineering by creating new knowledge and more powerful modeling, analysis and optimization methods to solve design problems. This can be used to increase the value of current technologies or enable new technologies and, hence, new value. Most engineers solve problems as part of their job and this often requires some element of research. As a result, almost everyone in the R&D system conducts some degree of research. The motivation for research may be induced top-down with specific targets to sustain improvement along a known trajectory or it may happen autonomously in a way that may disrupt the status quo and open up new opportunities. Most research does not need to be centrally managed, but research that is very dependent on others does require coordination with the different stakeholders. This includes research done within the R&D organization by individuals and groups, as well as research that requires collaboration with other entities in the R&D system. This creates a research pipeline of sustaining and disruptive research topics from many sources. Harvesting the value derived from the research pipeline is a team game. A pathfinding process is needed to sample the pipeline and help those that demonstrate realizable value become implemented and deployed. The role of the R&D manager is to create the culture and processes where all of this research thrives and value is realized quickly. This may require separately funding autonomous and disruptive research to maintain a healthy research pipeline.

Chapter 9, The R&D Organization: The R&D organization includes the employees and other assets in the manager's direct chain of command. The employees work on strategic planning, engineering research and development and manage the engagements with the rest of the R&D system. These functions, along with a culture of learning, value creation, diversity and inclusion form the foundation of the R&D organization as depicted in Figure 1.1. As a consequence, the R&D organization is much more than an organizational chart. It is also the strategy, values, capabilities, processes, reward systems and people needed to deliver value. Generally speaking, whether its reporting structure reflects that or not, the operation will exhibit the characteristics of a matrix and will rely on intrinsic networks to enable the dissemination of best-known methods

Figure 1.1.   The foundation of the R&D organization.

and autonomous innovation. Since the strategy influences the organization's design, a change in the strategy may result in a re-design of the organization and a healthy organization can quickly change to accommodate that. Not to mention, leading the R&D organization and directing the rest of the system requires different approaches that depend on the type of problem being addressed. Finally, as we have already noted, a culture of learning and continuous improvement is the foundation for winning. Inculcating that culture cannot be delegated and requires the manager to take personal responsibility for instilling the right behavior.

Chapter 10, The R&D System: The R&D System is the specific collection of entities, assembled and directed by the R&D organization to work together and deliver the engineering product or service (Figure 1.2). This includes the R&D organization along with its suppliers, customers and other partners. The R&D organization (led by the R&D manager) directs the co-development, collaboration,

Figure 1.2.   The R&D system.

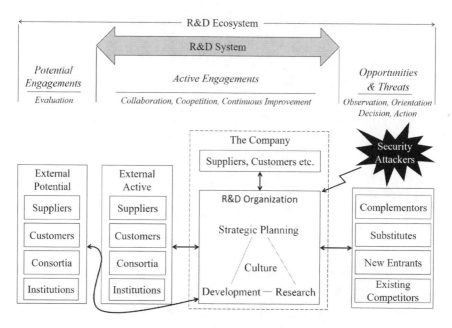

Figure 1.3.   The R&D ecosystem.

coopetition and continuous improvement across the R&D system. Although these entities do not directly report to the R&D manager, the manager must still provide the leadership to assemble and conduct the entire orchestra.

The R&D ecosystem shown in Figure 1.3 is larger and consists of the entire network of entities that are affected by each other even if they do not work together directly. This includes the entities that pose a threat or present opportunities like competitors, complementors and security attackers. It also includes entities such as suppliers, customers, consortia and institutions that may potentially become members of the R&D system. Each entity in the ecosystem affects each other, creating a constantly evolving relationship. Directing the R&D system requires comprehending the whole R&D ecosystem shown in Figure 1.3 to understand and act on the threats and opportunities posed by competitors, new entrants, substitutes, complementors and security attackers. Generally speaking, there is little or no direct engagement between the R&D organization and these entities and this is why they are not included in the R&D system. However, there may be indirect engagement with some of them via a consortium, industry standards bodies, industry forums, etc. or via other coopetition ventures. Directing the R&D system also requires evaluating potential suppliers, institutions, consortia and adding the appropriate ones to the R&D system. Finally, directing the R&D system means creating the culture, platforms, processes and governance needed to execute the co-development, collaboration and continuous improvement required to achieve the greatest velocity.

1.1.3 *Special Topics*

Chapter 11, Engagement with Academia: The R&D organization's engagement with academia spans activities that include: small one-time projects, education and hiring of new college graduates, continuous informal connections at conferences, major joint research projects, joint research goals within consortia and long-term interaction based upon national funding organizations. There is a great opportunity for both the R&D organization and a university to benefit if they engage. However, for the engagement to be beneficial, all parties must understand the objectives of the other parties. As described in previous chapters, the R&D organization's success is determined by delivering value to the customers. Academic success is determined by delivering educated students and by research that discovers truth.

Chapter 12, Information Security: The manager of the R&D organization is responsible for directing the engineering work and protecting all of the relevant information. Information security means protecting organizational, process, engineering, personnel, and product information while ensuring privacy means protecting information about individuals. The manager must drive the establishment of a culture where all employees are aware of the importance of information security to block threats at the source. With this in mind, the manager of the R&D organization must have a good working knowledge of information security technology and practices to properly allocate the resources needed and to establish the right balance of access control. This also requires conducting regular risk assessments and establishing the right legal and technical framework for managing intellectual property (IP) entering and exiting the company.

Chapter 13, Leadership: We dedicate this chapter to reinforcing the critical leadership behaviors that must be developed and applied to implement what we have discussed so far. Put succinctly, leadership is a combination of dreaming and doing. It spans the fuzzy front end of articulating a big and bold vision, translating that into a strategy that creates differentiable value and then motivating, funding and orchestrating all of the entities in the R&D system to work harmoniously together to deliver results with the greatest velocity. To be an effective leader, the R&D manager must balance the traits of curiosity, learning and having an open mind with a problem-solving mindset and a laser focus on results. Most of all, an effective leader knows that their success is based on the trust people have in them to do the right thing, the right way, for the right reason and they work hard to maintain that trust.

Chapter 14, Summary and Final Thoughts: Holistic engineering management covers a lot of territory. In this chapter, we provide a summary of the essential directives for the manager of the R&D organization from each chapter. Since people are the key to a holistic engineering R&D organization's success, we also provide our thoughts on how the manager should look after themselves and their employees to ensure everyone remains healthy, happy and full of energy. Finally, we touch on recent events in the world and their implications to some of the concepts we discussed in the book.

1.2 Important Terms

The R&D ecosystem that we cover in this book has many players and some standard terms are needed to avoid confusion. We have already defined such things as the R&D organization, the R&D system and the R&D manager in the preceding section. Their definitions and others can also be found in the glossary. However, other important terms and concepts are needed before the reader starts with the first chapter.

The R&D ecosystem:

- Suppliers: An entity that delivers a product or service to a consumer. In this book, it usually refers to entities that currently or potentially may deliver to the R&D organization. An internal supplier is from within the same company, an external supplier is from another company.
- Customers: An individual, group or company that receives and uses the product or technology. In this book, the customer usually refers to the entity that receives the engineering product or service delivered by the R&D organization.
- Consortia: A group of companies and other entities that pool resources to work together on specific topics of mutual interest such as standards and pre-competitive research. Examples include the Semiconductor Research Corporation (SRC) and the World Wide Web Consortia (W3C). Consortia can be the vehicle for coopetition between entities that normally compete with each other on topics that benefit each other or the broader industry.
- Institutions: A society or organization founded for a purpose. In this book, this refers to societies or organizations formed that further the objectives of engineering and technology. They may be public or private. Some are a source of research (e.g., universities and colleges) and others help with the development of the engineering community and set standards (e.g., the Institute for Electrical and Electronic Engineers (IEEE)). Others provide independent research

services (e.g., The Interuniversity Microelectronics Centre (IMEC)).

- Competitors: An entity that is engaged in commercial competition with the R&D organization or its company.
- Complementors: An entity that provides a product or service that increases the value of the product or service of another entity. A commonly referred to example is Microsoft and Intel.
- New Entrants: New competitors who enter an existing industry.
- Substitutes: Technologies that do not directly compete with a product but make it irrelevant. It becomes a substitute for the prior technology.
- Security Attackers: An individual or group that attempts to illegally obtain information or cause damage to the company.

Reporting Hierarchy and Networks:

When we refer to the structure of an organization, we mean the combination of the formal reporting hierarchy and informal or formal intrinsic networks of people shown in Figure 1.4 [8].

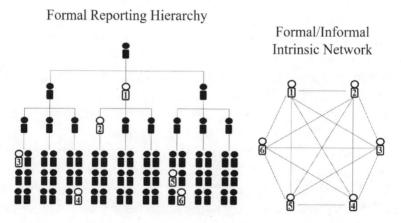

Figure 1.4.   The formal reporting hierarchy and the intrinsic network.

The exact reporting hierarchy varies depending on the need but it is the formal chain of command often represented in the form of an organization chart. Intrinsic networks are usually a collection of experts from different parts of the R&D organization's hierarchy that either form on their own or are created to address specific opportunities or threats. In some cases, the intrinsic networks will include members from entities outside of the R&D organization that are part of the R&D system.

Entities, Groups and Working Groups:

Since the terms to denote different levels of the reporting hierarchy and types of intrinsic networks can be company-specific, we have defined our terms. As shown in Figure 1.5, we use "entity" as a generic term for anything that may be a company, group of people, or an individual. A working group is an informal or formal network of people that is not part of the formal reporting hierarchy but typically reports its findings to a specific entity. A group is a term used for any type of organizational unit that is part of the formal reporting hierarchy. In this case, the R&D organization is also a group, but to differentiate it from

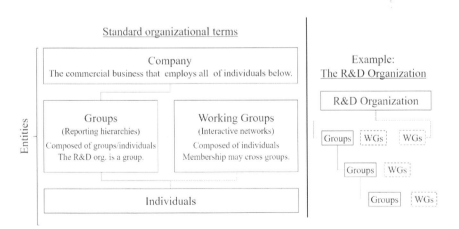

Figure 1.5. Entities, groups, working groups, individuals and organization.

other groups we call it the R&D organization. The company is a commercial business that employs people to achieve its goals. Entities working within the same company can generally share company information freely. Entities in different companies have more restrictions on how they work together.

The Workflow:

We call the series of connected engineering tasks a workflow. In several places in this book, we will dive deeply into various engineering workflows to provide examples and reinforce points. Workflows may be completed by an individual or a group. Workflows from different groups will often connect to other group's workflows to create cross-group workflows within the same company or across multiple companies. Regardless of their scope, the completion of every task in a workflow presents an opportunity to learn and, therefore, creates a potential learning cycle as shown in Figure 1.6. The workflow is the most common target for continuous improvement. We will also sometimes refer to a value stream or a value stream map. The value stream differs from a workflow because it represents the complete journey that chronicles the contributions of value

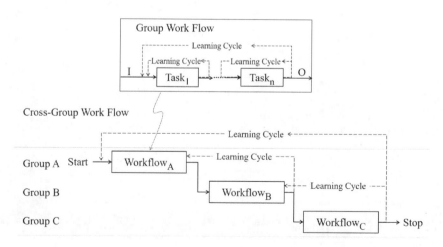

Figure 1.6.   Tasks, workflows and learning cycles.

Lines of communication vs. team size

Span	4	5	6	7	8
Lines	6	10	15	21	28

Group with Span of 6
1 manager, 6 direct reports

Figure 1.7.   Lines of communication vs. team size, highlighting a span of 6.

(and waste) from start to finish in a development process. There is overlap but we reserve the use of that term for when we are looking more broadly across an entire product development flow.

Lines of Communication:

Another way of looking at the interactions inherent across the R&D organization is achieved by mapping the possible lines of communication between members of the same group and between groups. This can provide a framework for understanding how information may flow between people. The number of people that directly report to a manager is called the manager's span. As shown in Figure 1.7, the lines of communication grow as the span increases. In the figure, we highlight a span of 6. The appearance is similar to that of the intrinsic networks that we described before but, in this case, it represents the lines of communication that exist within formal groups and across the reporting hierarchy. For such things as information security, it is instructive to show interactions using this approach.

Figure 1.8 shows how the lines of communication and hierarchy grow if we keep a span of 6 and grow the organization.

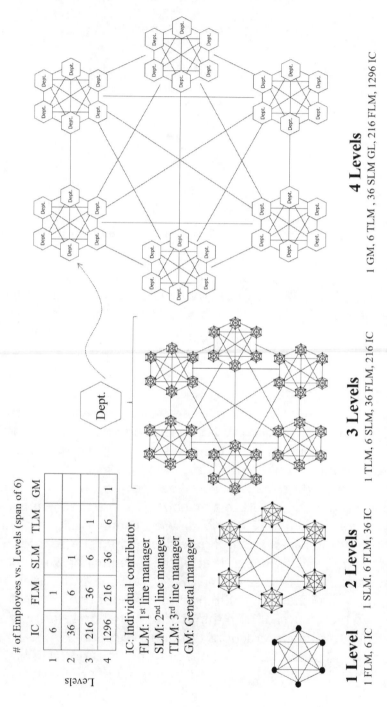

# of Employees vs. Levels (span of 6)

Levels	IC	FLM	SLM	TLM	GM
1	6	1			
2	36	6	1		
3	216	36	6	1	
4	1296	216	36	6	1

IC: Individual contributor
FLM: 1st line manager
SLM: 2nd line manager
TLM: 3rd line manager
GM: General manager

Dept.

1 Level
1 FLM, 6 IC

2 Levels
1 SLM, 6 FLM, 36 IC

3 Levels
1 TLM, 6 SLM, 36 FLM, 216 IC

4 Levels
1 GM, 6 TLM , 36 SLM GL, 216 FLM, 1296 IC

Figure 1.8.   The number of levels of hierarchy vs. the number of employees.

Seven or fewer employees necessitate one manager, 8 to 43 employees necessitate two management levels and so on. In this book, when we talk about the R&D organization, we will assume a span of 6 with 3 to 4 levels of reporting hierarchy.

1.3 A Simple Example of an R&D System

To provide a small sample of the challenges and opportunities that exist within the R&D system we can look at the journey of a request for a small enhancement in a design feature from an external customer. To keep it simple we will assume the R&D system is composed of the R&D organization, one external supplier, one internal supplier and one external customer. We also assume that the R&D organization has a span of 6 and three levels of hierarchy. In this example, a request is made by the customer, the solution is architected by the R&D organization, the specs are delivered to the suppliers who then design and release the collateral back to the R&D organization for integration. The R&D organization then releases the solution to the customer who deploys it to its users.

Figure 1.9 highlights that the work will touch the development functions of each entity. However, as we noted before, each entity in the R&D system may have its own set of development processes. Each one may have a unique process for capturing requests, prioritizing them against other requests, assigning resources and tracking progress. Some may follow a waterfall development model and others may use an agile model.

The underlying cultures may be different as well with some participants more open to sharing information and collaborating and others not. As a result, differences between the development models and cultures may cause unnecessary delays before any work is started.

Tuning each entity's development processes or culture to gain better alignment can benefit each member of the system. For example, if everyone used the same agile development

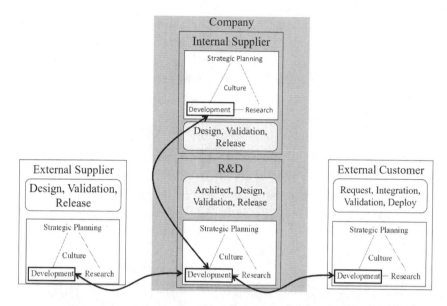

Figure 1.9.   Interaction of functions across the R&D system.

principles and aligned their development sprints then more work can be parallelized. The existing processes and culture in each entity often make even a small change difficult to achieve. So, instead, the differences are often bridged by padding the schedule with buffers to account for expected delays. This gives everyone a false sense of being on schedule when, in fact, they are operating much slower than what is possible.

If we drill down to the level of the workflows, it might look something like what is shown in Figure 1.10. The example shows the request coming from the customer on the far right (the dashed line) and going to the R&D organization where it is turned into a specification that is passed on to the different engineering groups that need to be involved (one external and internal supplier and the R&D organization). Each entity then uses its own set of workflows to design, validate and release the design for the next consumer in the flow as shown by the solid lines. The two suppliers release the collateral to be used in the R&D organization's engineering work, and the R&D

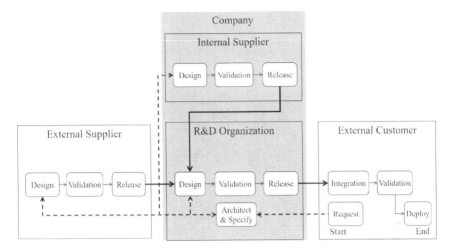

Figure 1.10.   Interaction of the workflows across the R&D system.

organization completes the design, validates the result and releases the result to the customer. The customer then validates and deploys the solution in its company.

In this case, even if the development model and process are the same, the workflows in each entity may be different. They may use a different set of engineering tools and methods to design and validate the work. The models they use to measure performance, validate functionality and meet other constraints may not agree with the models used by the other entities. It is even possible that the data format of the final design is slightly different than the format needed by the receiving engineering tools or models used by groups in the R&D organization. All of this does not matter too much while they are operating in their group, but once they hand-off the results to another entity, there may be problems with the integration that leads to rework or quality problems. One can try to catch these problems ex post facto by creating rules and systems to check incoming collateral for known incompatibilities. However, this is shooting behind the duck since it does not capture new issues. It also eventually leads to a bloated set of rules that

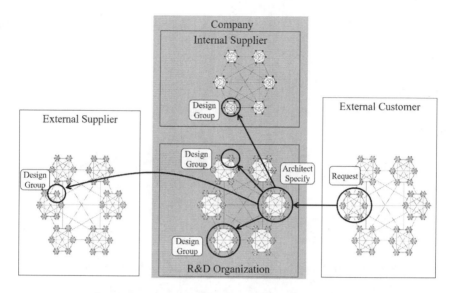

Figure 1.11.   Information flow for a request.

prove increasingly difficult to maintain. Eventually, the time spent in dealing with false positives and checking and reworking the collateral can start to far exceed the time spent doing the engineering the work in the first place. When that happens, people will find creative reasons to override or wave violations and this can lead to catastrophic results. Although each entity feels like they are being very productive internally, the system's overall velocity may be very low due to the rework and delays that result from differences in the workflows.

Figure 1.11 shows a third view of the same example that highlights the lines of communication that may exist within each group. Each group is composed of a nested set of networks, as described in Figure 1.8. Each network represents the possible interactions within each group. In some cases, the interactions are formal and in other cases, they may be informal. The figure depicts the flow of information originating from the customer's initial request, being received, architected and specified by a group in the R&D organization then being communicated to the different entities responsible for doing

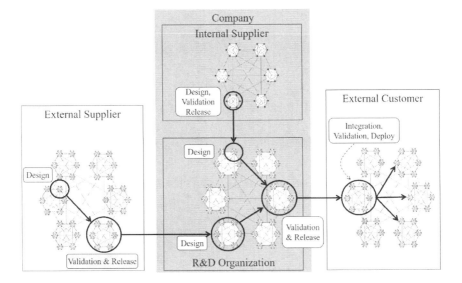

Figure 1.12. Information flow to satisfy a request.

engineering work. It is likely that once information is received by one group, the rest of its network may have access to it. It is also possible that adjacent groups may be asked for help (formally or informally) and also gain access to the information. Similarly, Figure 1.12 shows the flow of information once the engineering work starts. In that case, the design and validation results may be shared within the group's network and with other adjacent groups as needed.

The networks associated with each group and the informal networks that may sprout up enable innovation and problem-solving. However, this also means that sensitive information that is not meant to be broadly distributed can propagate through a supplier or other entities' networks quickly unless it is properly managed. Providing it to just one person in one group does not prevent it from spreading to the rest of the group and other groups. Also, information that one supplier has provided to the R&D organization may also propagate through the networks in the R&D organization. If this happens, it may find its way to one of its competitors. So, in addition to providing the means to protect the R&D organization's sensitive information when it is provided to another entity, it is also important to provide the means to protect the information received from other entities in the R&D system that work with the R&D organization.

One may still ask why we should care about all of this as long as we tightly control all of the information and everyone delivers their engineering collateral on schedule? The answer is that if we only do that, then all of the wastes we discussed get legitimized and baked into schedules. If we add up the time it takes to only work on the value-added activities, then we would find that the inherent latency of a workflow is very, very small compared to what it actually takes. This only gets worse as the number of entities in the system and the interdependencies grow. The inherent latency should be the target and not a schedule that is built to accommodate existing waste. Although one will never get to the ideal state, constantly striving towards it will lead to greater velocity than if we satisfied ourselves with just meeting the usual schedule.

References

[1] Service, T. (2014). Symphony Guide: Beethoven's Ninth ('Choral'), *The Guardian Newspaper*. September 9, 2014. Available from: https://www.theguardian.com/music/tomserviceblog/2014/sep/09/ symphony-guide-beethoven-ninth-choral-tom-service [May 2020].

[2] DeGusta, M. (2012). Are Smart Phones Spreading Faster than Any Technology in Human History? *MIT Technology Review*. May 9, 2012. Available from: https://www.technologyreview.com/s/427787/ are-smart-phones-spreading-faster-than-any-technology-in-human-history/ [May 2020].

[3] iFixit (2018). *iPhone XS and XS Max Teardown*. September 21, 2018. Available from: https://www.ifixit.com/Teardown/iPhone+ XS+and+XS+Max+Teardown/113021 [May 2020].

[4] Yang, D. and Wegner, S. (2018). *Apple iPhone Xs Max Teardown*. October 3, 2018. Available from: https://www.techinsights.com/ blog/apple-iphone-xs-max-teardown#costing [May 2020].

[5] Apple Corp. (2019). *Apple Supplier Responsibility 2019*, Available from: https://www.apple.com/supplier-responsibility/pdf/Apple-Supplier-List.pdf [May 2020].

[6] American Association for the Advancement of Science (1989). *Science for All Americans,* Chapter 3 (Project 2061).

[7] Boyd, J. R. (1976) *New Conception for Air to Air Combat.* Colonelboyd.com. p. 19. Available from: https://www.colonelboyd.com/boydswork [May 2020].

[8] Kotter, J. P. (2014). *Accelerate*, (Harvard Business Review Press) p. 63.

CHAPTER 2

Culture

> Culture eats everything for breakfast.
> — **Robert J. Aslett**

This chapter is about the values and beliefs employees have that are specific to the workplace and guide their work. We call this the culture of the R&D organization. Culture is the second chapter of the book because of its impact on everything we will talk about later. The phrase, "culture eats strategy for breakfast", is often attributed to Peter Drucker, and it speaks to the influence that the culture has on an organization's performance. We believe that this phrase does not go far enough and prefer the variant, "culture eats *everything* for breakfast". Providing autonomy to employees is critical to enabling the pace of innovation needed to increase velocity. A common culture that reinforces the R&D organization's values and strategy enables employee autonomy while ensuring fidelity to the mission. Conversely, autonomy without a common culture leads to chaos.

Everyone has a set of personal values and beliefs that guide them through life. We are not talking about those. We are referring to the set of common values and beliefs that are specific to the workplace and govern how people work together. Some of those may be the result of explicitly stated values reinforced by the leadership and some may never be written down and result from what employees learn by observing their leaders and peers. This is the muscle memory of an organization that guides the behavior of each employee such that when faced with ambiguous or incomplete information, they instinctively know what to do and are empowered to act. A strong culture means that employees agree on "how we do things here" and "what we value here". This helps overcome inherent flaws in any organizational structure, processes and operations and enables employees to act autonomously and still do the right thing. When an organization's actual culture aligns with the stated values and strategy then it is a powerful catalyst for rapid progress. Because of its overriding importance, the culture is at the heart of the R&D organization.

In reality, the correlation of the actual employee behavior to the stated values varies with each employee. The alignment (or misalignment) needs to be understood and managed since it has a major impact on the ultimate success of the organization. History provides examples of companies that have articulated the right values and reinforced them until they became ingrained in each employee. History also provides examples of companies that articulated the right values but reinforced them half-heartedly or, worse, role-modeled the opposite behavior. In these cases, there is a very poor correlation between employee behavior and the stated values, which often leads to catastrophic results.

Aligning to a strategy sometimes requires a change in culture within the R&D organization and other entities in the R&D system. Alignment on a common culture within the R&D organization is critical. Alignment with other entities in the R&D system does not have to be complete. However, better alignment in some areas can make a tremendous impact on the overall performance of the system. This is extremely difficult to accomplish. In those cases, the manager of the R&D organization will need to take the lead and work with others to make it so.

Finally, the manager of the R&D organization has the biggest impact on setting the culture and must lead by example. What a leader says does not resonate with employees as much as what they do. Employees understand that the latter is a far better indication of the leaders' actual values than the words in a presentation. Any perceived dissonance between what is said and what is done is evidence to the employee that the stated values are not taken seriously by the leadership and therefore do not have to be taken seriously by them.

2.1 What is Culture?

Culture defines "how we do things here" and "what we value here".

Employees in an organization with a strong culture have agreement on "how do we do things here" and "what do we value here." Former Intel CEO, Andy Grove [1], described the importance of culture in the following way,

> *"When the environment changes more rapidly than one can change rules, or when a set of circumstances is so ambiguous and unclear that a contract between the parties that attempted to cover all possibilities would be prohibitively complicated, we need another mode of control, which is based on cultural values. Its most important characteristic is that the interest of the larger group to which an individual belongs takes precedence over the interest of the individual himself. When such values are at work, some emotionally loaded words come into play — words like trust- because you are surrendering to the group your ability to protect yourself. And for this to happen, you must believe you all share a common set of values. A common set of objectives, and a common set of methods"*

When culture complements strategy it increases velocity.

As depicted in Figure 2.1, the organization's culture complements its vision and provides a foundation for how the strategic planning, research and development functions work together. When all of these elements share a common culture, then it is a powerful catalyst for rapid progress. When there is dissonance, success will be elusive because employees do not share the values needed to execute the strategy properly.

Figure 2.1.   Culture is at the heart for the R&D organization.

2.2 The Importance of Values

The right values must be instilled with everyone to drive the right culture.

At Intel, Andy Grove's views manifested themselves as a set of Intel values that had remained fairly constant throughout its history until recently. As former employees of Intel, we can attest to their reinforcement. It helped that every employee had these values handily attached to their employee badges. The most recent description of the Intel values published on the corporate web-site is shown in the following. In a case study at the end of this chapter, we will also show the previous set of values and explore the changes that were made.

Intel Values
   From the Intel Corporate Website May 2020 [2]

"<u>Fearless:</u> We are bold and innovative. We take risks, fail fast, and learn from mistakes to be better, faster, and smarter next time.

<u>Inclusion:</u> We strive to build a culture of belonging. We create a space where everyone can contribute to their fullest potential and deliver their best work. We welcome differences, knowing it makes us better.

<u>Customer Obsessed:</u> We listen, learn, and anticipate our customers' needs to deliver on their ambitions. Our customers' success is our success.

One Intel: We appreciate, respect and trust each other. We commit to team over individual success. We are stronger together. Innovators at heart, we bring fun to work every day!

Truth and Transparency: We are committed to being open, honest, ethical, and timely with our information and feedback. We constructively challenge in the spirit of getting to the best possible result. We act with uncompromising integrity.

Quality: We deliver quality and ensure a safe workplace. We have the discipline to deliver products and services that our customers and partners can always rely on."

Intel is a great example of a company's commitment to clearly articulate a set of values. However, they do not need to be written down to get the point across. A summary of Apple Inc. values were articulated at an investors conference call in January 2009 by, then Chief Operating Officer, Tim Cook (see the following). The description of Apple's values that exist, "regardless of who is in what job" left no doubt about the expectation he had for every employee and the importance he placed in inculcating these values across the company.

An Articulation of Apple Values
Tim Cook at the January 2009 Apple Investor Conference Call [3]

"We believe that we are on the face of the earth to make great products and that's not changing. We are constantly focusing on innovating. We believe in the simple not the complex. We believe that we need to own and control the primary technologies behind the products that we make, and participate only in markets where we can make a significant contribution. We believe in saying no to thousands of projects, so that we can really focus on the few that are truly important and meaningful to us. We believe in deep collaboration and cross-pollination of our groups, which allow us to innovate in a way that others cannot. And frankly, we don't settle for anything less than excellence in every group in the company, and we have the self-honesty to admit when we're wrong and the courage to change. And I think regardless of who is in what job those values are so embedded in this company that Apple will do extremely well."

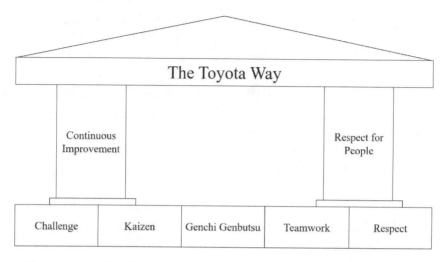

Figure 2.2.   The Toyota Way as summarized by Liker and Hoseus.

Another well-studied example of the importance of culture lies somewhere between these two examples. Toyota Motor Company has been singled out for its unique and powerful internal culture that led them to leadership in the automotive industry. The values were not originally written down. Instead, they were set by the example by the leadership, fellow workers, and through training based on the pillars of continuous improvement and respect for people. To the rest of the world, this became known as lean, lean manufacturing, or lean product development. As Toyota expanded to other countries, they saw the need to capture the essence of the Toyota Way more formally and unveiled it internally in 2001. The depiction here and in Figure 2.2 is courtesy of Jeffrey Liker and Michael Hoseus [4].

The Toyota Way 2001

"At the top level there are two pillars, continuous improvement, respect for people and "all Toyota team members, at every level, are expected to use these two values in their daily work and interactions."

Respect for people is a broad commitment. It means respect for all people touched by Toyota including employees, customers, investors, suppliers, dealers, the communities where Toyota has operations, and society at large. Respect for people has sub-categories of "respect" and "teamwork" shown in the foundation of the house.

Respect: We respect others, make every effort to understand each other, take responsibility and do our best to build mutual trust.

Teamwork: We stimulate personal and professional growth, share the opportunities of development and maximize individual and team performance.

Continuous Improvement is the second pillar. Toyota leaders believe people who are continuously improving are what have allowed Toyota to grow from a small loom company in a farming community to a global powerhouse. Continuous improvement is defined as, "We are never satisfied with where we are and always improve our business by putting forth our best ideas and efforts."

There are three categories under "continuous improvement" that complete the foundation of the Toyota Way House:

Challenge: We form a long-term vision, meeting challenges with courage and creativity to realize our dreams.

Kaizen: We improve our business operations continuously, always driving for innovation and evolution.

Genchi Genbutsu: We practice Genchi Genbutsu — believing in going to the source to find the facts to make correct decisions, build consensus and achieve goals at our best speed."

Intel, Apple and Toyota all have a history of technology leadership. Their values stress collaboration to anticipate what customers will value and the willingness to fail early, learn and adapt. Using the representative values of anticipation and adaptability we can construct the 2×2 matrix shown in Figure 2.3 that is probably common to most successful high technology companies and engineering organizations. As shown in the 2×2 matrix, a company whose strategy is based on technology

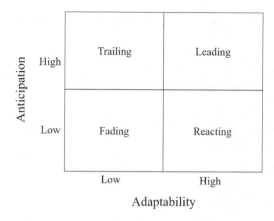

Figure 2.3.   The values of anticipation and fearlessness.

leadership needs a culture that enables *anticipating* the future (included in Intel's value of "customer obsession") and willingness to quickly fail and *adapt* (included in Intel's value of "fearless"). A company whose strategy is based on anticipating the future but participating at a lower cost may not invest in being the first to adapt. These companies have a clear vision of the future but purposely trail the leaders and their values should reflect that. Similarly, a company's strategy may be based on rapid adaption to present market conditions. They may choose not to invest in anticipating the future but are excellent at reacting. Once again, this could be an explicit strategy of the company and its values should reinforce that. Finally, there might be companies whose strategy is based on squeezing out profits from legacy technology that will gradually fade away. In those cases, there is little need for anticipation and adaptability and the values of the company should reflect that. Although we used companies as examples, creating the appropriate values that support the strategy also holds for engineering organizations within a company.

As we noted in the previous section, when the organization's culture aligns with its strategy, it is a catalyst for rapid progress. For Intel, Apple and Toyota the values of the company are tied to their strategy and how they believe success will be achieved. These values have been instilled in the employees through different means and have been a major reason for their continued technology leadership.

2.3 Stated Values vs. the Actual Culture

An organization's actual culture may not align with its stated values.

Even in organizations with a very strong culture, there is some variability among the employees on the understanding and practices that form the organization's culture. In reality, the organization's culture is a distribution of the relationship between the organization's actual culture and its stated values. As shown in Figure 2.4, there are a variety of possible relationships between the actual culture and its stated values. The solid line shows the case where almost all of the employees

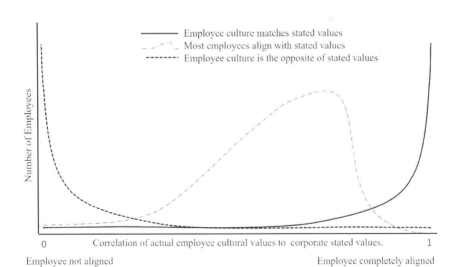

Figure 2.4.   The correlation of stated and actual cultural values.

think and act in alignment with the stated company values. For almost all employees, there is a high correlation between their own values and the company's stated values so that the actual culture is represented by the organization's stated values. The other extreme is the dashed line which shows that employee behavior is directly opposed to the stated values. The dot-dash line represents what is usually seen. In that case, most of the employee values aligned with the stated corporate values.

Dissonance between stated values and leadership actions lead to disaster.

Nicely articulated values, like the ones we mentioned, help describe the desired culture in the organization but are not sufficient to ensure the behavior matches the words. They have to be role modeled by the leadership every day and reinforced by governance, operations, performance management and reward systems. It is a never-ending task that requires persistence and investment. We will talk more about the manager's important role in this later in this book but to highlight what happens when this is not done let us consider an extreme example. Enron Corporation was a major energy company originally founded in 1985. In its 2000 annual report [5], Enron stated its mission as, managing "efficient, flexible networks to reliably deliver physical products at predictable prices". A big part of that was energy products. On page 53 of the report it highlighted four core company values:

Enron Values 2000

"Communication

We have an obligation to communicate. Here, we take the time to talk with one another ... and to listen. We believe that information is meant to move and that information moves people.

Respect

We treat others as we would like to be treated ourselves. We do not tolerate abusive or disrespectful treatment.

Integrity

We work with customers and prospects openly, honestly and sincerely. When we say we will do something, we will do it; when we say we cannot or will not do something, then we won't do it.

Excellence

We are satisfied with nothing less than the very best in every-thing we do. We will continue to raise the bar for everyone. The great fun here will be for all of us to discover just how good we can really be."

The 2000 annual report claimed an income of $1.36B on revenue of $100B but the company was bankrupt by the end of 2001. As explained in the Money magazine article of January 2002 [6],

"It seems hard to believe now, but Enron (ENE) used to be the envy of corporate America. In less than a decade, the Houston company transformed itself from stodgy gas-pipeline operation to natural gas and electricity trading powerhouse. Dazzled by sizzling earnings growth, giddy investors bid up Enron's shares 312% in two years to a high of $90.75 in 2000. Then someone turned out the lights. Beset by marketplace woes and man-agement mishaps, the stock already had tumbled 53% when chief executive Jeffrey Skilling stunned investors by resigning last August. After that, the bad news came at hyperspeed: $1.2 billion in shareholder equity zapped by risky hedging deals, a Securities and Exchange Commission probe, a last-chance merger with rival Dynegy called off and, finally, a bankruptcy filing. By the end of November, the stock had plummeted to 26[cents], obliterating $67 billion in market cap — a shocking fall for a company that just last year occupied the No. 7 spot on the Fortune 500."

Subsequent investigations exposed that those who raised concerns about financial results or practices were intimidated. It became clear that Enron's culture bore no resemblance to their stated core values, especially *Integrity* and *communication* [7].

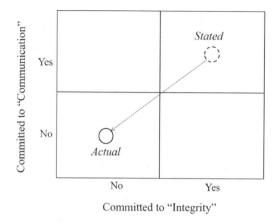

Figure 2.5.   Enron's stated values and actual culture.

"A culture of secrecy and a remarkable lack of transparency prevented any realistic assessment of the company's financial risk... Enron's numerous partnerships were shrouded in secrecy, tucked away off the balance sheet. They were used to shift debt and assets off the books while inflating earnings. The chief financial officer ran and partly owned two partnerships, a clear conflict of interest."

As depicted in Figure 2.5, the reality of Enron was that the actual culture of the company bore no resemblance to the stated value of communication that stressed, "Here, we take the time to talk with one another ... and to listen" and the stated value of integrity that stressed, "We work with customers and prospects openly, honestly and sincerely." Most distressingly, the company's failure was devastating to the 12,000 employees who lost their jobs, and most of their retirement savings.

Another example of the dissonance between company culture and stated values occurred at Wells Fargo in 2016. As the news reports revealed, the company culture did not support its stated values. However, contrary to the Enron case, Wells Fargo took steps to recover and correct its behavior. Wells Fargo defines its values in the following way [8].

"Five primary values guide every action we take:

What's right for customers.

We place customers at the center of everything we do. We want to exceed customer expectations and build relationships that last a lifetime.

People as a competitive advantage.

We strive to attract, develop, motivate, and retain the best team members — and collaborate across businesses and functions to serve customers.

Ethics.

We're committed to the highest standards of integrity, transparency, and principled performance. We do the right thing, in the right way, and hold ourselves accountable.

Diversity and inclusion.

We value and promote diversity and inclusion in all aspects of business and at all levels. Success comes from inviting and incorporating diverse perspectives.

Leadership.

We're all called to be leaders. We want everyone to lead themselves, lead the team, and lead the business — in service to customers, communities, team members, and shareholders."

On October 4th in 2016 NPR [9] reported that,

"Former Wells Fargo employees describe toxic sales culture, even at HQ." "Former employees of Wells Fargo tell NPR that a toxic high-pressure sales culture at the bank drove some workers to deceive customers and open unauthorized accounts — even in the blank's own headquarters building in San Francisco."

And the NPR account further stated that,

"Wells Fargo's embattled CEO John Stumpf has been spending long hours defending himself in congressional hearings trying to explain the scandal engulfing his bank. 'Wrongful sales practice behavior,' Stumpf told the Senate banking committee,'

goes against everything regarding our core principles, our ethics and our culture'".

On May of 2017 Vanity Fair [10] reported,

"How Wells Fargo's cutthroat corporate culture allegedly drove bankers to fraud."

On December 28, 2018, in a press release from the Attorney General's office [11],

"Attorney General Josh Shapiro today announced that Wells Fargo Bank N.A., the nation's biggest bank, will pay $575 million to resolve claims that the bank violated state consumer protection laws by: (1) opening millions of unauthorized accounts and enrolling customers into online banking services without their knowledge or consent, (2) improperly referring customers for enrollment in third-party renters and life insurance policies, (3) improperly force-placing and charging more than 850,000 auto finance customers for unnecessary and duplicative insurance policies, (4) failing to ensure that customers received refunds of unearned premiums on certain optional auto finance products, and (5) incorrectly charging customers for mortgage rate lock extension fees.

Later the release said,

"The states alleged that Wells Fargo imposed aggressive and unrealistic sales goals on bank employees and implemented an incentive compensation program where employees could qualify for credit by selling certain products to customers. The states alleged that these sales goals and incentive compensation created an impetus for employees to engage in improper sales practices in order to earn financial rewards. Those sales goals became increasingly difficult to achieve over time, the states alleged, and employees who failed to meet them faced potential termination and career-hindering criticism from their supervisors."

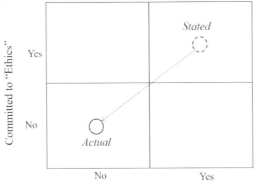

Figure 2.6.   Wells Fargo's stated values and actual culture.

As depicted in Figure 2.6, Wells Fargo in 2016 was driven by objectives that created behaviors that did not align with its value of ethics that stressed "the highest standards of integrity, transparency, and principled performance" and the value of doing "what's right for customers.

The Enron and Wells Fargo examples are not engineering organizations, but the lessons still apply. The well documented corporate examples highlight that when it comes to values, what is said is often not the same as what is done. They are examples of a lack of role modeling by leaders, incentivizing behavior that ran counter to the stated values and having the case where some values dominate and trample the others. These cases are also a reminder that there are consequences when the company's actual culture is not aligned with the professed values. In most cases, the consequences are more subtle and may show up as frustration with such things as slow progress, miscommunication, unnecessary re-work and the inability to make decisions and see them through. Without the strong alignment between the right values and the organization's actual culture, the R&D organization will be fighting an uphill battle against their competition.

2.4 Setting the Culture

Changing a culture requires persistence

A more common example of the impact of culture for an engineering R&D organization might be when it finds that the market it serves has shifted and decides it must change the way it works. Consider the example of an engineering organization designing a small variety of high volume, high margin customized technologies that needs to shift to producing a high variety of low volume, low margin, quick turn-around technologies. In this case, the organization may need to move from a culture where each engineering group controls all aspects of product architecture, design and validation to a matrix model that relies on centralizing common components to reduce costs and deliver more products. The stated values of the engineering organization may change to express the importance of accepting the trade-offs that come with re-using engineering work and collateral from central groups, however, the culture is likely to be anchored in the past by behavior that continues to emphasize customization. Product teams may decry the poor quality of the engineering output delivered by central organizations that do not meet the performance they are used as a reason to open up and "improve" what they received. Central groups may find that they are given specifications that are still geared to a custom mindset and become overwhelmed supporting them. This can create a lot of bad tension in the organization and reduce productivity unless the culture starts to align with the stated values and strategic direction.

Making a change to the culture to meet the change in strategy mentioned above is very difficult and will take multiple cycles of reinforcement. For example, if the technology development cycle time is 1 year it may take three iterations and, therefore, three years to align behavior and start seeing the benefit. In the meantime, everyone in the organization is watching the leadership to see if they crack and revert to previous behavior. This is especially true if the previous culture was strongly instantiated in the organization. It may have been the right culture at the time and strong alignment on that culture previously led to success.

There are several approaches one can use to drive such a transformation and the topic warrants its own book. Fortunately, many have been written and there are good resources available. A well-known framework for organizational transformation was defined by John P. Kotter [12–13] who listed eight leadership requirements for success shown in the following. We encourage the reader to read the source material and learn more.

1. Create a sense of urgency
2. Build a guiding coalition
3. Form a strategic vision and initiatives
4. Enlist a volunteer army
5. Enable action by removing barriers
6. Generate short term wins
7. Sustain acceleration
8. Institute the changes

Although there are different approaches and each transformation model may provide a slightly different take on the steps to follow, we have found something common in every successful transformation. It is the unrelenting persistence from the manager and leadership team to articulate, role-model and reward the new behavior. This is not a quick process. For large organizations, it can take years to inculcate the new behavior fully. This puts large organizations in transition at a disadvantage when competing against incumbents or small start-ups that are inherently more agile. As we will discuss later, steps can be taken in the organization design to trade-off efficiency, cost, or other items to temporarily deal with misaligned culture and remain competitive but ultimately the organization needs to adapt to the new culture.

The act of transitioning to a new culture also presents its own risk in the short term. The performance of the organization will likely degrade while the transformation is happening.

Old customers will not be happy with the changes and the new customers will not yet see the value that will be provided. This is the valley of death when the organization moves from doing the wrong thing well to doing the right thing poorly (hopefully, on the way to doing the right thing well). This can lead to second-guessing and a call to move back to the comfort of the old way of doing things. This is when most transformation efforts fail. When they fail to get across the valley of death, it is largely because the leadership team did not truly understand the new vision and did not provide the commitment needed on their part to see it through. As a result, they either declared victory too soon or quietly gave up.

The stated values must not contradict one another.

Let's now look at a specific example of how the alignment between the actual culture and values may manifest itself in a way that Andy Grove described earlier. Although the most recent version of the Intel values does not explicitly include it anymore, for many years *Results Orientation* and specifically the part that referred to "focus on output" was the most strongly embedded value in all employees. In a large company like Intel that must sometimes evaluate and compare the performance of many people across different functions the value of *Results Orientation* was very important and "focus on output" defined a means of quantifying the relative performance of people based on measurable results supported by data. Although there were exceptions, the majority of Intel employees used this to help judge how to prioritize their work and evaluate others. When done correctly, it created a positive feedback loop that relies on data to make judgments and reward people for real results.

On the other hand, there was also the stated Intel value of *Risk-Taking* and specifically, the part that referred to "encourage and reward informed risk-taking". Thorough analysis and debate (being informed) before taking a risk was a value that was easily adhered to for narrow engineering decisions with a small number of participants in the same team where decision making was relatively simple. However, when more organizations were involved, the analysis became more complex and took longer. At some point, it conflicted with the understanding that the company also valued and rewarded decisive leaders. As a short cut, managers would sometimes commit to a risky plan without the proper analysis and debate needed to be properly informed to look decisive. In some cases, they did this in the hope that an upstream partner would fail before they did and take responsibility for a slip in a schedule that they also could not meet.

These two examples also highlight the fact that for many employees, the interpretation of the stated values by their management can often seem to contradict each other. How does a person properly demonstrating the value of *Risk-Taking* (and who ultimately failed) compare to someone who was conservative but delivered a technology that exemplified the value of *Results Orientation*? Of course, a simple answer is that if judged correctly by management they both produce valuable results if the risk taken in the first case generated useful knowledge. The unfortunate reality is that an employee's risk-taking may not be properly rewarded in these cases. When employees see that they learn quickly that they should "focus on output" by delivering in volume. Although *Results Orientation* is a great value, if certain aspects of it dominate the culture at the expense of the other values, then the R&D organization is not getting the intended behavior. The entire set of values should complement one another and be applied in a balanced way for each to be effective. Also, the leadership must be honest with themselves and their employees and avoid token values that, in reality, are subordinate to other values.

Adjustments to cultures across the R&D system can benefit everyone.

Although setting the culture within an R&D organization or company seems daunting enough, when we expand the scope to include operating in the larger R&D system, we need to consider the added complexity of optimizing work across multiple organizations or companies that may have very different cultures. We cannot expect that all of the participants in the R&D system will promote the same values and share a common culture. Some differences are rooted deeply in the history of the company. For example, there may be suppliers whose engagements are predominantly driven by sales growth, who are most interested in protecting their niche position or ones that are driven by new research and development opportunities. Other differences relate to their business strategy or how they develop and deliver their technology or services. Those differences may seem to make sense for each one considering their strategy and the value they believe they provide. However, when considered within the context of the bigger R&D system, some may combine to create something analogous to an endothermic chemical reaction that siphons a lot of energy from the surrounding R&D system and leaves the surrounding environment colder and slows overall velocity. The ideal result is analogous to an exothermic chemical reaction where the interaction of the participants creates energy and feeds the collective performance and velocity. Later in this chapter, we look at a fictional case study of where a tweak to the cultures of engineering groups in a consumer and a supplier benefits both companies.

The manager of the R&D organization must lead by example.

As already mentioned, the manager has the most important role to play in establishing the right values to enable the organization's strategy and then leading by example to align the behavior to those values. Employees understand that what leaders do is a far better indication of the leader's true values than the words written on a slide in a presentation or other form of communication. This applies to how they act and how they communicate but, more importantly, what they reward and what they decide to start, to continue or stop. Any perceived dissonance between what is said and what is done is evidence to the employee that the stated values are not taken seriously by the leadership and therefore do not have to be taken seriously by them. This was certainly true in the cases we described earlier where the behavior of the leaders, actual measures for success and the way people were rewarded ran counter to the stated values of the companies. However, it also applies in more subtle ways such as when the manager of an organization states they value continuous improvement but only rewards reactive efforts to overcome mistakes or poor planning and not the proactive efforts others may have made to prevent them from happening in the first place. All of these examples pile up over time to define for the employees what are the actual values of the leadership regardless of those stated elsewhere.

Although leading by example is the most important element in setting a culture, there are different ways to do this. Like the values themselves, there is not one personality trait or approach that everyone should adopt as long as there is fidelity between actions and values. In the case of Intel, the corporate values are written down and can be attached to every employee's identification badge. At Toyota the culture has been codified to some degree but is mostly passed down through observation, mentoring and training. At Apple, Tim Cooke can succinctly express the values of the company verbally in a way that leaves no doubt about what he wants to see. All of these approaches will work if there is consistency between what is written down or said and the behavior of the leadership.

Like the different means by which values can be articulated, there is also not one set of values that makes sense for every organization. As we noted earlier, a set of values is tied to the organization's vision and strategy. Because of that, there will be differences between organizations. An organization with a vision and strategy designed to win business in a market that values stability and predictability may have a different set of values then an organization with a vision and strategy to win business in a market that values rapid change. Therefore, it is very important for any organization first to define the right values given their vision and strategy.

2.5 Summary

Culture is the muscle memory of an organization that guides the behavior of each employee such that when faced with ambiguous or incomplete information, they instinctively know what to do. A strong culture means that employees agree on "how we do things here" and "what we value here". It is at the heart of *holistic engineering management* because it enables the autonomy needed for rapid decision making while maintaining continuity. When the culture aligns with the stated values and strategy of the R&D organization then it is a powerful catalyst for rapid progress but when there is significant dissonance the consequences can be catastrophic. As a result, alignment on a common culture within the R&D organization is critical. Alignment with other entities in the R&D system does not have to be complete but there are cases where better alignment in a couple of areas can make a tremendous impact on the overall performance of the system and benefit everyone. In those cases, the manager of the R&D organization will need to take the lead and work with others to make it so. Finally, the manager of the R&D organization and the leadership team has the biggest impact on the organization's culture and must lead by example. What a leader says does not resonate with employees as much as what they do. Any perceived dissonance between what is said and what is done is evidence to the employee that the stated values are not taken seriously by the leadership and therefore do not have to be taken seriously by them.

2.6 Key Points

- Culture defines "how we do things here" and "what we value here".
- When culture complements strategy, it increases velocity.
- The right values must be instilled to drive the right culture.
- An organization's actual culture may not align with its stated values.

- Dissonance between stated values and leader actions lead to disaster.
- Changing a culture requires persistence.
- The stated values must not contradict one another.
- Adjustments to cultures across the R&D system can benefit everyone.
- The manager of the R&D organization must lead by example.

2.7 Case Studies

2.7.1 *Changing the Values at Intel Corp*

Background

For most of its history, Intel Corp. had the same set of stated values. Although some wording changed, they more or less stayed the same until May of 2020. A snapshot of what was on the corporate website in September 2019 is shown as follows:

> "*Quality:* There's a strong commitment at Intel to quality and continuous improvement. This commitment extends to all levels of the company. We strive to achieve the highest standards of excellence; do the right things right; continuously learn, develop, and improve; and take pride in our work.
>
> *Risk Taking:* Risk taking recognizes that some failures are unavoidable. Some experiments will generate favorable outcomes, while others will lead to disappointment. But even the disappointments can be turned into gains. We strive to embrace a growth mindset in everything we do, foster innovation and creative thinking, embrace change and challenge the status quo, listen to all ideas and viewpoints, learn from our successes and mistakes, and encourage and reward informed risk taking.
>
> *An Inclusive, Great Place to Work:* At Intel, we believe a productive and challenging work environment is vital to our success. This requires an open, candid, and respectful approach to working with one another. Building trust and maintaining dignity are critical in our very diverse global workforce and

environment. Ultimately, we want every employee to look forward to coming to work each day. We strive to create an inclusive work environment that fosters diversity; treat one another equally, with dignity and respect; be open and direct; promote a challenging work environment that develops our workforce; work as a team with respect and trust for each other; win and have fun; recognize and reward accomplishments; manage performance fairly and firmly; and be an asset to our communities worldwide.

Discipline: Our employees pride themselves on their ability to make and meet commitments, a quality we're able to attain by clearly communicating our intentions and expectations. We strive to conduct business with uncompromising integrity and professionalism; ensure a safe, clean, and injury-free workplace; make and meet commitments; properly plan, fund, and staff projects; pay attention to detail; and keep Intel information secure. "Discipline is all about planning. A lot of people think that it's presentation skills, but it's really preparation skills."

Customer Orientation: Intel's concept of customer orientation goes beyond conducting business in the marketplace. It's absolutely crucial that we listen and respond to our team members within Intel in a cooperative and supportive manner. We listen and respond to our customers, suppliers, and stakeholders; clearly communicate mutual intentions and expectations; deliver innovative and competitive products and services; make it easy to work with us; and excel at customer satisfaction.

Results Orientation: We strive to set challenging and competitive goals, focus on output, assume responsibility, and execute flawlessly. To help drive that process, we encourage each other to assume responsibility and to confront and solve the inevitable problems that arise along the way. Our well-known practice of constructive confrontation has served us well in recognizing issues early and dealing with them quickly and efficiently in a problem-solving mode."

In January of 2020 Intel's new CEO explained the transformation that the company was continuing to go through from a PC-centric to a data-centric company [14],

> *"Legendary Intel CEO Andy Grove talked about "strategic inflection points," periods of fundamental change in a business, an industry, or even the world. At such times, he said, clarity about what you uniquely do and why you do it is essential to defining your path forward.*
>
> *The emergence of data as a transformational force is such a moment — for the world and for Intel. We already create technology that enriches lives, serving every industry, every sector of society, and nearly every person on earth, every day. But now nearly everything looks more like a computer and therefore what it means to have "Intel Inside" is changing.*
>
> *....*
>
> *This profound evolution in computing opens a much larger opportunity with implications for every aspect of our business. As Intel's leadership team, we must reimagine the world through this change, and move forward boldly, not constrained by history. This includes the iconic Intel culture that helped define not just our company, but Silicon Valley itself. In some regards, that culture is more differentiating than ever, particularly the value we place on brilliant engineering, integrity, truth, and transparency. But we also know that we must continue to evolve our culture. We must approach every day with a growth mindset.*
>
> *...*
>
> *Our ambition is to play a much larger role in our customers' success."*

In May 2020, the Intel values published on the corporate website went through a major change and were articulated as:

> *"<u>Fearless:</u> We are bold and innovative. We take risks, fail fast, and learn from mistakes to be better, faster, and smarter next time.*

Inclusion: We strive to build a culture of belonging. We create a space where everyone can contribute to their fullest potential and deliver their best work. We welcome differences, knowing it makes us better.

Customer Obsessed: We listen, learn, and anticipate our customers' needs to deliver on their ambitions. Our customers' success is our success.."

One Intel: We appreciate, respect and trust each other. We commit to team over individual success. We are stronger together. Innovators at heart, we bring fun to work every day!.

Truth and Transparency: We are committed to being open, honest, ethical, and timely with our information and feedback. We constructively challenge in the spirit of getting to the best possible result. We act with uncompromising integrity.

Quality: We deliver quality and ensure a safe workplace. We have the discipline to deliver products and services that our customers and partners can always rely on."

Questions:

(1) What differences do you see between the two sets of Intel values?
(2) Why do you think the changes were made?
(3) Results Orientation was a value in the original list. Are those sentiments represented in one of the new values?

2.7.2 *Changing the Culture of an R & D Organization*

Background

Wendy was based in the United States and was the new manager of an R&D organization composed of 300 engineers equally spread across the United States, India and Germany. Wendy's organization was part of a large multinational company called that delivered electronic components to auto companies. Her company targeted luxury brand automobiles, where volumes were low, but profit margins were high. Wendy's group designed high-performance audio components for that market.

In Wendy's company, the components were custom-built for each car model. The non-recurring engineering and manufacturing costs were high but that was more than compensated for by the higher profit margin. The company had won several industry awards for quality for the products that Wendy's engineering group designed. Wendy's employees took great pride in the industry recognition and whenever a significant new award was received, they celebrated with a special event.

The group Wendy led was formed 20 years previously and had very little attrition. Its budget had grown almost every year. Most of the management and senior technical leadership had been in the group from the beginning when they established their set of values. The ones that were most ingrained in employee behavior were:

(1) Uncompromising quality.
(2) The customer is always right. Just say Yes.

Soon after Wendy assumed the leadership of the engineering organization, the company announced record annual revenue and profits. However, senior management also noticed that in the final quarter of the year, automakers started to migrate to suppliers who delivered a complete infotainment system that contained most of the components Wendy's company sold separately. The total cost to the automaker was lower than if they bought the components separately and assembled them themselves. Also, customer surveys had started to show that the quality of the car audio was becoming less important to consumers than the number of functions integrated into the larger infotainment system.

Wendy's company decided to go through a major strategic shift to re-position itself as a company that delivers complete infotainment systems. They also broadened their target market to include a wider range of automobiles from economy through luxury models. The company placed Wendy's group and other electronic component groups in a new system engineering group under a single general manager. The new group's mission was to deliver entire subsystems, like the infotainment system, that addressed the broader market.

The Problem

Although Wendy was convinced of the need to change to align with the new company strategy, her leadership team and employees were not. They did not agree that consumer tastes were changing. They felt they had several promising new technologies in the pipeline that would increase their quality leadership in the industry and felt that is where the focus of the company should be. Finally, since the company was still making a very good profit, they questioned the urgency of the need to act now.

Actions Taken

To create a sense of urgency, Wendy collected publicly available data on market analysis and forecasts, competitor margins. She combined it with relevant company information and the new corporate mission to paint a picture of the need for change for the company and in her organization. She communicated this via internal newsletters, videos, all-hands meetings at each site, email updates and other employee forums. She also held group sessions with her staff and individual one on one meetings with influential leaders in the organization to make her case for change. She set an aggressive goal to decide on and announce all of the changes to the organization needed to align with the new strategy in one month.

In parallel, Wendy tried to identify like-minded leaders in the organization to build a strong coalition of respected managers and individual engineers who could help guide the changes and influence others. She wanted the team to represent a cross-section of the organization. Over time, she hoped the team would expand to include more people from the rest of the organization as details started to be planned. The guiding coalition's first goal was to create a new vision statement, values and strategy for the organization that would align with the company's new strategy of delivering integrated systems.

Soon after it kicked-off, Wendy realized that some of the members of the guiding coalition she formed were not actually in agreement with the need for change. However, she did not replace them since she felt it was important to hear their perspective. Also, some of her direct staff told her that they would not participate because they believed it was more important for them to focus on the execution of their existing engineering work.

Nevertheless, Wendy forged on and ended up writing most of the vision, strategy and values herself. She went on to restructure her group and eliminated or merged some existing teams. However, she tried to give all of her previous direct staff similar roles to what they had before. Ultimately, Wendy announced the new vision, strategy, values and resulting re-structuring exactly one month after she started as she had promised.

The Results

Soon after Wendy announced the changes, it seemed like the momentum had been lost. The guiding coalition stopped meeting. Although people were operating in a new organizational structure, they were still behaving and making decisions as before. In some cases, elaborate informal processes were constructed by employees to work around the new structure and official processes so they could continue to work the way they did before. The result was that after one year operating in the new organization employee morale had reduced and they were still unable to effectively integrate components into the larger infotainment system they were now chartered to deliver.

Questions:

(1) What were the conflicting problems that Wendy faced?
(2) What should Wendy have done when she discovered the members of the guiding coalition were not all on board with the change?
(3) How could Wendy have held her staff accountable for making the changes a success?

(4) What were the repercussions of letting some of her direct staff skip participation in the guiding coalition?
(5) Should Wendy have given all of her direct staff roles in the new organization even though some disagreed with the new direction?

2.7.3 *Changing the Culture at Two Companies*

Background

Jim was the R&D manager of an engineering group in a very large technology company named ABC Inc. Jim's group designed and delivered critical technology for one of the six product groups in the company. There was one product group for each market segment and the other five product groups also had dedicated engineering groups that delivered similar technology designed for their particular needs. The engineers in each group used engineering tools and services from external suppliers to complete most of their work. These tools and services had a major impact on engineers' productivity and were critical for meeting very tight schedules and aggressive product targets. To move fast, each engineering group had the authority to make most of their own decisions which included selecting their suppliers and negotiating the price. This level of autonomy had been part of the company's culture since it started and the managers enjoyed the freedom to make their own decisions. As the company grew this led to redundant work in the different engineering groups but the extra cost and effort were considered worth it to maintain technology leadership in the different markets.

Mary was the R&D manager of an engineering group in XYZ Tools Inc. Her group designed and delivered engineering tools for companies like ABC Inc. Ever since the company was small, success had been built on a high degree of personalized service and making extraordinary efforts to accommodate every customer request was a part of their culture.

XYZ Tools had maintained the number one market share since the company was young but it was at risk of being passed. As a result, the company took a corporate goal to increase revenue by 15% and required each group, like Mary's, to sign up at least 20 new customers for the upcoming year. Employee bonuses were tied to achieving those goals and each group's progress was reviewed regularly by the CEO. One of Mary's primary targets for new customers was ABC Inc. Since each product group was large and acted independently, each one was considered a separate customer by her company and each one could bring in substantial revenue. Mary was willing to invest time and effort upfront in convincing ABC Inc. of the superiority of the technology and services her group provided. Since the six individual engineering groups in ABC Inc. made their own decisions, Mary approached each one separately.

Jim's engineering group at ABC Inc. was the first to meet with Mary and her group. They were impressed enough with the initial discussion to start an evaluation and Mary's team worked incredibly hard to meet the success criteria. After a couple of months of work, Jim's group decided to adopt the technology from Mary's team. The decision was based on the quality of the engineering tools, the service and on Mary's promise to deliver several enhancements that Jim's team requested.

The Problem

As shown in Figure 2.7, Mary's group was soon involved in five additional engagements across ABC Inc. Each one had its success criteria, enhancement requests and decider. Since revenue growth and design wins were the top priority for XYZ Tools, Mary and her team dedicated more and more of the group's resources to those evaluations and developing the new features each group requested. Mary soon noticed that even though each evaluation's success criteria were very similar, the different engineering managers at ABC Inc. still wanted her to repeat the effort for their

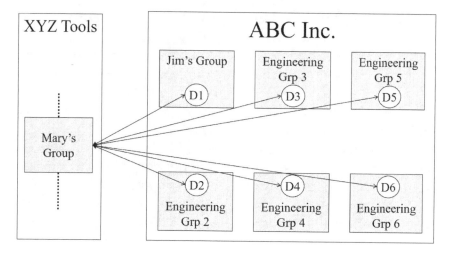

Figure 2.7.   ABC Inc. and XYZ Tools current state.

group. As a result, most of the year was going to be spent on supporting multiple redundant evaluations across ABC Inc. Also, all of the new features requested by the different groups were slight variations of each other. Accommodating all of the customized changes to their engineering tools caused their technology to resemble a Swiss army knife with each new feature increasing the technical debt that XYZ Tool engineers had to carry forward and validate with every subsequent release.

The complexity of validating the increasingly complex tool technology grew to the point where it started to impact the quality and reliability of the technology Mary's group was providing. Also, the effort to support multiple evaluations and develop a large number of customized features for each group siphoned resources from the support needed by the newly acquired customers in ABC Inc. Mary recognized that this was unsustainable and something needed to change. In parallel, Jim noticed that his group's requested enhancements were getting delayed and the service was declining. He felt that the underlying technology and services provided by XYZ Tools would provide them a competitive advantage but it was clear that something needed to change.

Actions Taken

In their next meeting together, Mary and Jim discussed the current state and agreed on the problem. They put together a proposal that they thought would benefit both companies. The proposal focused on optimizing the XYZ Tools Inc. technology and services for all of ABC Inc. and not the individual engineering groups. This meant centralizing the engineering feature decisions and issue management in one central body in ABC Inc. as shown in Figure 2.8. This required the ABC engineering groups to consolidate, filter and merge requests and speak with a single voice to Mary's group. This eliminated the redundant evaluations and feature enhancement requests so Mary's group could dedicate more time to quality control and user support.

The Results

The proposal required changing the culture of both companies in specific areas. It depended on other engineering managers in ABC Inc. relinquishing some of the autonomy they had and agreeing

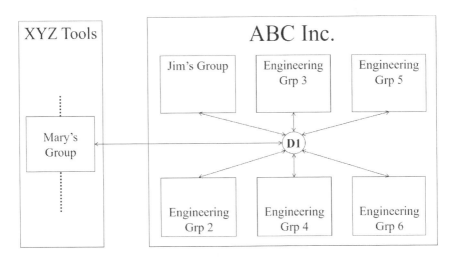

Figure 2.8. ABC Inc. and XYZ Tools proposed state.

to follow the decisions made by the central decision-making body. It also required that engineering groups in XYZ Tools resist the temptation to say yes to every request and, instead, work with the central decision body to converge on what would and would not be done. Instead of instinctively saying "yes" to every request, Mary's employees would sometimes need to say "no".

Both Mary and Jim called meetings with their respective management to outline the proposal. In both cases, they were met with stiff opposition. At ABC Inc., the other engineering managers felt that it impinged upon the autonomy they needed to act fast and enable their own unique engineering needs. They felt the quality problems with XYZ Tools were not their problem and ABC Inc. should instead look for an alternative supplier. At XYZ Tools Inc. Mary's management felt that by consolidating all of the decision making into one central body at ABC Inc. they will lose the opportunity to cherry-pick individual groups and gain some sales even if other groups at ABC Inc. decided not to use their technology. As a result, in both ABC Inc. and XYZ Inc. the proposal was rejected and Mary and Jim were forced to go back to the drawing board.

Questions:

(1) If the proposal was implemented it would likely have a short-term negative impact on both companies and a longer-term positive impact. Explain the negative and positive impacts that could be expected.

(2) How should Jim address the concerns about the loss of autonomy by the ABC Inc. engineering groups?

(3) How should Mary address the XYZ Tool's concerns about losing the opportunity to cherry-pick groups to target?

References

[1] Grove, A. S. (1995). High Output Management, 2nd Ed. (Vintage Books) p. 147.

[2] Intel Corp. *At Intel, Our Values Define Us*. Available from: https://www.intel.com/content/www/us/en/corporate-responsibility/our-values.html [May 2020].

[3] Lashinsky, A. (2009) *The Cook Doctrine at Apple*. Fortune, Jan 22, 2009.

[4] Liker, J. and Hoseus, M. (2008). *Toyota Culture — The Heart and Soul of the Toyota Way*. (McGraw Hill, New York) pp. 14–15.

[5] Enron Corporation (2000). Enron Annual Report 2000.

[6] Gibbs, L. and Pachetti, N. (2002), The Lessons for Investors; Hindsight, Shmindsight. There's much to learn when a Stock loses $67 Billion in Value, Money Magazine. January, 2002.

[7] Bloomberg. (2001) Enron: Let us count the culprits, Business Week December 16, 2001.

[8] Wells Fargo (2020). *Our Code of Ethics and Business Conduct*. Available from: https://www08.wellsfargomedia.com/assets/pdf/about/corporate/code-of-ethics.pdf [June 2020].

[9] NPR (2016). *Former Wells Fargo Employees Describe Toxic Sales Culture, Even At HQ*. October 4, 2016. Available from: https://www.npr.org/2016/10/04/496508361/former-wells-fargo-employees-describe-toxic-sales-culture-even-at-hq [May 2020].

[10] Mclean, B. (2017). *How Wells Fargo's Cutthroat Corporate Culture Allegedly Drove Bankers to Fraud*. Vanity Fair Magazine. May 31, 2017.

[11] Shapiro. J. (2018). *Attorney General Shapiro Announces $575 Million 50-State Settlement with Wells Fargo Bank for Opening Unauthorized Accounts and Charging Consumers for Unnecessary Auto Insurance, Mortgage Fees*. Office of the Attorney General, Commonwealth of Pennsylvania, December 28, 2018.

[12] Kotter, J. P. (1998). *Harvard Business Review on Change*, Chapter 1 "Leading Change: Why Transformation Efforts Fail. (Harvard School Press) pp. 1–20.

[13] Kotter, J. P. *8-Step Process*. Available from: https://www.kotterinc.com/8-steps-process-for-leading-change/. [May 2020].

[14] Intel Corp. *Intel 2019 Annual Report,* Available from: https://s21.q4cdn.com/600692695/files/doc_financials/2019/2019-Annual-Report.pdf [May 2020].

CHAPTER 3

Diversity and Inclusion

Companies that embrace diversity and inclusion in all aspects of their business statistically outperform their peers.

— Josh Bersin [1]

In the previous chapter, we discussed the critical role that culture plays in the R&D organization. We feel it is important to devote a chapter to an important topic heavily dependent on the culture. That is, creating an environment of inclusion that benefits from a diverse workforce.

Corporate diversity and inclusion can add value to the products a company sells. Similarly, a diverse workforce and inclusive culture can increase the value of the engineering R&D organization's product. The performance of the overall R&D system is also improved when the diversity of the R&D organization encompasses the diversity of its customers and suppliers.

Whereas diversity is measured by the variety of the workforce, inclusion is measured by the perceptions of the individuals within the workforce. While there are legal requirements for diversity, meeting the legal requirements and creating policies and procedures will only create an inclusive environment when the culture supports inclusive values. The manager must embody empathy and role-model inclusive behaviors. The manager has multiple types of communication, and relying upon one form for all cases is a lost opportunity. Group communication is needed to set tones and individual communication is necessary for sensitive exchanges. The manager must be flexible, but in all cases, must make sincere efforts to ensure that each member of the workforce feels truly appreciated.

Finally, diversity definitions and goals vary with time while inclusion varies with individuals. What is acceptable today may be incorrect tomorrow. As a result, it is important that the R&D manager continuously learn and improve.

3.1 Applying Diversity and Inclusion

We include diverse perspectives in the chapter on diversity and inclusion.

Although we have read a great amount of material and we have each had experience as managers attempting to create a diverse and inclusive workplace, we know that a key requirement is to listen to individuals. In keeping with that thought, we interviewed many diverse individuals from the engineering and academic community. The interviewees were informed that we were writing a book on engineering management. The interviews were one-on-one. Some interviews were in person; some were by phone. The basic structure of the interviews started with an open time where the interviewee could say anything they wanted about diversity and inclusion, and we stayed quiet and took notes. Then they were asked four basic questions:

(1) "Have you ever experienced being left out or ignored?"
(2) "What bad experiences have you had?"
(3) "What good experiences have you had?"
(4) "What can managers do?"

We have used their input for the content of this chapter.

3.2 Diversity, Inclusion and Value

Diversity and inclusion lead to greater value; the R&D system is diverse.

A diverse workforce provides multiple perspectives that can enhance the R&D organization's performance. An obvious case where diversity provides insight that improves the R&D

organization's product is the engineering of the user interface of technology. Incorporating ideas based on a wide variety of user experiences will lead to a more intuitive solution that is easier to use for a broader customer base. Another important part of diversity and inclusion that leads to greater value is customer relations. Managers often attempt to loosen up with informal comments and behavior before the R&D people get down to intense technical discussions with the customer. This is good, but managers need to listen to their workforce so they know what informal actions might be viewed as slightly offensive to the customer. These may not be offensive enough to ruin the discussion but are not productive. The manager may never know the customer was offended unless someone on their staff understands the customer's customs and beliefs enough to point out problems. As a side note, one universal rule that was provided by one of the interviewees was: "no jokes." When other interviewees were asked about this, they also confirmed that it was a good rule to follow.

3.3 Focus on the Individual

Diversity means including different types of people and their opinions.

When addressing a problem or situation the first step is a clear statement of the situation. Dr. Charlotte Acken often quoted Charles Kettering in her math classes and stated, "A problem well stated is a problem half solved." This is true in math, engineering, and life in general. With this in mind, the dictionary [2] definition of diversity is:

> *1: the condition of having or being composed of differing elements: VARIETY*
>
> *especially: the inclusion of different types of people (such as people of different races or cultures) in a group or organization*
>
> *2: an instance of being composed of differing elements or qualities: an instance of being diverse*

A diverse workforce can be measured by numbers and percentages, and while that is quantifiable, it is not the goal. The real goal is the inclusion of all of the members of the diverse workforce.

There are several features of diversity in the workplace. The obvious cases of diversity are different genders, races, religions and cultures. Less obvious, but often considered, are belief systems. Some ignored cases include disabilities and special needs. Diversity can be blatantly or subtly discouraged. Blatant examples may include hiring practices that prevent diverse candidates from being sought or hired or by undermining existing diverse employees. A subtle way of ignoring them is to acknowledge the presence of people with a diverse background but never take action on their ideas or comments. As noted earlier, diversity is not just a statistical goal, it can lead to a competitive advantage when it is coupled with an inclusive culture. Creating a diverse and inclusive workforce starts with the R&D manager's belief that it is the right thing to do.

Inclusion requires a conscious effort to ensure all individuals are heard, respected and treated with empathy.

The dictionary definition [2] of inclusion is:

> *the act of including: the state of being included*
>
> *...*
>
> *the act or practice of including students with disabilities with the general student population*
> *the act or practice of including and accommodating people who have historically been excluded (as because of their race, gender, sexuality, or ability)*

The dictionary definition goes on to provide examples,

> *"Academic libraries have traditionally struggled to address problems of equity, diversity and inclusion. The low representation of people of color in library staff has been a particular*

shortcoming, despite many initiatives to attract minority staff to the field."
— Lindsay McKenzie

"Tech workers say they are more interested in diversity and are more willing to work to promote inclusion in their workplace ..."
— Jessica Guynn

"Meaningful civic inclusion even now eludes many of our fellow citizens who are recognizably of African descent."
— Glenn C. Loury

True inclusion requires a level of trust on the part of all. It also requires a conscious effort to treat everyone as individuals, not just as symbolic members of a group. For example, while an employee has some experiences and viewpoints similar to some other people of the same ethnicity, they are still individuals. The R&D manager must know enough about the individual to be able to truly include them. Based on our interviews we found two opposing views held by some underrepresented individuals. Some said they did not want to be expected to speak for everyone like them. They said the managers must learn from more general sources, and let the individuals speak for themselves. Others said, yes, they were willing to speak for their group because then, at least, management gets input, and as long as that input is not ignored or misused, then they will speak up for their group. This emphasizes the point that a diverse group of people does not behave as a single monolith. Individuals will respond differently and it is dangerous to assume one way of thinking for an entire group of people.

The manager must provide the leadership needed to create a culture of inclusion. Policies and practices will not in and of themselves create an inclusive climate. The manager of an R&D organization accomplishes this with both personal interactions and public behavior.

Diversity is not just a statistical goal.

A diverse workforce is not just a mixture of genders, races, cultures, religions, and abilities. A diversity of experiences and viewpoints is the key to diversity. A person selected from an underrepresented group who has similar viewpoints to the dominate group, or whose differences of viewpoint are ignored, will provide numerical diversity only in a token way. A person added to the R&D organization's work environment to create diversity only adds value if their contribution is enabled by the culture and harvested by the organization.

Role models are important.

An important resource for everyone's career is a role model. The perception of a lack of diversity within an R&D organization is often the result of a lack of diverse role models. One interviewee said:

> *"I was not seeing women in senior management — except administrative and HR positions. All men in leadership are engineers, which creates a competitive environment for the women engineers- and a distance between women."*

It takes a special person to overcome the lack of similar role models. A senior executive gave this quote:

> *"I have managed global teams. My perspective is I am held accountable as a professional. Doesn't matter who or what I am. I don't want to talk about diversity and I didn't join such groups. I was generally the only female in the room. Picture 40 executives and I am the only women. Then about one and a half year ago I was told by the CEO that I needed to include diversity and inclusion goals to HR. OK, I need to get out of the mode that diversity and inclusion is just delegated to HR."*

This realization from both of these quotes is that for anyone (these examples are female engineers), role models are a valuable means for an R&D organization to get the most out of all of the employees. Employees are often motivated to pursue greater achievements by seeing successful engineers just like themselves.

Recently, at Portland State University (PSU), a survey of undergraduate women was taken to identify why so many switch majors. The most common feedback was the lack of female professors. So, the existence of diversity is important in and of itself. However, as another interviewee said, it is not enough:

> *"Diversity and inclusion are inextricably linked. Diversity without inclusion is tokenism."*

Inclusion is measured from the perspective of the person being included

The success of policies and procedures implemented to support inclusion must be measured from the perspective of the person being included. A key element to inclusion is perspective. From the manager's perspective, everything may appear to be fine for the employee. From the employee's perspective, they might feel completely left out. A simple example of this is a manager who holds a luncheon with a group of diverse employees but always checks their cellphone during lunch. Another example is a manager who asks for employees' inputs during the lunch but does not take any notes, or never addresses the comments in the future. The key for a manager is to focus on what the individual's perspective might be, rather than evaluating that perspective. As Mike Fromowitz [3] advised,

> *"Stop and think about how many different cultures you come into contact with at work. Our world's diversity is what makes marketing so fascinating. When you take time to understand this diversity, you show respect for other people's cultures. Remember, your audience is more than a collection of stereotypes and anecdotal evidence — you need to understand*

how they speak and what they care about to reach them on a personal level. If you do, they won't just remember your ad; they'll reward your brand with lifelong loyalty."

Modifying this from advertising to management, the manager of the R&D organization has it a bit easier because they know the makeup of their workforce. One "Aha" here, of course, is that the manager must know the makeup of their workforce. The key then is to get input from the employees themselves on how to become more inclusive. This step is often missed by the manager who believes they already know what to do. As a result, the culture of the workplace requires the trust of the employees to feel free to communicate their perspective. The manager will reinforce trust by acknowledging the input and seriously considering the appropriate choices. That not only includes avoiding automatic rejection of ideas but also automatic acceptance of such ideas. The input must be analyzed and utilized appropriately.

3.4 Policies and Culture

Policies will not create inclusion without an inclusive culture.

The manager must create the proper policies and procedures for an inclusive work environment and the management decisions regarding task assignments and priorities will also encourage inclusive behavior. However, to be successful the manager must demonstrate leadership. The manager must publicly include all of the employees in their interactions. And just as important, their actions must not overlook misbehavior. Two of the interviewees gave specific examples of where they reported improper behavior on the part of someone, and very soon afterward, that person was promoted. This sent a very clear message to them not to waste their time reporting misbehavior and to also leave that organization as soon as practical. On the positive side, three interviewees gave specific examples of managers going out of their way to help smooth and facilitate the person's success. These were specific cases of managers recognizing that policies and procedures did not substitute for proper and helpful behavior. In fact, policies and procedures sometimes got in the way.

Meeting the legal requirements for diversity is not enough.

The benefits of diversity do not automatically result in management and others embracing diversity. As a result, corporate policies supporting diversity are also driven by legal requirements. As an example, several US laws directly affect diversity policies. This list is a sample of the relevant laws:

- Title VII Civil Rights Act (1964)
- Pregnancy Discrimination Act
- Americans with Disabilities Act (1990)
- ADA Amendments Act
- Age Discrimination in Employment Act (1969)
- Equal Pay Act (1963)
- Employment and Reemployment Rights Act
- Civil Rights Act (1991)
- Rehabilitation Act (1973)
- Genetic Information Nondiscrimination Act (2008)

As one can see, there are many variations in local and international laws related to diversity. However, for the policies supporting the legal requirements to work, they require that the culture support the values underlying those policies. Specifically, from a managerial viewpoint, the benefits attributed to diversity must be understood beyond the simple adherence to the law. For diversity to add value to the R&D organization, the manager must internalize the importance of inclusion and not just treat it as a checkbox to fulfill a legal or policy requirement. All of the people interviewed for this chapter stressed that diversity and inclusion are not just a checkbox or the management topic of the year. Just as the intention to do right thing with respect to diversity does not replace its legal requirements, so too, the meeting of the legal requirements is not a substitute for real inclusion.

Managers need to be aware of their natural blindness to biases.

As David Rock and Heidi Grant Halverson wrote regarding bias [4],

> *"These are examples of common, everyday biases. Biases are nonconscious drivers — cognitive quirks — that influence how people see the world. They appear to be universal in most of humanity, perhaps hardwired into the brain as part of our genetic or cultural heritage, and they exert their influence outside conscious awareness. You cannot go shopping, enter a conversation, or make a decision without your biases kicking in.*
>
> *On the whole, biases are helpful and adaptive. They enable people to make quick, efficient judgments and decisions with minimal cognitive effort. But they can also blind a person to new information, or inhibit someone from considering valuable options when making an important decision."*

Also, as Scopelliti et al said [5],

> *"People exhibit numerous systematic biases in judgment (Tversky and Kahneman 1974, Kahneman et al. 1982, Nisbett and Ross 1980), many of which are due to unconscious processes (Morewedge and Kahneman 2010, Wilson and Brekke 1994). A lack of conscious access to judgment-forming processes means that people are often unaware of their own biases (Nisbett and Wilson 1977) even though they can readily spot the same biases in the judgments of others. Consequently, most people tend to believe that, on average, they are less biased in their judgment and behavior than are their peers. Most people recognize that other people are likely to be biased when judging an attractive person, for example, but think that their own judgment of an attractive person is unaffected by this type of halo effect. Because the majority of people cannot be less*

biased their peers, this phenomenon is referred to as the bias blind spot (Pronin 2007; Pronin et al. 2002, 2004)."

As stated previously, a problem well stated is a problem half solved. The first step is to acknowledge the problem. It is common to avoid admitting to a problem that is preventing inclusion rather than addressing it. One problem that prevents inclusion is racism. Robin Diangelo gives two categories of racism denial that are based upon the speaker claiming to be a good person therefore not a racist. One category is I am "color blind," which includes the following excuses, some of which are from her book [6]:

- I was taught to treat everyone the same.
- I don't see color.
- Everyone struggles, but if you work hard...
- I was picked on because I was white/I was poor (so I don't have white privilege).
- I don't care if you are pink, purple, or polka dotted.
- Race doesn't have any meaning to me.
- So-and-so happens to be black, but that has nothing to do with what I am about to tell you.
- Focusing on race is what divides us.
- If people are respectful to me, I am respectful to them, regardless of race.
- We are from the North; racism is a Southern problem.
- I am a liberal, conservatives are racists.
- I joined the NAACP, so I am not a racist.
- I contribute to the ACLU, so I am against racism.

The other category Diangelo gives is what she calls color celebrate, which is to embrace the difference as a credit. Some of the following examples are from her book:

- I have a friend who is Black.
- My brother-in-law is Black.
- I was a missionary in Africa.

- I was in the Peace Corps.
- I marched in the sixties.
- We adopted a child from China.
- I went to a diverse school/I lived in a very diverse neighborhood.
- I lived in Japan and was a minority, so I know what it is to be a minority.
- I live among [fill in the blank] so I am actually a person of color.
- I have a Native American ancestor.
- I send Christmas cards to Black people.
- My school was integrated.
- I can speak Spanish.
- I hire minorities for my yard/house work.
- I invite minorities to my parties.

One of our interviewees said there is a sociological term "conditioned oblivion" in which people cannot see their own privilege or bias because they have been conditioned to think that is the normal situation. The key learning is that the manager must not allow a self-evaluation to result in excuses or defenses against promoting inclusion. A person who thinks they are already doing great will not seek to improve. Therefore, a person must acknowledge their weaknesses first and then work to fix the situation. The goal is not to eliminate these types of biases, but to recognize them and eliminate the negative consequences.

3.5 Diversity and Inclusion Over Time

Diversity definitions vary with time, inclusion varies with individuals.

While there are many practices that organizations use to attempt to create diversity, there is not a standard set of practices that all managers can perform to create diversity. Also, the very definition of diversity changes with time. The managers must keep abreast of the world around them and their employees to ensure continuous improvement with respect to diversity and inclusion. The terms and solutions for diversity today are not the same as they were in the past nor will they be the same in the future.

As stated before, inclusion is based upon an individual's perception. Therefore, inclusive effort must be individualized. Some people feel included via public recognition; others feel included with private feedback. Also, what is perceived to be respectful and inclusive by one person or in one situation may be perceived to be the opposite to another person or in a different setting. In the 1950s, many people considered the term "Negro" to be respectful and inclusive (as evidenced by Dr. Marin Luther King's use of the word in many of his speeches). Today, most people consider the term, "Black" or "African American" to be more respectful and inclusive. Even in that case, many believe one of those words (or different variations of them) is more appropriate than the other. Therefore, a person wishing to be respectful and inclusive in 2020 will address someone differently than someone trying to be equally respectful and inclusive in the 1950s. This is where the manager needs a general attitude of sensitivity. As has been stated in this chapter in many places, the R&D manager must treat people as individuals. Find out what the person prefers. This was described to us first hand when discussing gender pronouns. One woman in a relationship said her partner insisted on the use of, "they", but that she usually used, "she". The correct term is very individual. One of our interviewees, a professor of sociology, also has the following rule, "No Jokes." The rule recognizes that what one person sees as funny hurts other people, and this changes with time.

The bottom line is that the R&D manager must know their employees and invest time in learning more about them and society. Above all, they must treat each person as an individual. One of the people interviewed gave some pretty clear and reasonable ideas for managers:

"— *Day to day treat people decently.*
— *Reassure people.*
— *Be sure to send a message to jerks and obnoxious people.*
— *formal position matters — group stuff is BS.*"

And more specifically, she said:

"1. *Just shut up and smile, don't ask stupid questions.*
2. *Pay attention.*
3. *Figure out what you need to do.*"

What attitude should a manager have if they follow her advice? Joe Torre was a very successful manager of the New York Yankees baseball team. He explained his success with a highly talented and diverse group of athletes in an interview with Fortune magazine [7]:

"I try to understand what motivates other people," Torre was explaining, pondering a bowl of oatmeal at a Manhattan restaurant two days into the new season. At 60, the dark features that seemed so menacing to opposing pitchers during his playing days have become the stock image of Yankee power–controlled, yet suggestive of deep water just beneath. "Some players may be critical of a decision I make, but I'm more into 'Why did they say it?' as opposed to what they said."

So, the R&D manager must listen, recognize differences in perspectives, decide what the right thing to do is, and then do it.

3.6 Proper Communication

Communication is essential for true inclusion.

Managers have many methods of communication available to them. Each method has appropriate applications and uses. One method is communicating to a big audience via a large group event or presentation. However, as one interviewee pointed out,

> *"I am sick and tired of people who want to throw a party and declare victory. Or people who target a few well-known groups. I am disappointed at looking at people group by group. Everyone deserves respect. And no one's civil rights depend upon the understanding or awareness of others."*

This has several instructions for the manager. First, a person has civil rights whether or not the manager understands or is aware of them. Second, a big diversity event is not a solution by itself. Third, do not lump people into groups, the key is to treat people as individuals. However, the most important message is that the manager's attempts at understanding and awareness, even if well-intentioned, do not excuse the need for corrective actions.

Communication needs to be bidirectional. One way of accomplishing this is via small staff meetings. All but one of the people we interviewed said they had often experienced being ignored in such meetings. A couple of people mentioned that when they made a point, it was ignored until a male colleague made the same point. Since the manager is in control of the meeting, they must be sure that points made by all are heard and acknowledged. Another means of bidirectional communication is a one-on-one meeting. This is a good place for the manager to listen and hear the person's specific views. This does not mean that the manager should interrogate them on what they feel or think. It should be a conversation where both parties participate and the R&D manager listens to the input from the employee.

Modern communication utilizes many mediums: email, texting, telephones, video conferencing, and more. Each can be very useful in the right situation. However, sometimes the most convenient method may create more problems than it solves. For example, email is a very convenient method of communication because one can send a message when it is convenient for the sender and the receiver can read it when it is convenient for them. However, for sensitive topics, email can lead to misunderstandings. In fact, this is true for most topics, a misreading can lead to a spiral of misunderstanding and hurt. Remember that inclusion is measured by the person to be included, not the manager trying to create an inclusive environment. When the manager hears of any misunderstanding, they must listen rather than defend or explain.

Also, addressing diversity isn't just about discussing diversity. One of the people interviewed said too much talking about diversity just splits people. Specifically, their complaint was "lots of talk and no action." On the other hand, another person interviewed pointed out how many people shy away from talking about race. She pointed out that some people use the statement "You cannot change history" as a cop-out to avoid discussing race. For a manager, this means dealing with individuals. Specifically, the interviewee said, "People in power must acknowledge their privileged position and allow a person to talk." One person said that "when you say you don't see color then you are saying you don't see me".

3.7 Summary

As we quoted under the title to this chapter, *"Companies that embrace diversity and inclusion in all aspects of their business statistically outperform their peers."* Therefore, managers must set the policies, goals, and procedures to create a diverse workforce. However, diversity by itself is not enough. Employees must also be included, and they must feel included. Diversity and inclusion are inextricably linked. Diversity without inclusion is tokenism (wasting a precious opportunity). Whereas diversity is measured by the variety of the workforce, inclusion is measured from the perspective of the person being included. As a result, the R&D manager must utilize the leadership skills to drive a culture of inclusion. This is a never-ending task. One interviewee said she could think of no examples of a good manager when it came to inclusion.

While many diversity and inclusion challenges seem difficult for an R&D manager, there are some simple things to keep in mind. Always, treat people as individuals, listen to their input, include them in decisions and live the value of inclusion. The R&D manager is still ultimately responsible for making key decisions but the decisions should role-model an inclusive behavior. The R&D manager is also responsible for reinforcing the right behavior in the organization. If misbehavior is occurring, the R&D management must put a stop to it. This also includes when their own words or actions are pointed out as offensive to someone. This makes a big difference to both the individual and the organization and shows that the R&D manager takes diversity and inclusion seriously.

3.8 Key Points

- Diversity and inclusion improve the performance of the R&D organization and the R&D system.
- Diversity means including different types of people and their opinions in the R&D organization.
- Inclusion is measured by the person being included.
- Inclusion requires empathy and a conscious effort to ensure all individuals are heard and respected.
- Policies and legal requirements will not create inclusion without an inclusive culture.
- Diversity definitions vary with time, inclusion varies with individuals.
- The R&D manager must be aware of their natural blindness to biases and continuously learn and improve.

References

[1] Bershin, J. (2019). *Why Diversity and Inclusion Has Become a Business Priority*. Published December 7, 2015. Updated March

16, 2019. Available from: https://joshbersin.com/2015/12/why-diversity-and-inclusion-will-be-a-top-priority-for-2016/ [May 2020].

[2] Merriam Webster. Available from https://www.merriam-webster.com/ [May 2020].

[3] Fromowitz, M. (2017). *Hall of Shame: More multicultural brand blunders.* Available from: https://www.campaignlive.com/article/hall-shame-multicultural-brand-blunders/1423941 [May 2020].

[4] Rock, D. and Halverson, H. G. (2015). *Beyond Bias, Neuroscience research shows how new organizational practices can shift ingrained thinking. Strategy + Business,* Available from: https://www.strategy-business.com/article/00345?gko=ed7d4 [May 2020].

[5] Scopelliti, I., Morewedge, C. K., McCormick, E., Min, H. L., Lebrecht, S., and Kassam, K. S. (2015). Bias Blind Spot: Structure, Measurement, and Consequences, Management Science 61(10):2468–2486. Available from: https://doi.org/10.1287/ mnsc.2014.2096 [May 2020].

[6] DeAngelo, R. (2018). *White Fragility.* (Beacon Press, Boston).

[7] Useem J and Munoz, L. (2001). A Manager for All Seasons Joe Torre gets the most out of his workers, makes his boss happy, and delivers wins. He may be the model for today's corporate managers. And he's not afraid to cry, Fortune Magazine. April 30, 2001.

CHAPTER 4

Value, Waste and Velocity

Understanding value exposes waste. Eliminating waste leads to greater velocity.

— **Robert J. Aslett**

Employees operate more effectively when they understand what constitutes value and how their work contributes to it. This helps them identify tasks that have value and those that do not. When coupled with a culture that motivates everyone to continuously learn and improve, a clear understanding of value enables everyone to reduce or eliminate time spent on wasteful tasks and increase the time spent on tasks that have value.

Generally speaking, an engineering task has value if it contributes to the company's vision in a way that customers will pay. However, determining who the customer is and what constitutes payment is not always straight forward. This is especially true for R&D organizations that deliver to customers in the same company. Also, all value is not equal. Decisions need to consider the entire R&D system and what provides the greatest value from the customer's perspective.

There are well-known sources of waste identified in product development and manufacturing that can act as a guide when analyzing value streams and workflows to eliminate waste. It is also important to keep in mind that management is a necessary waste and poor management behavior can also become a major source of unnecessary waste.

The rate of deployed value vs. time is what we call velocity. Continuously increasing velocity is required to enable the pace of innovation required to thrive. Although they often complement each other, greater velocity does not necessarily equate to greater efficiency, productivity, quality, or lower costs (or vice versa). In fact, it makes sense in some cases to temporarily accept less efficiency and productivity or more costs to speed deployment of value when the technology or market is new. These are the type of trade-off discussions that need to be directed by the manager of the R&D organization spanning the entire R&D system. It all starts with gaining agreement on what is value.

4.1 Value

The vision and what the customer will pay determines value.

As we noted, an engineering task has value if it contributes to the company's vision in a way that the customer will pay. Since the engineering R&D organization may deliver output directly to a paying customer or another internal entity the meaning of a paying customer varies. For example, most engineering work creates design information needed for someone else to build and then sell the technology. In that case, an engineering task has value if it produces the design information needed to build a technology that has value. With that in mind, we use the following definition:

> *An engineering task has value if it contributes to the company's vision in a way that customers will pay.*

Engineering tasks span strategic planning, engineering research, and development. Strategic planning includes tasks in the never-ending cycle of observation, orientation, decision and action we will discuss in Chapter 6. Development includes tasks in the iterative cycles of planning, modeling, prototyping, analysis, optimization and testing to converge in a final design. We will discuss this more in Chapter 7. Engineering research includes tasks related to the creation of new knowledge and more powerful modeling, analysis and optimization methods to solve engineering problems. We will discuss this more in Chapter 8.

A customer is an entity that receives the output of the engineering work. The engineering R&D organization may deliver directly to a paying customer that is external to the company. In that case, the customer gives a clear indication of the value of the engineering output by paying for it. The customer may be another entity within the same company. For example, the customer that receives the engineering output may be the company's factory that must manufacture the technology or another engineering group that uses it for its work. In these cases, the value of the engineering output may be determined by the consuming internal entity, but in practice, it is often more complicated than that. Similarly, the customer of the engineering output could be a company initiative with no immediate customer or is done for the benefit of humanity. The discussion about value can be even more complicated in that scenario.

The company's vision is an aspirational view of the ideal state it is seeking. It may also include a mission that is a more actionable description of what is being done to achieve the vision. Although the company's financial results are an essential part of its success, the vision often goes beyond the description of financial results. For example, in 1909, Henry Ford articulated the decision to manufacture only one model of the automobile (the Model T) in the following way to employees [1]:

> *"I will build a motor car for the great multitude. It will be large enough for the family but small enough for the individual to run and care for it. It will be constructed of the best materials, by the best men to be hired, after the simplest designs that modern engineering can devise. But it will be so low in price that no man making a good salary will be unable to own one — and enjoy with his family the blessing of hours of pleasure in God's great open spaces."*

Henry Ford called this the "universal car". There is no mention of financial results in this vision. Today, it is common that companies include a desire to meet environmental or societal goals as part of their vision. Some shareholders may be attracted to a company based on those considerations in addition to the earnings per share. Other shareholders may be attracted by the long-term potential described by the vision. In that case, short term financial results and customers may not be as important as their belief in the vision. That is why we expanded the usual description of value to include "contributing to the vision of the company" in addition to "in a way that the customer will pay".

Bearing this in mind one can construct a simple 2 × 2 matrix like the one shown in Figure 4.1 as a guide. As we show in the 2 × 2 there are four possibilities: existing value, potential value, unaligned value and waste.

Existing Value: If an activity contributes to the company's vision and the customer will pay for it then it has clear existing value. However, since an organization's resources are finite, something with clear existing value still might not be pursued in favor of something of greater value.

Potential Value: If activities align with the company's vision but no customer is willing to pay for them it has potential value. The key question then is whether or not there is a sponsor within

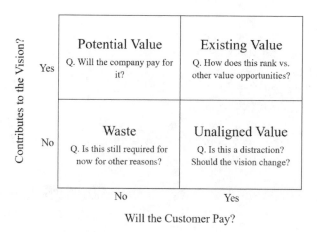

Figure 4.1. Types of value.

the company willing to fund the activity. This often is the case for research that has yet to prove itself but is funded as part of a corporate research budget.

Unaligned Value: If a customer is willing to pay for an activity but the result is not aligned with the company's vision then it is what we call unaligned value. The question to ask in that case is whether the effort will distract from delivering the vision. If so, it may not be worth pursuing. On the other hand, if the customer is exposing an opportunity that the company's vision or strategy is missing then it is also important to evaluate whether the vision needs to be updated.

Waste: If no customer is willing to pay for something and it is not aligned with the company's vision then it is called waste. However, as we will see later, just because it is waste does not mean that we can drop it all together since it still might be required for other reasons.

As an example, let us refer back to the vision for the Model T described by Henry Ford in 1909 and note the type of engineering tasks that fall into the upper right quadrant of Figure 4.1 (existing value). In addition to the target price set at $500 per automobile, there were several attributes that Henry Ford felt consumers would be willing to pay for [2]. Henry Ford described those as:

- Quality in material to give service in use.
- Simplicity in operation.

Table 4.1. Example of engineering activity value vs. consumer.

Consumer	An engineering activity creates value if...
Paying customer	The engineering output is aligned with the vision of the company and the customer will pay for it.
Internal entity	The engineering output enables other engineering or development organizations to deliver value.
Company objective	The engineering output enables a corporate strategy or objective and is funded.

- Power in sufficient quantity.
- Absolute reliability.
- Lightness.
- Control.

A sample of technology innovations for the Model T Ford [3] that contribute to the listed attributes include:

- Adoption of lightweight, high tensile strength steel alloy (Vanadium steel) that was used, "in many of the Model T's critical highly stressed parts including the crankshaft, forged front axle, and wheel spindles."
- Removable cylinder head, and cylinders. "Before the T, car engines typically had their heads and blocks cast as a single lump of iron, with separate cylinders, bolted to the crankcase, making them heavy and time-consuming to produce or repair."
- Integrated ignition-system magneto into the engine flywheel. "This setup eliminated the storage battery and helped ensure a car that started reliably with its hand crank, anytime."
- Fully enclosed powertrain. "The Model T featured a one-piece steel cover enclosing the underside of its engine, flywheel, transmission and universal joint. This cover dramatically helped contain the lubricants within and kept the car cleaner than many of its contemporaries."
- Three-point suspension for better use in rough terrain. "The three-point arrangement gave ample axle articulation on rough terrain while permitting the chassis to flex without twisting the engine."

These technologies were deemed to be required to meet the vision and attract the targeted consumers (based on what Ford thought the customer would pay for). Therefore, engineering tasks that were required to develop those technologies have clear existing value.

However, there were also some engineering activities that fit the company's vision but for which the customer may not pay.

This may include the many experiments that were pursued to explore new concepts that failed to materialize. These activities created useful knowledge in pursuit of the vision but customers would not necessarily pay for them. In these cases, the company's management demonstrates its belief in the potential value by making a conscious decision to fund the activities. In those cases, the customer was effectively the company itself. These tasks would meet our definition of potential value that occupies the upper left quadrant of the 2 × 2 shown in Figure 4.1.

There may be other engineering activities that do not align with the company's vision but for which the customer will pay. For example, by 1909 the Ford automobile company was producing several models. They were all making money for the company but only the new Model T aligned with the newly articulated vision. To avoid distractions and make it clear what the focus of the company needed to be, Henry Ford stopped production of all of the other models. So, in that case, the management decided very clearly that something that a customer was willing to pay for was not sufficient was worth producing since it did not fit the vision. These types of activities occupy the bottom right of the 2 × 2 in Figure 4.1.

Like the Ford example, when Steve Jobs returned to Apple in 1997, Apple was swimming in products and there were more than 70 Macintosh Performa (computer) models released between 1992 and 1997 [4]. All of the products were contributing some revenue but were diluting the vision of the company. They were also an example of what we call unaligned value. Steve Jobs recognized this and immediately eliminated 70% of new projects and reduced the number of product lines from 15 to 3. One year later, the iMac debuted.

As we shown in Chapter 2, this thinking was articulated well by Tim Cook in 2009 [5] when he said.

"We believe in saying no to thousands of projects, so that we can really focus on the few that are truly important and meaningful to us."

Of course, as we noted earlier, in the lower left of the 2×2, there are engineering activities that do not align with the vision and for which no customers will pay. We call that waste and in Henry Ford's assembly line, Tim Cook's Apple or any organization, there are many examples. Some of the waste is necessary and some of it is unnecessary. We will elaborate on that shortly.

The customer and how they pay depends on the charter of the entity.

In engineering workflows that span the R&D system, there will be many small supplier-consumer hand-offs needed to deliver the completed work to the paying customers. For example, the output from one engineering task may be handed off to another engineer for further work or integration into a larger system. In that case, there is a supplier-consumer relationship between the two engineers. This applies to larger groups as well. In essence, the consumer informs the supplier what they need and the supplier delivers that. For engineering organizations that do not directly engage with the final paying customer of the technology the internal recipient of their work often acts as a proxy for determining what the ultimate end customer will pay. The assumption is that if the downstream group in the development chain are requesting it then it must have value.

The requests are based on what the receiving group has determined to be of value. However, like the group before them, they may not have direct engagement with the paying customer and are basing their determination on the stated requirements of the next group they deliver to downstream. As a result, a long workflow may have many hand-offs of information before reaching the paying customer. As anyone who has played the telephone game as an icebreaker can attest, as the number of information hand-offs grows the accuracy of the message reduces. Therefore, trusting each consuming entity in the chain to accurately translate the needs of the eventual paying customer in their requests often leads to mistakes.

As noted in the discussion of the 2 × 2 other requests may stem from corporate strategic objectives, government regulations, or other tactical objectives in the company that have no direct relationship to a traditional paying customer. In this case, the paying customer is not an entity that will ultimately purchase the technology but management of the company or another entity that directly or indirectly funds the activities.

The bottom line is that there are three scenarios to consider based on the organization's charter and its objectives. These define how they should determine value. They are:

(1) Engineering output is deployed directly to a paying customer.
(2) Engineering output is deployed to another internal entity.
(3) Engineering output is deployed to meet a corporate objective.

Scenario 1: Output is delivered directly to a paying customer. Determining what engineering activities constitute value when the output is delivered directly to a paying customer is the simplest case to analyze. If the customer values what is delivered, they will spend money to purchase it. These activities can be split into two types. The first are those that provide a set rate of return. In these cases, the customer is willing to pay a certain amount for the output of activity regardless of the volume. For example, the customer may be willing to pay for the design of a simple and intuitive user interface (UI) and use it everywhere. However, the customer will not pay more for another one. The second are activities where the customer is willing to pay incrementally more money for incrementally more output. For example, a customer may be willing to pay incrementally more for each additional software feature that improves their productivity. As a result, a good question to ask when evaluating what the customer is paying for is: if we diverted more people to this activity to increase the output, would the customer pay even more? This helps clarify and prioritize different value-adding activities.

As we mentioned with the 2 × 2 discussion, if the activities that the customer will pay for also align with the vision of the company than the activities have clear existing value. The case where there is alignment on the vision but no customer is willing to pay for it requires a management decision on the potential value. The case where the customer is willing to pay but there is no alignment with the vision also requires a management decision on whether the distraction is warranted or if the vision needs to be tuned to comprehend the new business opportunity.

Scenario 2: Output is deployed to another internal entity. As we mentioned, determining the value provided by engineering R&D tasks become more difficult when they enable another engineering or product development organization within the same company. In those cases, there may not be a direct connection between the engineering output and the paying customer. Also, it typically does not cost the consuming internal engineering and product development organizations anything more to ask for more. This can result in a long list of poorly prioritized requests. As a result, internal engineering R&D organizations often struggle to define and measure how their work contributes to what the end customer is willing to pay. It is not as simple as blindly providing exactly what the consuming engineering or product development organizations request. Ideally, the internal engineering R&D organization is still able to collaborate with the paying customer at the end of the development chain. When that is possible, it becomes a balance between providing what the next consuming internal entity needs and working directly with the end customer to get to the best engineering solution.

Ideally, all participants work together in close collaboration and total transparency so decisions can be made jointly with all of the relevant information. In reality, there will be tension since each participant represents a different perspective. In that environment, the engineering R&D organization doing the work has the responsibility to know the roadmap and plans of the other internal entities in the development chain and the end customer who ultimately pays as well as their own. They also must understand the root of the paying customer's requests, assess the value from their perspective, propose solutions and constructively challenge what they believe may be incorrect assumptions made by other internal entities. By doing this, they can make sure the engineering output of their and the other organizations best satisfies the true need of the end customer and the next internal entity in the value stream. When there is a disagreement that cannot be reconciled it should be escalated to the appropriate management body for resolution. As we will see later, when we talk about the Value Stream Map (VSM) this type of debate helps identify and eliminate waste across the whole product development flow.

Scenario 3: Engineering output is deployed to meet a corporate objective. In addition to the traditional paying customers who purchase technology and generate revenue for the company, there are other important customers that do not directly pay money but are critical to the company's success. For example, the management of the company may provide funding for activities that do not have an external customer. This could include such things as reducing the company's carbon footprint, providing critical services in a national emergency, and meeting government regulations. All of these examples require some form of funding to succeed and it often comes from the company itself. So, similar to the previous scenarios, if the company is willing to fund the tasks, then they create value.

Table 4.1 below summarizes the three cases we discussed. They offer a guide for others to follow but each R&D organization needs to go through their journey of defining value and constructing a similar framework. There is a large measure of subjectivity to the decision of what constitutes value and it is rarely as straight forward as the table suggests. The most important thing is that the leaders in the organization have a robust conversation about what constitutes value for them since it not only provides a means to get everyone aligned but it often exposes inconsistencies in people's assumptions and perceptions that need to be resolved.

Everyone in the organization needs to understand how value is defined.

Whether or not the organization delivers directly to a paying customer or to other entities within the company they all share the common challenge that as you move down into the hierarchy of any organization the connection between the day to day work and the corporate vision and mission can get muddied. As we will soon see the real power of having a deep understanding of the value the organization delivers is that each workflow in the organization can be properly tuned by employees to eliminate waste and increase velocity. This requires the active participation of everyone to determine what tasks contribute to value.

The examples we have given are quite simple and the reality is that the entire R&D system can be very complex. However, the debate that results from trying to determine the value provided by an organization and the engineering tasks is very important. Without it, eliminating waste and increasing velocity can be very ineffective.

4.2 Waste

Time spent on tasks that do not create value is waste.

There are several ways to measure waste. One can identify and rollup unnecessary expenditures on such items as capital, services and the workforce to show a financial impact. One can look at efficiency by measuring output vs. the input as well as the productivity of the individual engineers. These are all important, however, since we stress the importance of learning faster than the competition, we focus on wasted time. As we will discuss in Chapter 5, the ability to complete learning cycles faster than your competition is the key to success. Also, shifting time spent on wasteful tasks to tasks that create value increases the rate of deployed value (velocity). In a large R&D system, wasted time is the root cause of many issues that show up in poorly performing financial, efficiency and other metrics. With this in mind, we use the lean terms for waste: necessary waste and unnecessary waste.

Tasks that are a necessary waste must be done but should be minimized.

Ideally, the engineers and all of the processes and workflows are perfect and we do not need to test results, write status reports, or hold project reviews. However, engineers, the workflows and processes are not perfect, and therefore holding such things as a project review is usually required in some form to expose potential problems. In lean terminology, this type of activity is referred to as necessary waste and although you may not be able to eliminate this type of task, it should be completed effectively in the least amount of time as possible. If an operations review can be done effectively in 15 minutes there is no reason to consume any more time on it. If a status report communicates the essential information in a couple of lines that take five minutes to write it is better than a tome with reams of irrelevant information that takes hours to prepare.

Categorizing tasks as value-added vs. necessary waste can be difficult even when an organization has a good grasp of the value it provides to the company. For example, testing engineering work after it is completed is a form of necessary waste. However, many people may argue that more testing adds value since it will give customers more confidence in the quality of the product. What is forgotten in that argument is that the customer is paying for quality not testing and the most effective way to ensure quality is to build it into the development process such that testing is not needed. Of course, management itself is also a necessary waste. Ideally, employees do not need to be managed because they have a culture, processes and workflows that allow autonomy while ensuring consistency with the organization's strategy and goals. When managers internalize that key understanding then they become much better managers.

Tasks that are an unnecessary waste can be eliminated

Unnecessary waste is not otherwise required and should be eliminated. For example, time that elapses when one task is idle because it is waiting for another task to complete is unnecessary waste. The time that elapses searching for information or re-learning something that was forgotten are other examples of unnecessary waste. Similarly, re-work to fix a defect discovered by a customer is unnecessary waste. Although they must be fixed, the ideal state is that the defect should never have existed in the first place. Figure 4.2 depicts the relationship between value, necessary waste and unnecessary waste.

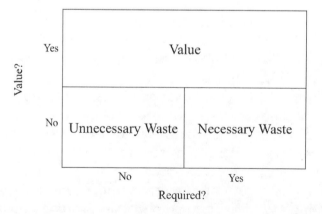

Figure 4.2.   Value, necessary waste and unnecessary waste.

4.3 Sources of Waste

As further elaboration one can look at well-documented sources of waste for manufacturing, software development and management.

The Seven wastes in a manufacturing system: Taiichi Ohno was the father of the Toyota Production System that became the basis for what was originally termed lean manufacturing. Ohno described seven wastes in a manufacturing system [6] and they have been paraphrased in the mnemonic device TIMWOOD shown in Table 4.2.

The Seven wastes of software development: Mary Poppendieck translated Ohno's seven wastes in a manufacturing system into equivalent wastes of software development [7] which we have paraphrased and shown in Table 4.3.

The eight wastes of management: Since this is a book on management, we have also added our list of 8 wastes of management and included them in Table 4.4 below. Often, bad

Table 4.2. The seven wastes in manufacturing.

Transportation	Moving parts and products unnecessarily.
(Idle) Inventory	Having more than the minimum stock necessary for a pull system.
Motion	Making unnecessary movements looking for parts, tools, docs, etc.
Waiting	Operators standing idle, equipment fails, needed parts fail to arrive, etc.
Over-processing	Performing unnecessary or incorrect processing steps.
Overproduction	Producing ahead of what's needed by the next process or customer.
Defects	Inspection, rework and scrap.

Table 4.3.   The seven wastes of software development.

Partially done work	Started but unfinished work.
Extra features	Features or functions in software that are not used.
Relearning	Finding or relearning knowledge that was previously known.
Hand-offs	Knowledge transfer associated with handing off work to a different team.
Task Switching	Switching between tasks that require resetting mental state.
Delays	Delays that prevent related work from being done at the optimum time.
Defects	Rework required to fix errors that should not have existed in the first place.

Table 4.4.   The eight wastes of management — Aslett, Acken and Yerramilli.

Creating Ambiguity	Leaving things up for interpretation
Creating Churn	Increasing the rate of unnecessary change
Creating an Untrusting Environment	Stifling risk-taking and creativity
Inaction	Unnecessarily stopping progress
Perpetuating Gratuitous Development	Not stopping activities of less value
Under/Poorly communicating	Starving employees of the rationale needed to align and adapt
Acting Arrogantly	Inhibiting constructive debate and driving away valuable employees
Coasting	Promoting the continuation of the status quo

management actions and decisions have a greater impact than good actions and decisions. As noted in the article "Bad is Stronger than Good: Evidence-Based Advice for Bosses" [8] and the book *The power of Bad* [9], it takes many "good" actions to overcome the negative effects of "bad" actions. As a result, management actions that create waste often require a much greater effort to fix than it took to create the problem in the first place. The 8 wastes of management are:

Creating Ambiguity: Creating ambiguity leaves things open for interpretation and prevents the autonomy needed by employees and others to make the right decisions and act. The manager creates ambiguity by not having a clear strategy, values, objectives and definition of success for the organization. They contribute to it by establishing irrelevant metrics that create waste, by not properly prioritizing work and by being unresponsive to others in the R&D system.

Creating Churn: Churn creates unnecessary change in the organization and wastes resources. The manager creates churn when they are impulsive and act too soon with incomplete or premature information. They may also contribute to it by being indecisive and not deciding quickly and effectively once information is complete. Once a decision is made, they can contribute to churn by being fickle and too easily reversing a decision once it is made.

Creating an Untrusting Environment: Creating an untrusting environment stifles risk-taking and creativity and slows down innovation. The manager creates an untrusting environment when they do not enable a psychologically safe work environment for employees to express themselves freely. This under-utilizes the full capacity of people and leads to poorer quality decisions. They contribute to it by demotivating employee's autonomous actions without a credible reason and rewarding the wrong behavior or the wrong people.

Inaction: Inaction unnecessarily stops progress and creates unnecessary waste. The manager causes delays when they do not identify and empower a single decision-maker for projects and organization processes. They contribute to delays when they are unwilling to decide on and deploy resources or when they do not constructively confront the source of late deliverables to them.

Perpetuating Gratuitous Development: Perpetuating gratuitous development steals resources from activities of higher value and reduces the organization's velocity. The manager perpetuates gratuitous development by not establishing a clear definition of value for the organization that is strongly linked to the customer. They contribute to it by not stopping projects with a lower value than others.

Under/Poorly communicating: Under or poorly communicating starves employees of the rationale they need to align with the strategy and values and adapt to change. The manager is guilty of under or poorly communicating when they do not give the proper information at the right time, using every available communication platform. They contribute to the problem by not giving a full rationale for the vision, strategy and values of the organization and important decisions and changes. When managers under-communicate employees fill in the blanks with rumors.

Acting Arrogantly: Acting arrogantly inhibits constructive debate and drives away valuable employees because they cannot freely express themselves. It leads the organization into a death spiral of decreasing value. The manager acts arrogantly when they do not address conflicts of interest such as favoritism or systematic negative biases such as "not invented here" (discounting external work), confirmation bias (discounting evidence that does not conform to existing beliefs) and self-serving bias (blaming others when bad things happen). These contribute to it by not hiring a strong staff, micromanaging those that are left and being unwilling to hear or internalize concerns from subordinates.

Coasting: Coasting promotes the continuation of the status quo and ignores the need for continuous improvement. It causes the organization to lag further behind the competition. The manager is coasting when they encourage a "no news is good news" or "don't rock the boat" attitude. They contribute to it by not investing in employee development, not inculcating the skills of a learning organization and not driving continuous improvement every day by everyone.

4.4 Shifting Time Spent on Waste to Value

Waste elimination creates more value. It does not eliminate people.

Although it will be tempting for some to look at waste elimination as a means to drive downsizing of organizations, that is not our purpose. We take the view that the best path is to eliminate waste to create more value with the resources one has. This means that the elimination of unnecessary waste and the minimization of necessary waste incrementally provides more time for value creation, as depicted in Figure 4.3.

Finding waste in the R&D system requires everyone to participate.

Big sources of unnecessary waste are easy to see and often get the attention of senior management. Many more sources of

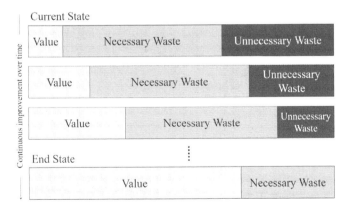

Figure 4.3. Increasing value over time.

waste do not rise to that level individually and never get a management initiative. Instead, workers just learn to live with them by creating workarounds or absorbing the expected waste in their planning and project duration estimates. These small wastes accumulate and can become debilitating over time. At that point, there may be a large process improvement initiative that tries to redraw processes and streamline the workflow. Even if this does provide some benefit, it is usually short-lived as the organization reverts to creating workarounds as new problems inevitably arise. To break the cycle of crises and management improvement initiatives, waste elimination must be something the organization does every day.

To identify waste in a workflow one needs to understand how the work is being done so a complete mapping of all of the tasks can be completed. However, in a complex environment, no single person knows how all of the work (and workarounds) is being done. Although there may be documents that capture the methodology in great detail, it is likely the actual work is done differently to work around issues not comprehended in the documentation. In work flows that include handoffs between different teams, it is also very likely that some teams misunderstand what their suppliers or consumers are doing. Therefore, the knowledge about what is really happening is distributed across many different individuals that work in the system every day. Getting a complete picture of how the work is getting done and where waste resides requires everyone to participate to comprehend and agree on the current state, identify and implement incremental improvements and then reconvene regularly to repeat the process. A relentless focus on waste elimination is required by everyone, every day to create more value. This must be a part of the culture.

The Value Stream Map (VSM) is an important tool for identifying waste.

When we refer to a value stream in this book, we will be using the lean usage of that term and not how it is also used to architect a business. In our case, the value stream is the complete series of activities needed to engineer a solution for the final customer. The value stream often starts with a hypothesis or customer request and ends with the deployment of an engineering solution. The customer may be a paying customer, another R&D organization, or the company itself.

A Value Stream Map (VSM) is an extremely useful tool for visualizing and analyzing workflows to identify how much time is spent on value-added activities and how much is spent on waste. It uses the definitions of value and waste we discussed in this book to identify opportunities to reduce waste and shift the time saved to value-added tasks. A detailed example of how this can be applied in practice is provided in the first case study at the end of this chapter.

When analyzing a VSM it is important to understand that all value is not equivalent and not all waste elimination has the same impact on the deployed value. Bearing in mind Amdahl's law, a local reduction in waste in one area may have less impact than another on the overall velocity across the complete value stream. Also, adding more time for some local value-added tasks may not have the same impact on the total deployed value as another. So, when evaluating the priorities for waste reduction and increasing value-add activities one needs to keep that in mind. Although everyone should be relentlessly eliminating the many, many incremental wastes as they are identified in some cases the order of addressing them can be optimized.

4.5 Velocity

Velocity measures deployed value vs. time.

Efficiency is an important metric. It is a measure of how well an input is utilized for an intended task or function. It is usually expressed as a ratio output/input where the input and output are often measured in tangible terms such as money and energy. Therefore, the units of measurement of the inputs and outputs will differ depending on the topic, field and industry.

Efficiency = Output/Input

Efficiency is a very important metric for comparing organizations against benchmarks and identifying where resources (people, machinery, etc.) may be under-utilized. However, one can be very efficient and create no value if the result is something the customer does not want.

Productivity is also an important metric. In our context, it is generally used to measure the efficiency of workers or a group of workers in a way that can be benchmarked with other entities. One measure of productivity may be the average number of components one factory worker can assemble per hour or per dollar of labor cost. For agriculture, productivity is often measured as the number of bushels of a crop per acre per year, assuming the same level of quality. Like efficiency, the units of measurement of the production and inputs will differ depending on the field and industry.

Productivity = Production/Input

Productivity is also a very important metric for comparing organizations against benchmarks with respect to the raw capacity of the organization. However, once again, if the customer will not pay for what is being produced (or the price varies), then the efficiency and productivity metrics do not tell the whole story. When we speak of velocity, we are speaking very specifically in terms of the deployed value vs. time or:

Velocity = Deployed Value/Time

The value is determined by the customer and we use the term "deployed value" to highlight that it is not good enough to release the output of the work, it must be deployed and being utilized before the value is realized. Increasing velocity implies that either the deployed value is increasing over the same time frame, the time it takes to deploy the same value is reducing or some combination of the two. Value and time are the critical variables and velocity is independent of the resources or cost associated with generating the value.

The R&D system must continuously increase its velocity to stay competitive. This is because for most technology there is only a fleeting moment of competitive advantage brought about by the initial engineering innovation. Once the innovation is understood by the competition firms, it can easily get overrun by competitors that move faster to iterate and improve on the idea. The only guarantee of continuing success is to be able to always move faster than the competition. Since the competition is always improving so must the R&D system we describe in this book.

Although efficiency and quality are not explicitly a unit of measurement for velocity, we know that eliminating time spent on wasteful activities has the effect of improving those as well. For example, a focus on reducing the spent on the necessary waste of testing often leads to innovations that build quality into the workflows. The result is better product quality. Similarly, reducing the time spent on other waste means that more time can be spent on value-added activities. This means that the output of the organization will increase while the input stays the same. This improves efficiency. Similarly, the productivity of each engineer will increase. Creating initiatives focused on improving quality and efficiency often is the result of a Value Stream Mapping exercise that has highlighted the related waste. The goal may have been to reduce waste but the side effect is often better efficiency, quality and productivity as well.

Velocity, efficiency and quality trade-offs are sometimes required.

Overall, a culture that enables learning cycles to experiment and continuously improve will increase velocity and improve efficiency, quality and productivity. This ideal state occupies the upper right quadrant of the 2 × 2 matrix shown in Figure 4.4. However, this will not always be the case since there may be temporary trade-offs required in velocity to enable efficiency and vice versa. For example, in a competitive environment, it may

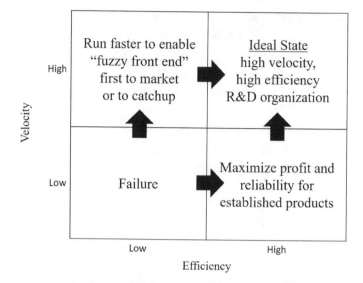

Figure 4.4. Velocity and efficiency trade-offs.

be necessary to move faster than the capacity of the R&D organization can handle to meet a competitive threat. In this case, a decision may be made to rapidly increase the rate of deployed value (velocity) even though it will cause some inefficiency. This may happen for a new technology where market leadership is the goal for the first version of the technology and efficiency is the goals for subsequent versions. In this case, a trade-off may be made to increase the velocity for the first release by brute force while sacrificing efficiency so success in the market can be achieved. For the following versions, velocity may be sacrificed to ensure reliability, lower cost and higher profit margin.

In a similar vein, the manager of the R&D system needs to be careful while setting goals for velocity, efficiency and quality. Expecting continued improvement in efficiency while addressing an interrupt like the ones just mentioned to increase velocity may not be realistic. It is important to reiterate the importance of improvement across all of those metrics but if a temporary trade-off is being made with efficiency or quality to enable greater velocity, then the expectations communicated to the employees need to be adjusted as well. It is not necessary to reduce an efficiency or quality objective that the organization has taken but it will be good to publicly communicate that for a short time a trade-off is being made and progress may be slower than desired. If the manager does not do this then the organization will be conflicted and may not fully commit to the immediate need for increased velocity. It is a difficult balance to maintain but one that that manager must communicate clearly.

4.6 Summary

Employees operate more effectively when they understand what constitutes value and how their work contributes to it. This helps them identify tasks that have value and those that do not. When coupled with a culture that motivates everyone to continuously learn and improve, a clear understanding of value enables everyone to reduce or eliminate time spent on wasteful tasks and increase the time spent on tasks that have value.

Generally speaking, an engineering task has value if it contributes to the vision of the company in a way that customers will pay. However, determining who the customer is and what constitutes payment is not always straight forward. This is especially true for R&D organizations that deliver to customers in the same company. Also, all value is not equal. Decisions need to consider the entire R&D system and what provides the greatest economic value from the perspective of the customer.

There are well-known sources of waste identified in product development and manufacturing that can act as a guide when analyzing value streams and workflows to reduce waste. It is also important to keep in mind that management is a necessary waste and poor management behavior can also become a major source of waste.

The rate of deployed value vs. time is what we call velocity and continuously increasing that is required to enable the pace of innovation required to thrive. Although they often complement each other, greater velocity does not necessarily equate to greater efficiency, productivity, quality or lower costs (or vice versa). In fact, it makes sense in some cases to temporarily accept less efficiency and productivity or more costs to speed deployment of value when the technology or market is new. These are the type of trade-off discussions that need to be directed by the manager of the R&D organization spanning the entire R&D system. It all starts with gaining agreement on what is value.

4.7 Key Points

- The vision and what the customer will pay determines value.
- The customer and how they pay depends on the charter of the entity.
- Everyone needs to understand how value is defined.
- Time spent on tasks that do not create value is waste.
- Tasks that are necessary waste can be minimized.
- Tasks that are unnecessary waste can be eliminated.
- Waste elimination leads to more value. It does not eliminate people.
- Finding waste in the R&D system requires everyone to participate.
- The Value Stream Map (VSM) is a tool for identifying waste.
- Velocity measures deployed value vs. time.
- Velocity, efficiency and quality trade-offs are sometimes required.

4.8 Case Studies

4.8.1 *Increasing Velocity Using a Value Stream Map*

Background

Company ABC develops and sells software applications that model and simulate the performance of very large central air conditioning units. Most of its customers are manufacturers who provide these systems for the construction of large office buildings and other facilities. One of the most important factors in purchasing such systems is the consumption of electricity since this drives up the total cost of ownership. Engineering applications that can accurately estimate the electricity consumption over many operating conditions and construction scenarios allow the manufacturers to optimize the design and installation to minimize electrical consumption. As a result, improvements in the modeling and simulation applications can provide them a competitive advantage. They are more than willing to pay more to companies like ABC Inc. for applications that provide improved modeling and simulation accuracy.

Three groups in company ABC work together to develop and release the software associated with the application. Recently, they got together to do a value stream map of the development flow across the groups. To keep it simple, they chose the beginning of the coding of a new or enhanced feature as the starting point for the analysis and the deployment of the updated application to the customer as the endpoint. Since the customers are willing to pay for improvements in the accuracy of the simulation software, they agreed that tasks that improve the accuracy of the simulation add value. Figure 4.5 is an abstracted representation of the result of their analysis.

Unit Development & Re-work #1: The unit development team does the coding of the individual feature. It is responsible for unit-level testing before it hands off its output to the integration team. In the picture, 1 day is spent on the coding and 1 day is spent on the testing. Since the coding work implements the improvement in simulation accuracy, this amounts to one day of value-added work. The 1 day of effort for testing is captured as necessary waste. So, for the first pass of coding and testing, there is 1 day being spent on value and 1 day being spent on necessary waste. In this example, the participants noted that the testing usually discovers 1 problem that requires rework at this point. This is denoted by the "Re-work #1" loop. The result is

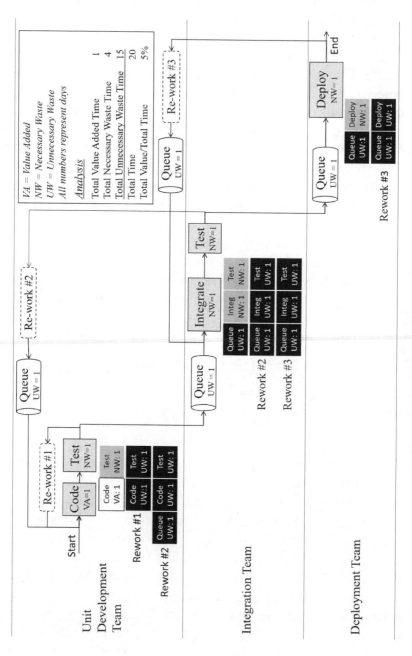

Figure 4.5.   Original value stream map.

that coding and testing activities are repeated. However, this time they were both classified as unnecessary waste since, if the code was implemented correctly in the first place, that extra work would not have to be done. By the time the code is passed on to the integration team, there had been 1 day of value-added work, 1 day of necessary waste and 2 days of unnecessary waste.

Integration & Re-work #2: When the code was delivered to the integration team there was a delay of 1 day as it sat in a queue awaiting a person to become available to do the integration. This was determined to be an unnecessary waste. It then took 1 day to do the integration. This was classified as a necessary waste since it was required but did not create additional value. It then took 1 day to do integration testing. As before this was classified as necessary waste. One pass through the integration flow yielded 2 days of necessary waste and 1 day of unnecessary waste. The participants noted that integration testing usually discovered one problem that requires re-work of the original code. This was denoted by the "Re-work #2" loop. The result was that the problem's description waited in a queue for a resource to become free (one day of unnecessary waste) and then went back to the coding and testing workflow (2 days of unnecessary waste). When the fixed code was delivered to the integration team it sat in the queue and then was integrated and tested again (3 days of unnecessary waste). So, by the time the integrated code was delivered for deployment, there had been 1 day of value-added work, 3 days of necessary waste and 9 days of unnecessary waste.

   Deployment & Re-work #3: When the integration team's results were delivered to the deployment team it sat in a queue for one day (unnecessary waste). It then took one day to deploy the results. This was classified as necessary waste since it was required work but did not, itself create additional value. So, one pass through the deployment flow took 1 day of necessary waste and one day of unnecessary waste. As before, the participants noted that the deployment usually uncovered one problem in the integration work that needed to be fixed. This is denoted by the "Re-work #3 loop". The result was that the problem was submitted to the integration team where it sat in a queue for 1 day for a resource to become free (unnecessary waste). It then went through the integration and test activities again (2 days of unnecessary waste). When the fix was delivered back to the deployment team it waited in the queue and then was deployed again. So, by the time the new feature was deployed without any defects, there had been 1 day of value-added time, 4 days of necessary waste and 15 days of unnecessary waste.

The Problem

When the teams summed up the time spent on value-added tasks, necessary waste and unnecessary they were surprised to see that only 5% of the total time used to develop and deploy the new feature was spent on value-added work. They saw that if they reduce the time spent on unnecessary and necessary waste and spend more time coding improvements in accuracy, they could provide more value in less time for the customers.

Actions Taken

In this case, the large number for unnecessary waste was largely due to the rework needed to fix defects found downstream and the wait time for the developers upstream to work through their queue and address them. The teams observed that the biggest reduction in waste would result from reducing the amount of re-work needed. As shown in Figure 4.6, they decided the best way to do this was by eliminating the hand-off between the unit development team and the integration team and adopting continuous integration so the code was continuously tested as it was written. This was achieved by merging the unit development and integration teams. By adopting continuous integration quality assurance was made part of the coding process and

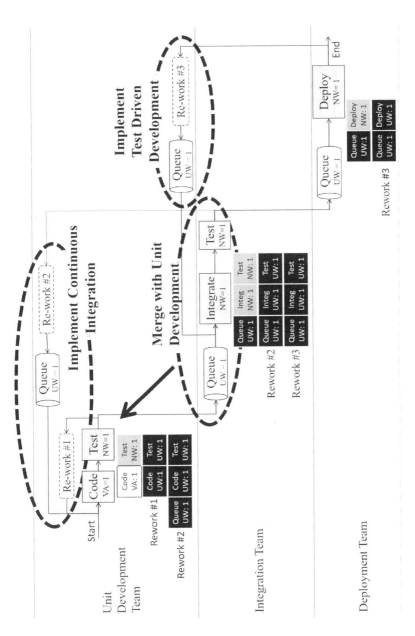

Figure 4.6. Identified improvements to value stream map.

separate testing was no longer needed. Although this slightly increased the time it took to code a new feature it eliminated the rework loop and associated queues. It also had the benefit of increasing the size of the coding team. As more efficiency was gained via continuous integration more people could be assigned to do coding work where the value was being generated. For the first round of improvements, they decided to keep the deployment team separate. However, they eliminated the rework loop associated with that step by adopting a test-driven approach that incorporated customer test cases into the continuous integration environment so mistakes encountered at the time of deployment were avoided.

Results

The new value stream map is shown in Figure 4.7. With the changes implemented, value-added tasks now consume 25% of the total time. In the case of ABC Inc., these time savings were used to increase the number of incremental improvements delivered in the same amount of time as before. In other cases,

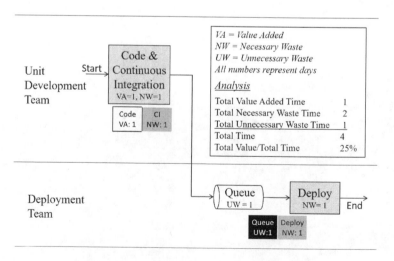

Figure 4.7.   The new value stream map.

the improvements were used to respond more rapidly to specific customer enhancement requests.

Questions:

(1) Which type of wastes applies to the example given?
(2) What could be done to further reduce waste?
(3) What work may be required to implement continuous integration?
(4) What challenges need to be overcome to combine the unit development and integration teams?
(5) What challenges need to be overcome to implement test-driven development?
(6) Can you dive deeper into the workflow and draw a value stream map for the unit development team's tasks before and after the changes?

References

[1] Ford, H. (2017). *My Life and Work*. (Create Space Independent Publishing Platform).
[2] *Ibid*, pp. 34–35.
[3] Brooke, L. (2008). Top 10 Ford Model T Tech Innovations That Matter 100 Years Later, Popular Mechanics. September 25, 2008. Available from: https://www.popularmechanics.com/cars/a3658/4284734/.
[4] Thomke, S. and Feinberg, B. (2012). Design Thinking and Innovation at Apple. Original January 2009, Revised May 2012 (Harvard Business School). Available from: https://www.hbs.edu/faculty/Pages/item.aspx?num=36789.
[5] Lashinsky, A. The Cook Doctrine at Apple, Fortune Magazine. Jan. 22, 2009. Available from: https://fortune.com/2009/01/22/the-cook-doctrine-at-apple/.
[6] Koenigsaecker, G. (2013). *Leading the Lean Enterprise Transformation*, 2nd Edition. (CRC Press) p. 11.

[7] Poppendieck, M. (2006). *Implementing Lean Software Development*, 1st Edition. (Addison-Wesley) pp. 73–82.

[8] Sutton, R. (2010). Bad Is Stronger Than Good: Evidence-Based Advice for Bosses, Harvard Business Review. September 8, 2010. Available from: https://hbr.org/2010/09/bad-is-stronger-than-good-evid.

[9] Tierney, J. and Baumeister, R. F. (2019). *The Power of Bad. How Negativity Effect Rules us and How we can Rule it*. (Penguin Press).

CHAPTER 5

Learning Cycles

He who can handle the quickest rate of change survives.
— **John R. Boyd [1]**

Engineers in the R&D organization and other entities are responsible for tasks in workflows and processes that span the R&D system. The completion of each task generates information on the effectiveness of the engineering decisions and the operation of the workflows and processes. When this knowledge is used to improve subsequent engineering decisions or incrementally improve the workflows and processes, then a learning cycle has been completed. Learning cycles are a critical part of enabling continuous improvement in the R&D system.

Whether it is in business, manufacturing, or engineering the reason for short learning cycles is aptly summed up in the quote from Colonel John Boyd in the subtitle. Although, we prefer a slight variation in the wording to, "Those who can handle the quickest rate of change survive". Translated into our frame of reference, the R&D system with the shortest learning cycles explores more ideas, makes course corrections sooner, improves development processes faster and delivers more value than the competition. It also means that the role of the engineer is much more than designing new or improved technology; it also includes continually improving the workflows and processes they use.

Learning cycles are not a new concept. They are just applying the scientific method of *observation — hypothesis — test* to engineering workflows and processes. It is a philosophy that is embedded in lean thinking and agile development practices. Machine learning, which enables systems to learn from experience automatically, is an example of "the scientific method on

117

steroids" [2]. Like machine learning, learning cycles are a means to conduct controlled experiments during execution and compare the results with expected outcomes. Changes that meet the expectation can become part of a revised standard. In this manner, learning cycles offer a means to continuously challenge and improve standards with the people responsible for their execution.

Many problem-solving and continuous improvement models utilize learning cycles and they can be applied to strategic planning, engineering research and development. More than anything else, implementing learning cycles requires a way of thinking and a culture where everyone in the enterprise is involved and view themselves as scientists conducting experiments to improve the workflows. This requires the conviction by everyone that this is a critical part of their job description and valued by their management. Enlightened leaders recognize that shorter learning cycles lead to better technology and by enabling everyone to improve their workflows they create a competitive advantage for the company.

5.1 The Importance of Learning Cycles

Short learning cycles have reduced software development times.

In the last 50 years, the use of software has grown dramatically. As this happened, so did problems with productivity and quality related to the rigid nature of the waterfall development model. Frustration spurred new thinking and in 2001 practitioners of several novel approaches to software development met and [3]

> *"What emerged was the Agile 'Software Development' Manifesto. Representatives from Extreme Programming, SCRUM, DSDM, Adaptive Software Development, Crystal, Feature-Driven Development, Pragmatic Programming, and others sympathetic to the need for an alternative to documentation driven, heavyweight software development processes convened."*

Among other principles the manifesto included [4]:

> *"At regular intervals, the team reflects on how to become more effective, then tunes and adjusts its behavior accordingly."*

One way this gets done is in a SCRUM. In 1995, Ken Schwaber [5] introduced SCRUM development as:

> *"SCRUM assumes that the systems development process is an unpredictable, complicated process that can only be roughly described as an overall progression. SCRUM defines the systems development process as a loose set of activities that combines known, workable tools and techniques with the best that a development team can devise to build systems. Since these activities are loose, controls to manage the process and inherent risk are used. SCRUM is an enhancement of the commonly used iterative/incremental object-oriented development cycle."*

In SCRUM, the iterative/incremental development cycle referred to is termed a sprint and these are linked together throughout the project. Figure 5.1 below describes what a typical sprint looks like and how each one takes the learnings and feeds it into the next one [6].

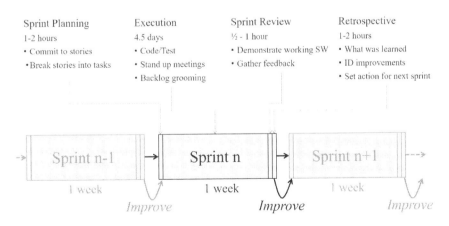

Figure 5.1.   A representative 1-week SCRUM sprint.

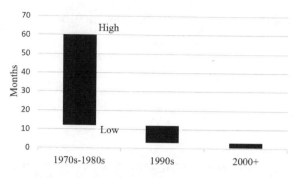

Figure 5.2.   Software development time, 1970s to 2000+.

Each sprint provides an opportunity to complete a learning cycle and the retrospective step provides the means to capture opportunities for improvement for the next sprint [7]

> *"The retrospective allows learnings and insights to be captured while experiences are still fresh, and before, negative patterns have a chance to harden in place. The goal is simple: to identify one or maybe two specific things to improve, and to create an action plan to implement those changes."*

As shown in Figure 5.2 [8], agile principles and new development technology tools have combined to shorten software development cycle times dramatically over the last 50 years.

5.2 Learning Cycles in the Engineering R&D System

Learning cycles exist at all levels of engineering and run concurrently.

The opportunities for learning cycles exist across the R&D organization at different levels of engineering and management and they often run concurrently with each other. Figure 5.3 depicts the hierarchy of the detailed engineering workflows, broader processes and overarching strategy of the R&D organization.

The workflow is the sequence of tasks used to get specific engineering deliverables accomplished. Their learning cycles are typically very short but still operate with a discrete cadence.

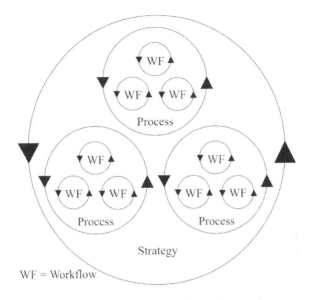

Figure 5.3. Concurrent learning cycles.

The process typically spans multiple workflows to set priorities, allocate resources, choose between options, determine schedules, etc. The learning cycles for a process may be longer since each step in the process may require more than one workflow to complete.

The strategy sets the direction and design of the R&D organization. This includes establishing the correct values, capabilities, structure, processes and other attributes needed to deliver the right mix of value.

The learning cycles for a strategy vary depending on the topic but can be quite long since they rely on getting feedback from the broader eco-system. We discuss this in much more detail in Chapter 6.

Learning cycles for workflows exist within and across entities.

Whether an individual or a group in the R&D organization conducts research, strategic planning, development, or a

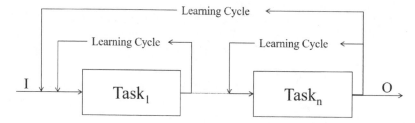

Figure 5.4.   A generic workflow with learning cycles.

combination of one or more of those functions, they are executing a series of specific tasks as part of a workflow as depicted in (Figure 5.4).

Whenever an individual completes a task or series of tasks, it is not uncommon for them to decide to try something different the next time. These learning cycles occur intuitively at the individual level. In a start-up, with a small engineering team located in the same room (or garage), the learning cycles can be very short since everyone is in constant face-to-face communication with each other. When an enterprise grows in size and workflows become dependent on other groups in the same organization or company then the learning cycles are no longer so intuitive and no one may exist who understands how the complete set of connected workflows operate in detail. Figure 5.5 is a simple example of how the work of designing an integrated circuit may look. In this case, there is the same basic workflow used by one team to create the circuit schematic for the sub-circuit, another team to integrate it into a larger circuit block and then the third team to integrated that block into the full chip. Although the workflows look the same, one group may accidentally set different conditions for the circuit simulations they do or they may use a different circuit simulation tool that does not generate the same results as the other groups. It's also possible that a change in how Group A does its circuit simulations might make the work of Group B more productive, efficient, or complete faster. As a result, learning cycles exist across the groups as well as inside them. We can go

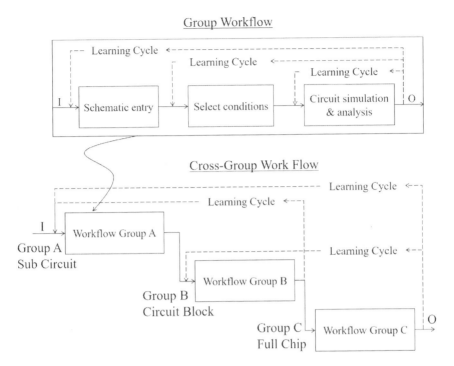

Figure 5.5. An example of connected circuit design workflows.

one step further and replace Group A with a group from an external supplier. When we do this, then the workflow spans multiple companies.

Learning cycles for processes exist within and across entities.

In addition to the detailed engineering workflows, the entities also have processes for planning, research, development and other functions that set priorities, allocate resources, choose between options, determine schedules, etc. Figure 5.6 shows a generic concurrent engineering process that we will discuss later in Chapter 7. In this simplified diagram, the process is meant to iteratively converge on the best design choice as the engineering requirements become more certain. It is a process often used to reduce risk when there is a lot of initial ambiguity.

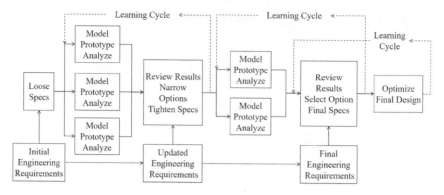

Figure 5.6.   An example of a concurrent engineering process.

In the first step, loose specs are provided based on the available engineering requirements. Three separate design options are chosen and modeled, prototyped and analyzed. When they are reviewed, one option is rejected, the specs are tightened based on updated engineering requirements and the process is repeated. Ultimately the choices are narrowed to the best one with the latest engineering information and the design is executed. In the figure, a learning cycle exists after each review that can be used to evaluate if the model/prototype/analyze step can be improved based on the results seen. This is repeated in the next step. After the design is implemented a retrospective can be done to see what can be improved with the final review, selection and specification step or the entire process altogether.

As with the previous example of the workflows, the processes of the different entities in the R&D system may need to connect. In the example shown, there may be engineering technology or collateral needed for each prototyping step from multiple suppliers. The prototyping step should also include close interaction with the customer as well. If the development processes of the suppliers and customers do not align with the concurrent engineering process, then there will be delays in the concurrent engineering process in the R&D organization. In this case, the learning cycles that traverse the R&D system and related suppliers and customers can be used to continuously improve the interaction. This is also true for all of the strategic planning, research and other processes that exist in the R&D organization that require interaction with other entities in the R&D system.

5.3 Enabling Learning Cycles

Learning cycles are a key part of continuous improvement.

In machine learning, an algorithm is trained by evaluating whether it provides the correct answer using a set of existing data. Once trained against known data, the resulting algorithm is optimized for speed and latency to make a prediction based on new data. This is a compute-intensive operation that benefits from a high volume of existing data. In our case, the scale is much smaller and timescales longer but there are similarities. Like machine learning, our learning cycles should also begin with a thesis or expected result. For example, the thesis may be that an incremental change to the workflow will save a certain amount of time. Each learning cycle either confirms the thesis or disputes it. If the thesis is confirmed, then the change becomes a part of the new workflow and if it is not confirmed, then the knowledge is archived but the change is discarded. In this way, learning cycles drive continuous improvement.

Learning cycles work best using intrinsic networks.

Later in this book, we will discuss the role intrinsic networks play in enabling innovation and information sharing across the R&D system. These networks usually consist of people with similar skills or knowledge from different parts of the R&D system. They are created bottom-up through the employees' autonomous actions or they are created top-down by management to deal expeditiously with a particular problem or opportunity. The networks may be composed of people within the same

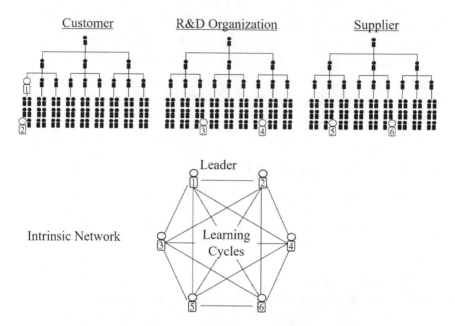

Figure 5.7.   An example of an intrinsic network.

group or spread across multiple groups as the need dictates, as depicted in Figure 5.7. Because they are largely unencumbered by processes, this type of network is ideal for getting the most out of each learning cycle and driving continuous improvement. For engineering work that is part of a sprint, intrinsic networks often form as part of the sprint retrospective step. In other cases, they may need to be created to make sure the right people are engaged when workflows or processes cross group boundaries.

Everyone must become a scientist and conduct experiments.

There are many different workflows and, hence, learning cycles embedded in the R&D system. They are far too numerous for one person or even a group of dedicated people to continuously analyze and improve. The only way of enabling continuous improvement across such a vast array of interdependent workflows is to empower everyone in the organization to improve the workflows they use and make this part of the culture.

Toyota has a strong culture of continuous improvement. In a summary of a study of the company published in 1999 Steve Spear and H. Kent Bowen captured the essence of the Toyota Production System [9],

> *"We found that, for outsiders, the key is to understand that the Toyota Production System creates a community of scientists. Whenever Toyota defines a specification, it is establishing sets of hypotheses that can then be tested. In other words, it is following the scientific method. To make any changes, Toyota uses a rigorous problem-solving process that requires a detailed assessment of the current state of affairs and a plan for improvement that is, in effect, an experimental test of the proposed changes. With anything less than such scientific rigor, change at Toyota would amount to little more than random trial and error — a blindfolded walk through life."*

A view of how the same idea worked at Apple was articulated by Steve Jobs, when asked about Apple's process for innovation [10],

> *"The system is that there is no system. That doesn't mean we don't have process. Apple is a very disciplined company, and we have great processes. But that's not what it's about. Process makes you more efficient. But innovation comes from people meeting up in the hallways or calling each other at 10:30 at night with a new idea, or because they realized something that shoots holes in how we've been thinking about a problem. It's ad hoc meetings of six people called by someone who thinks he has figured out the coolest new thing ever and who wants to know what other people think of his idea. And it comes from saying no to 1,000 things to make sure we don't get on the wrong track or try to do too much. We're always thinking about new markets we could enter, but it's only by*

saying no that you can concentrate on the things that are really important."

As Steven Spear, Kent Bowen and Steve Jobs note, and what we observed, continuous improvement works when everyone in the enterprise is involved and everyone views themselves as a scientist constantly running experiments and challenging each other to improve. Since learning never ends, it requires the conviction by everyone in the organization to sustain this forever. This is not easy since it takes effort to create the training, technology, management systems and culture to work smoothly. The incremental nature of the improvements also makes it extremely easy for organizations and their leaders to become impatient or complacent. However, enlightened leaders recognize that shorter learning cycles lead to better products. The resulting improvement in the workflows leads to shorter execution time, which enables moving faster than the competition when the needs in the market become clear. For outside observers, this may give the impression that the engineering manager is a visionary, but in reality, it is just the result of relentless cycles of improvement in the strategic planning, development and engineering research workflows and processes.

Learning cycles complement and co-exist with standardization.

Henry Ford democratized the automobile with the Model T Ford. The vision was simple [11],

"a motor car for the great multitude" that was built on such a way that, "no man making a good salary will be unable to own one."

The Model T Ford's lack of variety was famously captured by the statement that, "You can have it in any color you want as long as it's black". However, this belied that continuous improvement that occurred during its lifetime. When discussing the Model T, Henry Ford wrote [12],

"Our biggest changes have been in methods of manufacturing; they never stand still. I believe that there is hardly a single operation in the making of our car that is the same as when we made our first car of the present model. That is why we make them so cheaply."

Perhaps the greatest manufacturing innovation was the creation of the assembly line. The concept was based on the principle that [13],

"a man shall never have to take more than one step, if possibly it can be avoided, and that no man need ever stoop over"

The assembly line was a source of continuous experimentation that started in 1913. This was when the Model T was reaching its peak in production. There were many and frequent changes. An example of this for one part was given by Henry Ford [14],

"We had previously assembled the flywheel magneto in the usual method. With one workman doing a complete job he could turn out from thirty-five to forty pieces in a nine-hour day, or about twenty minutes to an assembly. What he did alone was then spread into twenty-nine operations; that cut down the assembly time to thirteen minutes, ten seconds. Then we raised the height of the line eight inches — this was in 1914 — and cut the time to seven minutes. Further experimenting with the speed that the work should move at cut the time down to five minutes. In short, the result is this: by the aid of scientific study one man is now able to do somewhat more than four did only a comparatively few years ago. That line established the efficiency of the method and we now use it everywhere."

Similar to manufacturing, using standard engineering tools, methods and workflows are critical when working in the complex R&D system to engineer technology with many interdependences. However, as the example points out, learning cycles and continuous experimentation complement standardization and make the standards better. This is because learning cycles are a means to conduct controlled experiments during execution and compare the results with expected outcomes. Changes that meet the expectation can become part of a revised standard. Those that do not can be recorded and then discarded. In this manner, learning cycles continuously challenge the standards and improve them.

5.4 Executing Learning Cycles

There are several ways to execute learning cycles.

As we noted at the beginning of this chapter, learning cycles are not a new concept. Learning cycles are used in different problem-solving, continuous improvement, strategic planning and customer collaboration models. Each approach has a particular strength depending on the application. The people involved in the workflows and processes must understand how these and other approaches can be used and then decide for themselves what makes the most sense to use. We introduce some of the ones we have found useful and that we will refer to later in the book. They are:

(1) PDSA, A3, AAR and SCRUM for workflow improvement.
(2) The OODA Loop for continuous strategic planning.
(3) LAMDA and the Design Thinking for technology development.

PDSA, A3, AAR and SCRUM for workflow improvement.

Plan-Do-Study-Act (PDSA)/Plan-Do-Check-Act (PDCA)

Dr. W. Edwards Deming was a leader in the theory and application of continuous improvement for product quality. He was a major contributor to the post-World War II improvement in Japanese manufacturing quality and his management philosophy help transform manufacturing. In 1982, Dr. Deming summarized his influential work on manufacturing quality and management in "14 Key Principles for Management" [15] which included the exhortation to:

> *"Improve constantly and forever every process for planning, production and service"*

During his career, Dr. Deming formalized a structured problem-solving model for continuous improvement known as Plan-Do-Check-Act (PDCA), Plan-Do-Study-Act (PDSA) or the Deming cycle depicted in Figure 5.8. For simplicity, we use the PDSA acronym described below [16].

> *"The cycle begins with the Plan step. This involves identifying a goal or purpose, formulating a theory, defining success metrics and putting a plan into action. These activities are followed by the Do step, in which the components of the plan are implemented, such as making a product. Next comes the Study step, where outcomes are monitored to test the validity of the*

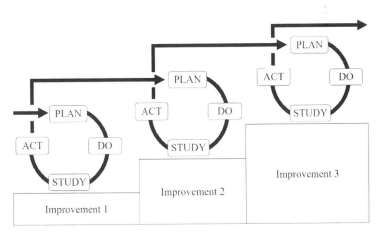

Figure 5.8.   Dr. W.E. Deming's Plan-Do-Study-Act (PDSA) model.

plan for signs of progress and success, or problems and areas for improvement. The Act step closes the cycle, integrating the learning generated by the entire process, which can be used to adjust the goal, change methods, reformulate a theory alto- gether, or broaden the learning — improvement cycle from a small-scale experiment to a larger implementation Plan. These four steps can be repeated over and over as part of a never- ending cycle of continual learning and improvement.

The A3 Plan

One tool that usually goes hand in hand with PDSA is the A3 process [17]. The output of this process is a single-page docu- ment that gets alignment on the problem, current state, root causes, proposed countermeasures and expected results. In addition to providing an efficient framework for visualizing and communicating improvement efforts, the accumulated A3 documents provide a valuable database of knowledge in the organization. An example is shown in Figures 5.9 and 5.10. The A3 document is split into two figures to make it readable.

The fictitious example starts with a simple problem statement (in the section titled "Problem") that the market share is drop- ping every quarter. This gets to the point without implying a solution and it is accompanied by data. The root cause analysis (in the section titled "RCA") summarizes the work that was done offline to dig into the reasons for the problem. This can be done using different methods. In this case, the key indicators high- lighted deficiencies in customer satisfaction, quality as well as the development and customer integration time. The analysis was done to understand the root cause of these poor results and a value stream map was completed to understand the complete development flow. This led to two specific root causes: (1) too many hand-offs/owners and (2) system testing occurs too late. The ideal state (in the section titled "Ideal State") provides a clear direction and ultimate target that needs to be achieved and the countermeasures (in the section, "Countermeasures/Alternatives")

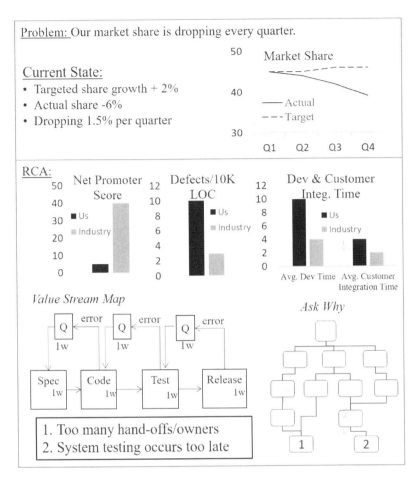

Figure 5.9. Example A3 document (first half).

defines the changes that will be made in the next iteration of the development cycle. The actions and associated dates are then provided (in the section titled "Next State") to create the executable plan with owners. Once the countermeasures have been implemented and the results measured, the process will be repeated and the A3 document will be refined to identify additional needed changes.

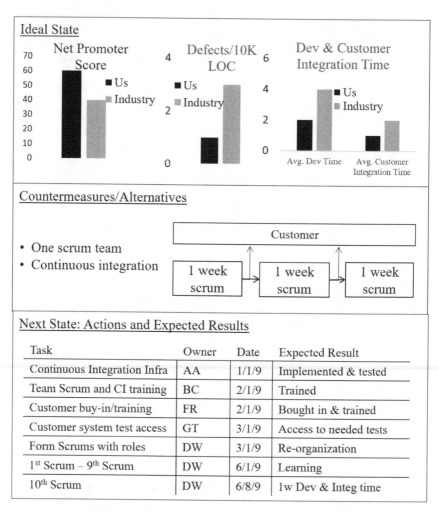

Figure 5.10.   Example A3 document (2nd half).

The After-Action Review (AAR)

In the 1970s the US Army began applying the concept of the After-Action Review (AAR) as "a method for extracting lessons from one event or project and applying them to others" [18]. This became an essential part of the US Army's Opposing Force (OPFOR) brigade that challenges troops in mock combat during training. The rationale and template for an AAR is described as [19],

> *"the units perform a collective self-examination in which the more general question, "How did the unit do?," is broken down into three more specific questions:*
>
> 1. *"What happened during the collective training exercise?" In other words, AAR participants attempt to specify the facts (i.e., the important actions and outcomes) of the simulated battle.*
> 2. *"Why did it happen?" Given the facts of the exercise, the participants attempt to explain the causes of particularly important actions and outcomes.*
> 3. *"How can units improve their performance?" Given that the previous two questions are answered, the participants determine appropriate actions to solve problems identified in their performance. Example actions include changes to unit standard operating procedures (SOPs) or increased training on basic drills.*
>
> *These questions are addressed during AAR sessions, which are conducted immediately after the end of a short exercise or during logical breaks in longer exercises."*

The AAR is an example of how one can capture learnings and decide on countermeasures that can be applied the next time. As Table 5.1 shows the AAR process also engaged the students much more directly in the analysis and identification of problems [20]. This type of engagement is an important element of continuous improvement.

SCRUM: The standup meeting, sprint review and retrospective

We described the SCRUM sprint earlier. There are three learning cycles embedded in each sprint in the SCRUM development model. The daily stand-up meeting is the opportunity for the team to discuss the status of the work in progress at a specific

Table 5.1.   Traditional performance critiques vs. AAR.

Characteristics of feedback sessions	Performance feedback method	
	Traditional performance critique	AAR
Soldier participation	Passive members of an audience.	Active participants in a discussion.
The main topic of discussion	Errors committed.	The sequence of events.
Direction of communication	One-way (from leader).	Two-way.
Atmosphere	Defensive.	Open to suggestions.
Instructional style	Traditional lecture.	Guided discovery and learning.
Source of information: why it happened?	Exercise leader and controllers.	Participants & members of opposing force (OPFOR), exercise leaders and controllers.
Source of Information: what happened?	Subjective judgment.	Objective performance indicators.

time each day and make adjustments. These meetings are very short and deal mostly with issues that need attention. The sprint review occurs near the end of the sprint and includes the customer. The goal of this review is to educate the customer on the current state and agree on any changes in the product backlog for subsequent sprints. Finally, as mentioned earlier in the book, the sprint retrospective provides the opportunity to capture improvements to development practices that can be implemented in subsequent sprints.

The combination of these learning cycles resolves issues rapidly, enables the development and priorities to be continuously adjusted with the customer involvement and continuously improves the development process. As a result, each sprint is not only a development activity but also a small experiment.

It is important to note that the objective is not to tear up the entire development process after one sprint but, instead, focus on a small number of digestible changes that can be implemented in the next sprint. Over the course of many sprints, these improvements lead to significant improvements in the overall productivity of each sprint.

The OODA loop for continuous strategic planning.

Colonel John Boyd was a jet fighter pilot and an influential military strategist in the US Air Force. In the 1960's he developed the Energy-Maneuverability Theory and championed the move to light-weight, maneuverable jet fighters that led to the highly successful F15 and F-16 generation of fighters. When referring to air to air combat he wrote [21],

> *"In order to win or gain superiority — we should operate at a faster tempo than our adversaries or inside our adversaries' time scale ... He who can handle the quickest rate of change survives"*

The key learning & decision process that had to be mastered was what he termed the Observation-Orientation-Decision-Action (OODA) loop shown in Figure 5.11. The OODA Loop recognizes that any model of reality is incomplete and needs to be continuously refined and adjusted. As Boyd noted in 1995 [22] when he presented his diagram,

> *"Note how orientation shapes observation, shapes decisions, shapes action, and, in turn, is shaped by the feedback and other phenomena coming into our sensing or observation window.*
>
> *Also note how the entire "loop" (not just orientation) is an ongoing many-sided implicit cross-referencing process of projection, empathy, correlation, and rejection."*

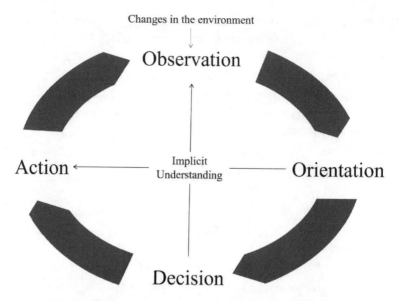

Figure 5.11.   A depiction of Boyd's OODA loop.

Since this loop exists for both adversaries in a conflict, the goal is to operate at a faster rate than the opponent by either speeding up your own loop, disrupting your opponent's loop or both. The Orientation step is where the most progress can be made to speed up one's own loop or, conversely, where the most damage can be done to their opponent's loop to confuse or slow them down.

The OODA loop provides a model for learning and adapting in an ambiguous and dynamic environment where inherent biases and new information must be continuously analyzed to synthesize a new reality and act. As noted, it originated as a model for combat operations but it has also been embraced for military strategy, law enforcement, corporate decision making and other fields.

In this model, each action needs to be tested to see if the result matched the hypothesis to understand what to adjust.

For jet fighters and other systems, this may be automated to fit within a very small timescale. For the analysis of something like a strategic action where feedback is not always automated, other means exist.

LAMDA and Design Thinking for technology development.

The Look Ask Model Discuss Act (LAMDA) Model

Dr. Allen Ward studied the Toyota Product System and described the learning cycles that happen during product development as Look, Ask, Model, Discuss and Act (LAMDA). He based it on his belief that [23],

> *"the aim of product development process is to create profitable operational value streams, and further, that creating usable knowledge is the key to doing so predictably, efficiently, and effectively. To create knowledge requires learning"*

LAMDA encourages direct observation of the work being done. It also encourages modeling multiple options, discussing them with the customers and other stakeholders for refinement, and conducting the experiments needed to test the hypotheses. Like other approaches, it is a continuous process that builds on the accumulation of knowledge over each cycle as depicted in Figure 5.12 [24]. Over time, the organization collects a lot of knowledge and establishes a deep understanding of the trade-offs associated with various product choices. Like in the case of the PDSA cycle, an A3 or Knowledge Brief can be used to succinctly capture the problem, current state, plans and results.

Design Thinking for technology development

Design thinking is an approach that employs learning cycles to look at design problems and opportunities holistically in close collaboration with everyone involved in producing and consuming the product. Similar to the other approaches already

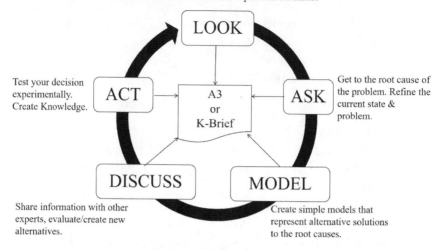

Go to where the work is being performed and directly observe how it is done. Define the current state & problem statement.

Test your decision experimentally. Create Knowledge.

Get to the root cause of the problem. Refine the current state & problem.

Share information with other experts, evaluate/create new alternatives.

Create simple models that represent alternative solutions to the root causes.

Figure 5.12.   The Look-Ask-Model-Discuss-Act (LAMDA) model.

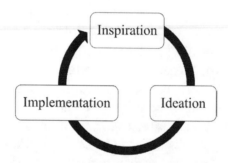

Figure 5.13.   The Design Thinking model.

covered, a key tenet is "direct observation". This is coupled with a very inclusive process of brainstorming and prototyping with the end customer to converge on the best answer incrementally. As described by Tim Brown in his Harvard Business Review article in 2008 and depicted in Figure 5.13 [25,26]:

"Design projects must ultimately pass through three spaces (see the exhibit "Inspiration, Ideation, Implementation").

We label these "inspiration," for the circumstances (be they a problem, an opportunity, or both) that motivate the search for solutions; "ideation," for the process of generating, developing, and testing ideas that may lead to solutions; and "implementation," for the charting of a path to market. Projects will loop back through these spaces — particularly the first two — more than once as ideas are refined and new directions taken."

Design Thinking is an important part of the "Apple Way" as described in the 2012 Harvard Business School article on, Design Thinking and Innovation at Apple [27]:

"Given the sleek appearance of iPods, iPhones, the iPad and Mac computers, and all these products' prominence in media depictions, it's tempting to attribute their popularity to Apple's ability to tap into a zeitgeist — a sense of what is popular, fashionable, trendy at the moment. But there is more to coolness than fashion. In fact, Apple goes beyond superficial trends and gets to the essence of customer experience such that its "design" seems to happen from the inside out, while the outside continues to be deeply appealing and, ineffably, "cool."

Larry Tessler, an engineer from Apple, described how this was applied with respect to his work on the early implementations of the user interface [28].

"In the user interface design, it was a lot of trial and error. We tried different things and found out what did and didn't work. A lot of it was empirically driven. I kept bringing new stuff in and saying, "What about this?" and [we would] set up tests so that different people could try it. For example, if you have a scroll bar . . . which way should [the arrows] be? When you scroll toward the bottom of a document, the document moves up, so there's some reason to think of a down arrow, and some reason to think of an up arrow. A really good question is, "What do people expect? When people see an arrow, which way do they think it will move?" What I found mattered much more than whether the arrow went down or up

was where the arrow was: if the arrow was at the top, they expected to see more of what was above, whereas if it was at the bottom, they expected to see more of what was below."

5.5 Summary

The R&D system is composed of people responsible for one or more tasks in a set of workflows that span R&D organization and the other entities required to get the work done and deliver value. The completion of each task in a workflow generates information on the engineering decisions and its operation. When this knowledge is used to improve subsequent engineering decisions or incrementally improve the workflows, then a learning cycle has been completed. Learning cycles are a critical part of enabling continuous improvement in the R&D organization and to the broader R&D system.

The R&D system with the shortest learning cycles explores more ideas, makes course corrections sooner, improves development processes faster and delivers more value than the competition. It also means that the role of the engineer is much more than designing new or improved technology it also includes continually improving the workflows and processes they use. Using learning cycles to reduce the time it takes to deliver value is a foundation of holistic engineering management.

Learning cycles are not a new concept. They are just applying the scientific method of *observation — hypothesis — test* to engineering workflow and processes. It is a philosophy that is embedded into lean thinking and agile development practices. Many problem-solving and continuous improvement models utilize learning cycles and they can be applied to strategic planning, engineering research and development.

Learning cycles are a means to conduct focused and controlled experiments during execution and compare the results with expected outcomes. Changes that meet the expectation can become part of a revised standard. In this manner, learning cycles offer a means to continuously challenge and improve standards with people responsible for their execution.

More than anything else, implementing learning cycles requires a way of thinking and a culture where everyone in the enterprise is involved and views themselves as scientists conducting experiments to improve the workflows. This requires the conviction by everyone that this is a critical part of their job description and valued by their management. Enlightened leaders recognize that shorter learning cycles lead to better technology and by enabling everyone to improve their workflows they create a competitive advantage for the company.

5.6 Key Points

- Learning cycles exist at all levels of engineering and run concurrently.
- Learning cycles for workflows exist within and across entities.
- Learning cycles for processes exist within and across entities.
- Learning cycles are a key part of continuous improvement.
- Learning cycles work best using intrinsic networks.
- Everyone must become a scientist and conduct experiments.
- Learning cycles complement and co-exist with standardization.
- There are several ways to execute learning cycles.
- PDSA, A3, AAR, SCRUM can be used for workflow improvement.
- The OODA loop can be used for strategic planning.
- LAMDA and Design Thinking can be used for development.

5.7 Case Studies

5.7.1 *Manufacturing, CAD and Design Learning Cycles*

Overview

In 1965 Gordon Moore observed that the number of transistors on a microprocessor was doubling every year while the cost of computing was declining. He later revised that observation in 1975 to the number of transistors doubling every two years. This observation drove long-range planning and targets in the semiconductor industry for decades.

The Problem

To stay on the trajectory defined by Moore's Law, the critical dimensions of the silicon process technology had to shrink with every new manufacturing generation. To accomplish that, the complexity of the manufacturing process increased (as measured by the number of mask layers). If we take a general industry assumption that each mask takes about one day to process in the fabrication facility, then the cycle time required to complete a chip grows as depicted in the generic example shown in Table 5.2 [29].

Actions Taken

To keep pace with Moore's Law, companies like Intel made major improvements in semiconductor manufacturing technology every two years. The time it takes to define an experiment, manufacture the test silicon in the fabrication facility, test and sort the output and then analyze the results is what some people call an information turn and what we call a learning cycle in this book. Figure 5.14 depicts a simple picture of the steps involved in each learning cycle.

Due to the increasing number of fabrication steps and layers of interconnect needed to connect more transistors the amount of information that needs to be analyzed and the time taken to generate test silicon will grow with each generation. As a result, to stay on the curve defined by Moore's Law,

Table 5.2.   Growth in manufacturing masks as transistor dimensions shrink.

Process node	Mask layers	Avg. cycle time (Days)
28 nm	40–50	40–50
10/14 nm	60	60
7 nm	80–85	80–85
5 nm	100	100

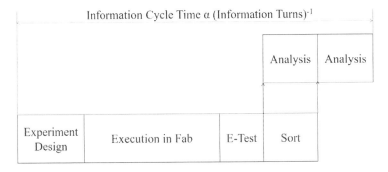

Figure 5.14. Typical Si manufacturing learning cycle.

improvements are needed to increase the number of wafers that can be run through a processing step every day to offset the growth in complexity.

In addition to the advances needed in manufacturing process technology, a new and improved microprocessor design needed to be executed and made ready for manufacturing at the same two-year beat rate defined by Moore's Law. This was needed to take full advantage of the increase in available transistors on the die and deliver performance improvements required by the customers. This meant changes in the circuit design and physical implementation and integrating more functions using the design rules for the new manufacturing process.

To design the new microprocessor in the same length of time (to stay on the two-year cadence) either the design engineering productivity or the number of engineers had to double every two years. This is required to address the growth in the numbers of transistors, increasingly complex design rules and more aggressive power, performance and other specifications. This meant that the Computer Aided Design (CAD) tools and design methods had to improve in parallel with the advances in manufacturing process technology and microprocessor design.

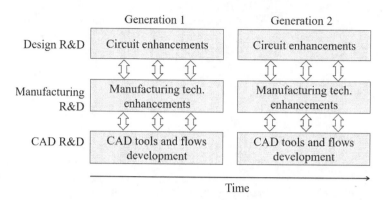

Figure 5.15.   Co-optimization on a two-year cadence.

Actions Taken

To come together and give the optimum result, all three of these components (manufacturing process technology, microprocessor design, CAD solutions) had to be "co-optimized" together simultaneously as shown in Figure 5.15.

Results

Establishing a 2-year cadence for overall improvement spawned a series of shorter learning cycles within the development of each new generation that would repeat like clockwork. This allowed suppliers, co-development partners and other collaborators to synchronize their work to each learning cycle with clear goals in mind. It also permitted customers of the semiconductor companies to plan their products several years in advance and take advantage of the standard beat rate of manufacturing and microprocessor architecture improvements.

Questions:

(1) What types of learning cycles were embedded in the 2-year cycle?
(2) What did Intel have to do to maintain a 2-year cadence?

(3) What were the advantages of Intel's development cadence?
(4) How can Intel stay on the same cadence as complexity grows?

References

[1] Boyd, J. R. (1976). *New Conception for Air to Air Combat*. p. 19, 24. Available from: https://www.colonelboyd.com/boydswork [Jan 2020].

[2] Pedro, D. (2017). *The Master Algorithm*. (Penguin Books, UK) p. 13.

[3] Highsmith, J. *History: The Agile Manifesto*. Available from: https://agilemanifesto.org/history.html [June 2020].

[4] Agile Alliance. *Principles Behind the Agile Manifesto*. Retrieved from https://agilemanifesto.org/principles.html [June 2020].

[5] Schwaber, K. (1995). Scrum Development Process, Object-Oriented Programming, System, Languages & Applications Conference, p. 1.

[6] Sims, C. and Johnson, H. L. (2011). *Elements of Scrum*. (Dymaxicon) pp. 81–96.

[7] *Ibid*, p. 91.

[8] Cockcroft, A. (2013). Velocity and Volume (or Speed Wins), Presentation at FlowCon 2013.

[9] Spear, S. and Bowen, H. K. (1999). Decoding the DNA of the Toyota Production System, HBR. September-October Issue, p. 111.

[10] Burrows, P. (2004). The Seed of Apple's Innovation, Business Week. December 12, 2004.

[11] Ford, H. (2017). *My Life and Work*. (Create Space Independent Publishing) p. 37

[12] *Ibid*, p. 10.

[13] *Ibid*, p. 40.

[14] *Ibid*, p. 41.

[15] Deming Institute. *Dr. Deming's 14 Points for Management*. Available from: https://www.deming.org/explore/fourteen-points [June 2020].

[16] Deming Institute. *PDSA Cycle*. Available from: https://deming.org/explore/p-d-s-a [June 2020].

[17] SixSigma (2017). *Lean Six Sigma A3 Process Keeps Problem Solving Organized*. Available from: https://www.6sigma.us/six-sigma-in-focus/a3-process-problem-solving/ [June 2020].

[18] Darling, M. Parry, C. and Moore, J. Learning in the Thick of It, Harvard Business Review, July–August, 2005.

[19] Morrison, J. and Meliza, L. (1999). Foundations of the After Action Review Process, U.S. Army Research Institute for the Behavioral and Social Sciences. Special Report 42, July 1999, p. 1.

[20] *Ibid*, p. 8.

[21] Boyd, J. R. (1976). *New Conception for Air to Air Combat*. p. 19, 24. Available from: https://www.colonelboyd.com/boydswork [Jan 2020].

[22] Boyd, J. R. (2018). A Discourse on Winning and Losing, Air University Press, Curtis E. LeMay Center for Doctrine Development and Education Maxwell AFB, Alabama.

[23] Ward, A. C. (2007). *Lean Product and Process Development*. (The Lean Enterprise Institute) p. xii.

[24] *Ibid*, p. xiii

[25] Brown, T. (2008). Design Thinking, HBR. June, 2008.

[26] *Ibid*.

[27] Thomke, S. and Feinberg, B. (2012). Design Thinking and Innovation at Apple. Original January 2009, Revised May 2012 (Harvard Business School). Available from: https://www.hbs.edu/faculty/Pages/item.aspx?num=36789.

[28] Moggridge, B. (2007). *Designing Interactions* (MIT Press). p. 89.

[29] Lapedus, M. (2017). Battling Fab Cycle Times, Semiconductor Engineering, February 16, 2017 Available from: https://semiengineering.com/battling-fab-cycle-times/ [June 2020].

CHAPTER 6

Strategic Planning

> Strategy without tactics is the slowest route to victory. Tactics without strategy is the noise before defeat.
>
> — Sun Tzu [1]

As we described in the introduction, the R&D manager leads their organization and also directs the R&D system within a larger ecosystem that includes existing and potential competitors that pose a threat and potential new suppliers, customers and others that provide opportunities for cooperation. To remain vital, the R&D organization must anticipate and respond to that evolving ecosystem. This is true even if the R&D organization is buried deep within a company and delivers only to other internal groups. The process of doing that is what we call strategic planning.

Strategic planning comprehends the competitive and cooperative forces in the ecosystem and determines the best mix of value for the R&D organization and the means to achieve it. Effective strategic planning results in actions that create differentiable and sustainable value for the customer. This is accomplished by continuously synthesizing new information, creating knowledge, forcing actions, and helping to implement them. As a result, the primary product of strategic planning is not the plan but the continuous loop of observation, orientation, decision and action.

Strategic planning differs from tactical planning in that it focuses on actions that materially affect the organization's direction. These are not easily reversed and set the organization's destiny. Strategic actions may be induced top-down by management or result from autonomous efforts of individuals or groups that do not align with the existing strategy. Managing the strategic planning processes that properly balance induced and autonomous actions and translates decisions into objectives that can be easily communicated to the organization is an important part of the R&D manager's job.

We address the strategic planning topic before we discuss the other core engineering functions and the design of the R&D organization itself because the strategy guides decision-making by the development and engineering research functions and has a major influence on the capabilities needed by the R&D organization and how they are structured.

6.1 Strategy and Strategic Actions

A strategy determines the best mix of value and the means to achieve it.

In his book, Competitive Strategy, Michael E. Porter wrote:

> *"Competitive strategy is about being different. It means deliberately choosing a different set of activities to deliver a unique mix of value... the essence of strategy is in the activities — choosing to perform activities differently or to perform different activities than rivals. Otherwise, a strategy is nothing more than a marketing slogan that will not withstand competition" [2]*

Therefore, the role of strategic planning for an engineering organization is to define the best mix of value that it can provide and describe the unique engineering methods, processes and technology that will be needed to create competitive differentiation. As we will discuss later, we feel it is also the responsibility of the strategic planning function to drive the actions needed and then help with the implementation as well.

A strategy may make the pie bigger for everyone.

In their book, *Co-opetition* [3], Adam Brandenburger and Barry Nalebuff used the concept of game theory to go beyond the usual win-lose paradigm and show that cooperation (even with competitors) can sometimes lift all boats. In their view,

"Business is cooperation when it comes to creating a pie and competition when it comes to dividing it up."

They went on to write [4],

"Even if it's a good game, think about creating a better one. Changing the game is the essence of business strategy"

For an engineering organization, this means that strategic planning also explores opportunities to collaborate with others to benefit the industry as a whole by doing such things as expanding the total addressable market or by reducing costs for everyone. This means taking advantage of opportunities to work with suppliers, customers and even competitors to expand the pie and change the game.

A strategy is built on assumptions. These define the risk.

A strategy is based on assumptions about the current state of the ecosystem and how it will react to a change. As a result, each activity required to deliver the mix of value proposed in a strategy is based on certain assumptions and if any one of those turns out to be false then the strategy, as defined, will fail. For example, if a strategy is based on signing exclusive licensing deals with a supplier then there is an explicit assumption that this is possible. If that ultimately turns out to be false then the overall value proposition will not be reached and the strategy will fail. Other more subtle assumptions may also be tied to that. For example, the possibility of an exclusive licensing deal may be based on the assumption that the supplier will see value in the agreement for themselves. This in turn may be based on an assumption about the relevant market conditions. If the assumed market conditions turn out to be false then that the deal will not be possible. Consequently, the risk associated with a strategy correlates to the size and number of assumptions that must be true for it to succeed.

Strategic actions materially affect the R&D organization's direction.

What further separates strategic planning from other types of planning is the motivation for the plan and the consequences of the actions. The motivation comes from the need to address relevant market, competitive or cooperative forces and the consequences are usually a change in direction for the organization that is difficult to reverse. In his book, Strategy is Destiny, Robert Burgelman [5] wrote,

> *"Strategy is therefore concerned with the external and internal forces that have the potential to materially affect the company's destiny."*

He went on to write that strategic actions differ from tactical actions in the following way.

> *"Strategic action is consequential; it involves resource commitments that cannot be easily undone and moves the company in a direction that is not easily reversible. This view provides a criterion for distinguishing between tactics and strategy. Actions are tactical if their outcome does not significantly affect subsequent degrees of freedom to act."*

Similarly, strategic planning for an engineering organization addresses the evolving forces in its ecosystem that require a significant change in direction. These typically are not decisions that can be made by individual groups within the R&D organization since they may require shifting resources between them.

The origins of strategic actions may be induced or autonomous.

Strategic actions can originate from the current strategy of the company or R&D organization. Robert Burgelman called these *induced strategic actions*. Other strategic actions can originate bottom-up within the R&D organization and are not aligned with the current strategy. Robert Bugelman called these *autonomous strategic actions.*

Induced Strategic Actions: The existing beliefs in a company or engineering organization drive the official strategy. This is influenced by the current ecosystem, past experiences and expectations for how everything will evolve going forward. Within the company, the different R&D organizations create initiatives in support of the strategy. They compete for resources and priority with other organizations using well-known standard processes.

Autonomous Strategic Actions: Independently, individuals or groups may start small initiatives in response to an emerging external environment. They typically have very little to do with the current strategy and therefore, are not funded very well. Most will not see the light of day because the failure rate is very high. In companies where the induced strategic actions are proving to be very profitable, it can also be very hard to justify doing anything else and, therefore, these can die on the vine. This is what Clayton Christensen famously called the innovator's dilemma [6]. However, important disruptive technologies can result from this path that ultimately change the official strategy of the company.

6.2 Strategic Planning Scope

Strategic planning balances induced and autonomous strategic actions.

Induced strategic actions by their nature are well known and managed through standard processes in the R&D organization or the company. However, strategic planning also must harvest the autonomous strategic actions bubbling up from within the R&D organization. There may be many autonomous strategic initiatives that start on their own at a low level, percolate for a while, and then reach a point where a decision is needed to make them part of the R&D organization's strategy.

To enable this, some level of strategic dissonance should be tolerated as long as it is motivated by an attempt to create value and not hidden from view. This requires a culture of "disagree and commit". This means that, after sufficient debate, people in the R&D organization commit to following the strategy but feel empowered to continually test and improve it. In other words, thoughtful disagreements can lead to thoughtful commitments. This is critical for innovation to thrive since it allows progress to be made on the stated strategy while allowing experimentation and further improvement. As Robert Burgelman wrote [7],

> *"Successful firms are characterized by maintaining bottom-up internal experimentation and selection processes while simultaneously maintaining top-driven strategic intent."*

The strategic planning network spans the entire ecosystem.

As shown in Figure 6.1, the strategic planning function is directed from inside the R&D organization. It interacts with all of the participants of the existing R&D system as well as potential new

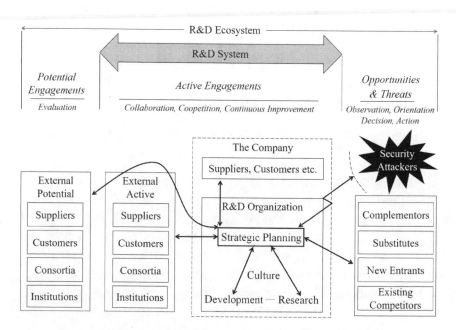

Figure 6.1.   Strategic planning and the R&D ecosystem.

suppliers, customers and other cooperators. It does not directly engage with existing competitors but must divine their direction and anticipate new forms of competition. Inside the R&D organization, close interactions exist between the engineering research and development functions. There may also be interactions with other institutions and industry consortia where suppliers, customers and competitors may co-exist. Although the process of strategic planning needs an owner, the participants can be organized in different ways to make the best use of the skills in the organization.

6.3 Continuous Strategic Planning

Strategic planning operates like an OODA loop.

The famous hockey player, Wayne Gretzky, was quoting his father's advice to him when he said, "Skate to where the puck is going, not where it has been." It is good advice and since then the phrase has been repeated by Warren Buffet, Steve Jobs and many other leaders. But how does the manager of the R&D organization know where the puck is going? Wayne Gretzky answered that question in the documentary "In Search of Greatness". As summarized by Ben McGrath of the New Yorker [8], it was the result of a lot of data collection and analysis that started when he was very young.

> *"Wayne Gretzky describes his habit, as a four-year-old, of watching hockey games on television with a pen and paper in hand. On the paper, he would draw a rink: the ice surface, as viewed from above. Then, while staring at the action on the screen, and without diverting his eyes to the paper, he would trace the movement of the puck. When play stopped, he would look down at his scrawls and observe what he recalls, in retrospect, as patterns: areas of the ice where the flow seemed to concentrate, and others that his pencil scarcely touched."*

As Wayne Gretzky explained in the documentary, this data collection and analysis continued when he was an adult playing in the NHL [9].

"When I was a player, I could tell you every single thing about every player that was in the league... Each and every night I watched every game. We really were enticed by preparation and what was going to transpire on the ice; who you are going to play against, what system they are going to play against you. You have to understand that every night the system you are going to play against is completely different and you have to be prepared for that."

Wayne Gretzky knew where the puck was going to be because he studied the game. If the manager of the R&D organization does not have a way to systematically study the game, then they will not know where the puck will be. Also, even if they have a way to that, but learn slower than the competition, they will not get to the puck first.

The main learning cycle of strategic planning is similar to the OODA loop we described in Chapter 5 and also shown in Figure 6.2. In this example, the observation step represents the process of continuously sensing changes in the R&D ecosystem and the operation of the R&D system. The orientation step represents the

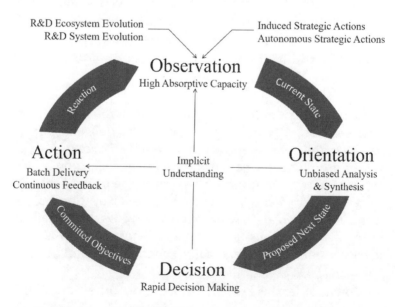

Figure 6.2.   The OODA loop for continuous planning.

analysis and synthesis of this information into a proposal and set of alternatives. In some cases, the decision to proceed is obvious and does not need a formal decision. This is shown in the figure by the line called "implicit understanding". The decision step leads to a reformulation of the overall strategy and associated objectives and key results. The action step includes the implementation of the strategy and the reaction to it. This is important since, as we noted before, strategic planning does not end with the delivery of a plan. It is a continuous loop that includes making sure the strategy is implemented and then learning from the result.

> *The primary product of strategic planning is not the plan but a continuous process that creates knowledge, makes decisions and helps implement the actions.*

As noted earlier when we first presented the OODA loop [10],

> *"In order to win or gain superiority — we should operate at a faster tempo than our adversaries or inside our adversaries' time scale"*

With this in mind keeping the cycle time of each loop as short as possible is important. For a one-person operation, the OODA loop is performed in a single brain. No meetings are needed and analysis is performed, decisions are made and action is taken in seconds or minutes. For a small team with no external dependencies completing the OODA loop may happen over coffee with the rest of the team. The complexity arises when R&D is distributed across multiple teams with external dependencies. This is the case we are concerned about in this book. In this scenario, the OODA loop is not a license for adding bureaucracy and slowing everything down. Done correctly, the OODA loops should still complete in minutes in some cases. Therefore, it is paramount that the R&D manager continuously improve the strategic planning process to eliminate waste and get as close as possible to the ideal state of making decisions at the same speed as a one-person operation.

Since the observation and orientation tasks are continuous it is up to the management of the R&D organization to decide on a cadence for formal decisions requiring management engagement. As noted, some decisions will be obvious based on past experience and will not need a formal review and decision meeting. It is important to understand what types of decisions need a formal review, the manager's approval or can be left to a different leader in the R&D organization to decide on and communicate. In general, decisions should be made at the lowest acceptable level of the R&D organization's management hierarchy.

The R&D manager is responsible for enabling robust strategic planning.

The role of the R&D organization's manager is critical. The R&D ecosystem evolves at a rapid pace and disruptive technologies and methods quickly evolve under the radar of most companies. As a result, they are missed by those doing well and seemingly on a comfortable improvement trajectory. This type of complacency ultimately leads to failure. The R&D organization manager must embrace their role in driving the strategic planning process and continue the momentum even when things seem to be going quite well.

We will now dive into the details of each component of the OODA loop for strategic planning. We discuss some models and tools that can be used in each step. However, there are many to choose from and whether a Porter force diagram is used to analyze the external forces in the industry or some other equally good model is up to the manager. The primary objective is to generate the right information to have a fruitful conversation. The tools and models can vary.

6.3.1 *Observation*

Observation senses change in the ecosystem. It requires conversations.

The goal of observation is to discover relevant change to the current state that requires action. As shown in Figure 6.3, a

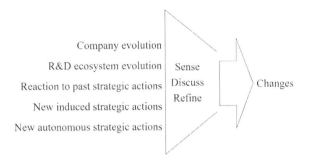

Figure 6.3. The observation function in the OODA loop.

current state is assumed from earlier iterations and the focus is on identifying the changes to the R&D ecosystem that present threats or opportunities. These include strategic actions from within the R&D organization that follow the trajectory of an existing value proposition (induced), changes that are coming from a different source with a new value proposition not included in the existing strategy (autonomous), changes in dynamics and capabilities across the R&D system, the unfolding R&D ecosystem and any reaction from recent strategic actions.

This type of observation is not easy. Many points of information may be irrelevant and those that appear relevant may be contradictory. Observation requires conversations between the leaders and content experts in the R&D organization and strategic planners (if they not already the same) to understand and categorize information. Decisions may also be clouded by conscious or unconscious biases. Also, observation is not only about finding new information. Old information may become relevant again in light of changes in the environment. Connections may be discovered between new and old information that provides new insights. Finally, the type of information sought is influenced by the analysis that is performed. As a result, knowledge, experience and intuition is needed to sort, filter and connect new and old information. The process of sensing, discussing and refining the observations is continuous and open-ended but the sampling of the output of this process can be done on a specific cadence.

Former Intel CEO, Andy Grove, once said that if you want to know someone's strategy, you need to watch what they do and not what they say. Toyota is famous for its philosophy of direct observation that exhorts its supervisors and managers to go beyond the charts and graphs and go to where the work is being done to see for themselves what is happening. When this is done, revelations sometimes occur that would not have been apparent by reading a chart. As Dyer, Gregersen and Christiansen note, observation is a critical trait of the innovator's DNA [11],

> *"Innovators carefully, intentionally, and consistently look out for small behavioral details — in the activities of customers, suppliers, and other companies — in order to gain insights about new ways of doing things."*

6.3.2 *Orientation*

Orientation analyzes and synthesizes new information.

The goal of the orientation task is to analyze information obtained through observation, understand the relative strengths, weaknesses, opportunities and threats across the ecosystem and then synthesize that into a recommendation. In most cases, due to the type of questions owned by the strategic planning function, they will require a formal review and decision. Similar to the case for observation, the orientation task is a continuous process and the output can be sampled continuously or on a specific cadence.

As depicted in Figure 6.4, new information, past experiences, assumptions and the culture of the people in the R&D organization affect the analysis and synthesis. In some cases, experience with a similar event may prove useful in bypassing the analysis and decision-making tasks and move directly to action. In other cases, the feeling that one has "been there, done that" may obscure a nuance that makes this time different. Also, the R&D organization's culture may bias the analysis in a direction that

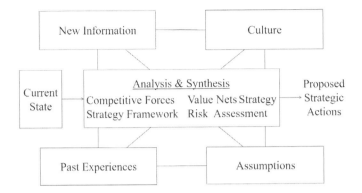

Figure 6.4. The process of orientation in the OODA loop.

people feel more comfortable with and miss disruptive events. Finally, assumptions about the ecosystem and the organization's capacity to deliver determine the perceived risk and incorrect assumptions will give a false sense of security. It is important to recognize and mitigate the bias that results from these during the analysis and synthesis of the new information.

The orientation task is the most complex step since it needs to analyze and synthesize information that spans the entire ecosystem and address biases that may cloud judgment. The following are three examples of the type of analyses that can be done in the orientation phase. They are complementary and, together, form a more complete picture of the analysis needed:

(1) Understanding competitive forces using Porter's model.
(2) Changing the game using the Value Net Analysis.
(3) Assessing the strategy vs. the current reality.

Now we will address them in more detail.

Understanding competitive forces using Porter's model

There are competitive forces inherent in each industry that must be considered when formulating a strategy. These need to be

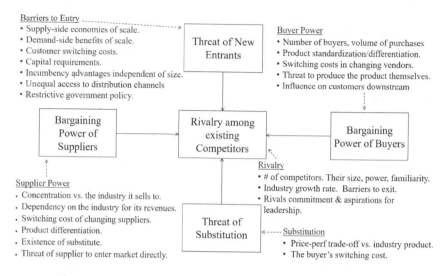

Figure 6.5.   Michael E. Porter's five competitive forces.

understood since the industry's composition will define the R&D ecosystem and strongly influence what strategies will succeed in the R&D organization. We will start with a review of Michael Porter's five forces [12] shown in Figure 6.5, which captures the degree of competition in an industry and the levers available to adjust the forces. The relative strengths of the forces will be different for each industry and change over time. It is part of strategic planning's function to understand them and propose strategies to shape them.

A *new entrant* may be a start-up or a company with an engineering function in an adjacent business that enters the market that the R&D organization serves. In this case, the goal of strategic planning is to be aware of all of the potential entrants and propose the proper response. There are often barriers to entry that work in favor of the status quo and block new entrants. If the cost of adopting new technology is high then a customer may be reluctant to entertain an alternative. Similarly, if the investment required for a new entrant to develop a competing technology is large, then they may not like to take the risk. However, for the technology industry, the threat of new entrants is ever-present and existing companies build strategies that raise the barriers to entry. Incumbents often integrate hardware and software capabilities with services and deliver them as a single package that is impossible to separate. This increases the customer switching cost, the supply-side economies of scale and capital requirements in the hopes of dissuading others from entering the industry.

A substitution is different from a new entrant. In this case, it describes something Clayton Christiansen termed a disruptive technology [13]. A disruptive technology starts with a unique value proposition but evolves in a way that enables it to supplant a different product or service. An example is the cell phone that has largely supplanted simple point and shoot cameras. The threat of substitution is high when the price and performance of a disruptive technology become competitive with the existing one since it brings along additional value associated with its original purpose. The threat increases further if the cost of switching to the new product is also low. More than any other case, this threat is the most difficult to determine. By the time it is clear that a specific disruptive threat exists, the threatened engineering organizations typically do not have time to react. Therefore, understanding the inherent risk posed by substitute technologies from adjacent industries is important. Not to mention, by being diligent strategic planning may also identify opportunities to disrupt other industries with their solutions.

A supplier provides the engineering technology, information, or services that are used by the R&D organization to do the engineering. Suppliers are in the business of making money just like their customers. They also have their analysis of the forces in the industry and want to maximize their leverage. If a supplier is in a powerful position, they may demand a larger price for their services and lower the customer's profit margins. However, this may also encourage the customer, if they are capable, to enter that business themselves. Some suppliers may be held in check by this possibility. Not surprisingly, the power of a supplier increases if there are not many other suppliers or potential substitutes. To counteract this, customers often nurture or seed other suppliers to increase competition.

The customer (*buyer*) purchases the product of the R&D organization. The customer may be an internal group in the same company, an external company, a distributor, or individual consumer. Similar to the case of the supplier the customer is often in the business or making money and many of the same dynamics apply but with the opposite effect. If a customer is in a powerful position, they may demand favorable pricing, additional services, or other benefits to commit to a purchase. The power of a customer increases if there are many suppliers and few customers. If the customer values low cost over differentiation, its power can be further increased by driving standardization across its supplier base to lower switching costs. As we already noted, the customer's power also increases if it is a legitimate threat to build the suppliers' products itself. In the technology industry, this is not unusual.

In the center of the picture is a box that represents the *rivalry* among existing competitors. It is a measure of the intensity of the competition within the industry. If there are many similar competitors with similar products battling for the same customers, then the intensity of the rivalry is high. If the competition is based on price then the profit potential is low. On the other hand, if the competition is based on new capabilities that open up new usages or markets then the rivalry can increase the profit potential.

An important point to consider is that an industry may also have a large barrier to exit (as opposed to a barrier to entry). In this case, once engaged in the industry it is hard to get out. For example, when technology is provided to a company that uses it in its product line, there will be an expectation of continued support for the duration of the product's lifespan. If the supplier decides to exit that business, it may not be possible without sustaining a very negative reaction from its customers.

Changing the game using the Value Net Analysis.

We now look at a different view based on game theory that combines competition and cooperation (aka coopetition) using the net value model from Brandenburger and Nalebuff [14]. In addition to the competitive forces, there are opportunities for cooperation with suppliers, complementors, customers and even competitors to expand the market or drive benefits for the entire industry. The value net model [15] depicted in Figure 6.6. looks at the industry through the lens of game theory. In a game, there are players. The players in this game are the subject company, its suppliers, customers, competitors and those who offer a product that makes the company's product more valuable to customers. They are termed complementors. It's important to keep in mind for this model and for the previous five forces model that the same company can be a supplier, competitor customer and complementor in different scenarios. Also, each player has their value net diagram. Suppliers of the R&D organization also have suppliers, competitors and complementors and other customers. These two facts make this a very dynamic analysis with many different ways to play the game.

Figure 6.6. The value net model and the R&D organization.

A key part of the value net exercise is to fully understand who is a competitor. A competitor is not just another company or engineering organization that provides the same technology. It also can be an entity that provides a substitute for the product (something Porter captured as the threat of substitution in the five forces model). An entity that makes it less attractive for a supplier to provide resources to support the development of a technology is also a competitor. When viewed from that perspective, the list of competitors can become very large and diverse.

Similarly, fully comprehending who is complementor is a key part of the value net analysis. A complementor provides a product or service that increases the value of the R&D organization's technology in the customers' eyes. The Microsoft operating system and Intel microprocessors are often cited as an example of this. Also, an entity that makes it more attractive for a supplier to provide resources to enable the R&D organization's technology development is a complementor.

In the value net model, five elements can be used to change the game:

Players: Becoming a new player in an existing industry is a big way to change the game and has an obvious upside if successful. Bringing other players can also significantly change the game. As the Porter model shows, bringing in new customers reduces the power of the other customers. Similarly, bringing in new suppliers reduces the power of other suppliers. Adding complementors makes sense by definition. Even adding competitors may prove advantageous in situations where obtaining a major business contract requires a 2nd source.

Added Value: How well the R&D organization will do depends on the relative added value of all of the players. Finding a different trade-off between product performance, features and cost can change the game. An example is trading off performance to reduce costs to enter a lower market segment. Creating virtuous cycles of growth with complementors that create additional value and wall off competitors has also happened several times in the technology industry.

Rules: Contracts can be constructed to change the game. For example, having a clause in a contract that permits a supplier to meet the price of any bid from another supplier reduces the chances of the existing supplier losing business. However, it also opens the door for competitors to bid down prices, knowing they do not need to deliver.

Tactics: Tactics are used to manage perceptions by either clearing things up, preserving obfuscation, or creating new obfuscation. For example, credibly communicating how you intend to react to a perceived threat can sometimes eliminate the threat. Conversely, hiding the true cost of technology development helps keeps the competition guessing and may slow their decision-making process.

Scope: An entity that is a player in one game may be a player in a different game too. In these cases, a strong added value provided to a customer in one game may be used to enter a new game that has the same customer. Applying tactics to create the perception of this type of linkage (or not) may also influence the strategy of competitors.

So, considering all of this, what happens in an ecosystem where there is no room for cooperation and where competitors are constantly catching up, leapfrogging each other, where imitation is common and no advantage seems permanent? In this case, the opportunity to change the game may be very limited and it once again boils down to velocity. As Brandenburger and Nalebuff point out, once it reaches that point,

> *"The game isn't about how good your products are; it's about how good you are at improving them. It isn't where you are; it's how fast you are moving, it isn't position; its speed. You never stand still; you're a moving target."* [16]

Taken together, the players in the competitive forces and co-opetition analysis form the more complete picture of the main participants in the ecosystem as shown previously in Figure 6.1.

Assessing the strategy vs. the current reality

As shown in Figure 6.1, the R&D organization is at the center of the R&D ecosystem. Equally important to understanding the external forces is the need to deeply understand the R&D organization itself and its strategic position in the industry. This means having a clear understanding of the current strategy and what it takes to win, the unique competencies inside the R&D organization, the strategic actions already in flight, and finally the biases that exist that influence the selection of induced vs. autonomous strategic actions.

Stanford professor Robert Burgelman worked closely with Intel's former CEO, Andy Grove, for many years and during that time he was given access to many internal strategic discussions as Intel became a dominant player in the semiconductor business. Figure 6.7 is a model used in those discussions to help understand the evolution of the strategic position of the company relative to the industry over time [17]. It became known within Intel as the rubber band model and it is a good model to apply for the R&D organization in the context of its ecosystem.

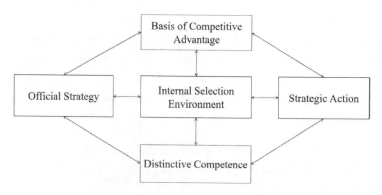

Figure 6.7.   Robert Burgelman's rubber band model.

In this model, the basis of competitive advantage closely relates to the conclusions from the analyses of the external forces we just described. It includes the players shown in the ecosystem and defines the basis for the competitive advantage for each position the R&D organization has staked out. Like everything else, this is dynamic and will evolve. There should be strong alignment between the distinctive competencies of the R&D organization shown in the bottom of the figure and the basis for competitive advantage shown at the top. The distinctive competencies include the underlying culture, the skills and related resources that help differentiate itself from the competition and support the basis for competitive advantage. The official strategy shown on the left of the figure captures the top-down communication of the strategy, values and priorities based on the senior management's beliefs. On the other hand, on the right of the figure, strategic action describes what people are working on. The internal selection process is the means used to manage that alignment of all of these components.

One goal of strategic planning should be to fully understand how the R&D organization fills out the rubber band model. By going through that exercise, disconnects and dissonances can be identified. Since autonomous strategic activities can become a source of innovation, the rubber band model should not always be in perfect alignment. However, there will also be cases where problematic delays in aligning to a new strategy will be highlighted using this simple model. In this respect, the role of strategic planning is to make sure the current state is comprehended so the right actions can be taken.

6.3.3 *Decision*

The goal of the decision task is to expeditiously decide on a new strategy or incremental adjustment and translate that into the proper set of measurable objectives and key results for the R&D organization. It must address initiatives on the induced and autonomous strategic paths.

Decide on what not to do as well as what to do.

A strategy requires a commitment to invest in new activities that will deliver the unique value proposition being pursued. In most cases, a changing value proposition and the need to invest in new activities require divesting from or reducing the investment in another. Therefore, identifying areas where investment should reduce or be eliminated is a fundamental part of any strategy. Also, a strategy does not only have to stipulate where investment should be reduced but also how that should be achieved.

With all of this in mind, we will focus on an example of new information requiring a strategic decision that will impact stakeholders from across the system. An example of the decision-making process is shown in Figure 6.8. it is divided into three parts that we will discuss in detail.

Prepare: Converge on a concise proposal

Assuming the observation and orientation tasks were completed satisfactorily, the result needs to be articulated to those that have a stake in the proposed strategy so a decision can be made on the new strategy or adjustment. There are a number of characteristics of a good proposal. They are:

Define the problem. If there is no agreement on the problem, then the rest will not matter because participants will not be

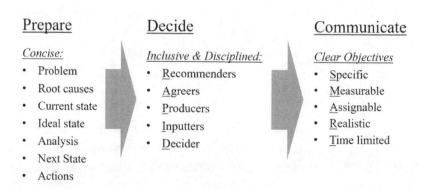

Figure 6.8. Preparing for the strategy planning decision meeting.

debating the same topic. Debating solutions before the problem is understood and agreed on is a major source of waste in most decision-making processes. In our case, this means that the people defining the problem need to synthesize all perspectives and weed out opinions to form a crisp system-level view of the problem. The problem then needs to be clearly stated in quantifiable terms. It should also be clear what will happen if no action is taken. Many people also fall into the trap of jumping too far ahead and framing the problem as the absence of the solution. However, the absence of a solution is not a problem. The problem statement should be agnostic to potential solutions and focus on data that explains the problem in tangible terms.

Articulate the root causes. The main problem statement should be analyzed and boiled down into its root causes. A simple exercise is to keep asking why until you get to the base reasons for the problem. When the root causes are clear then the required actions also become clear. If there is difficulty attaching an action to resolve a root cause then the actual root cause has not probably been identified.

Summarize the current state: Much like the problem statement, if there is disagreement on the current state, then the rest of the proposal will not matter. The current state should explain how things operate and what the state of the industry is today. This does not always match what is documented. This is where direct observation and conversations we discussed earlier is critical.

Articulate the ideal state: Describing the ideal state is critical for communicating the direction. Although the ideal state may never be achievable, it should be quantified. This enables employees across the R&D organization to exercise autonomy and make decisions that enable a step in the direction of the ideal state. The ideal state should ignore practical constraints; it is called "Ideal state" for a reason. This allows people to think more freely about the possibilities. The bottom line is that without a clear articulation of the ideal state the end result will most probably be close to the current state.

Show the analysis: The analysis that supports the problem statement, current state and feasibility of the ideal should be summarized. This leads to the discussion on what should be the next step.

Describe the next state: The next state represents the incremental step being proposed to get closer to the ideal state. The next step should not take too long to achieve since the purpose is to have short learning cycles and allow for adjustments. The next state's description includes details like the new workflow, technology and other important changes as well as a description of the measurable improvement that will be achieved. The next state is a realistic assessment of what can be achieved in that step given the resources, technology and other ingredients that will be provided. By starting with the ideal state and then bringing in constraints to define the next state a connection to the vision can be maintained.

List the required actions: These are the specific actions that define how the next state will be achieved. The actions are measurable with due dates and owners. They should include tasks associated with ending some work to enable investment in the proposal. These actions form the basis of the objectives and key results that will be cascaded to the different participants if the proposal is approved.

Finally, the big challenge is that all of this information should all fit on a single piece of paper. This helps crystalize the thinking and allows the easy sharing with all stakeholders. Keeping it to one page makes it easy to use as a discussion aid while it is being put together and then as a concise presentation later. When presented in a forum it can be expanded upon but its essence should fit on one page.

Decision: The decision meeting should be inclusive and disciplined

The purpose of this step is to make a decision. The discussion should include everyone truly required for the decision. Understanding who those people are is an important part of each decision. Since there may be many different stakeholders who need to attend from across the R&D organization, the time spent in the meeting is very expensive and the discussions should be short and to the point. As mentioned earlier this is not the time to brainstorm. That should have been accomplished before the meeting.

Several decision-making frameworks can provide a model for a decision meeting. For our example, we have used the RAPID model from Bain Consulting [18]. However, as before, the actual tool used is secondary to having a standard means to have a disciplined discussion and final decision. The RAPID model helps identify the key participants required for the decision. It defines who is recommending the action (R), who must agree (A), who are needed to implement the recommendation (P), whose inputs are additionally important (I) and who is the decider (D). This framework is very useful in identifying key stakeholders that need to be met before the decision meeting to provide education, obtain feedback, or brainstorm. It is also very useful for clarifying everyone's roles when the recommendation is discussed. Converging on the RAPID roles before the meeting is a forcing function that later makes the meeting itself proceed more quickly.

Another key element of the RAPID model is that the decision requested is published in advance, framed in a way that requires a Yes/No answer and has one decider. The decider is counted on to make a decision and therefore needs to be someone who has the authority to do so. We have observed that with the proper prework completed, major strategic decisions can be reviewed and dispositioned in 30 minutes or less.

Communicate: Translate decisions into clear objectives and key results

Once a decision is made it needs to be translated into executable form and communicated so relevant parties can align their priorities and goals. A format for this is the model developed by Andy Grove at Intel that has proliferated to other companies is called Objectives and Key Results (OKRs). Like other approaches, if they are not utilized properly, they can be ineffective and even detrimental to a company. However, when done properly they have proved to be very effective in focusing work on what is important and measurable. They have the advantage of being easily cascaded up and down the hierarchy of an organization as well as to other partners outside the organization. Quite simply, the Objective is what is to be achieved and the Key Results are how it will be achieved and by when. A key result is either accomplished or not. Although the strategy may take a long time to fully complete the OKRs should be graded often during execution to uncover problems. An excellent guide for setting clear objectives goes by the acronym SMART. They should be:

Specific: Clear and simple to understand.
Measurable: Meaningful progress can be tracked.
Achievable: Risk is acceptable and resources exist.
Relevant: Connected to the strategy and goals.
Timely: The schedule will maximize economic benefits.

This may sound trivial but there is a discipline required to create objectives and key results that are concrete, meaningful and motivate employees so everyone does not head off in the wrong direction. In his book, Measure What Matters [19], John Doerr gives a good example of OKRs that were written by Andy Grove in 1980 for his staff. They include his scoring (out of 1.0) for Q2, 1980.

"Corporate Objective
Establish the 8086 as the highest-performance 16 bit microprocessor family, as measured by:

Key Results (Q2 1980)
Develop and publish five benchmarks showing superior 8086 family performance (0.6)
Repackage the entire 8086 family of products (1.0)
Get the 8MHz part into production (0)
Sample the arithmetic coprocessor no later than June 15 (0.9)"

The objective is clear as are the key results. They are also simple to understand. They provided Andy Grove with the means to track progress and give his grade on the status every quarter. For our example, the decisions are translated into objectives and key results for the participants in the R&D organization. The initial objective and set of key results that come out of the decision meeting can be taken by others and cascaded down into their groups.

As a cautionary note, though, John Doerr also gives an example of goal setting and metrics that led to disastrous consequences for owners of the Ford Pinto [20].

> *"The Pinto was a firetrap, and Ford's engineers knew it, But the company's heavily marketed, metric driven goals — under 2,000 pounds and under $2,000 — were enforced by Iacocca "with an iron hand ...[W]hen a crash test found that [a] one pound, one dollar piece of plastic stopped the puncture of the gas tank, it was thrown out as extra cost and extra weight," the Pinto's in-house green book cited these three product objectives: "True Subcompact" (size, weight); "Low Cost of Ownership" (initial price, fuel consumption, reliability, serviceability); and "Clear Product Superiority" (appearance, comfort, features, ride and handling, performance). Safety was nowhere on the list."*

The lesson with the Ford Pinto is that setting clear objectives and measurable key results is a powerful means to align people but they can lead the employees astray if they miss an important component.

6.3.4 *Action*

Action communicates goals, aligns execution and collects reaction.

A colleague of ours once said, "a strategy is a hallucination without the commitment of resources". In other words, a strategy doesn't become real until you commit the resources to execute it. The goal of the action step is to ensure the strategy actions are implemented and to gather feedback on their impact. The work on a particular strategic action does not end until the reaction from the ecosystem is understood and compared to the original objectives. A positive and negative reaction is equally valuable since they provide information that can be used to improve subsequent efforts.

Translation into action

In recent years the move to self-managed teams with a high degree of decision-making autonomy has resulted in much greater agility. However, it means clearly communicating a system-level strategic decision is now even more important than before. Providing a clear message to all of the employees about the strategy will make it easier for individuals to make the right decisions within the R&D organization.

As explained earlier, OKRs are a very good way of capturing the objectives and key results of a decision. The OKRs that result from the decision meeting form the basis for action and are used by others to create their OKRs in support. The strategic planners that have been championing a proposal are a great resource when defining them. They provide content expertise and should work with leaders across the system to ensure fidelity to the initial intent.

Customers also need to be informed of decisions so they can prepare. In some cases, they will need to absorb new technology, in other cases, they may need to plan for the end of life of a technology they are using. That is not an easy conversation in many cases, but it is far better to do it as soon as possible so they can make adjustments to their plans.

Collecting reaction

Each strategic decision produces a hypothesis that expects a specific reaction from the R&D ecosystem. There is an expectation about the mix of value that will be created and how it will be done. Whether the value was achieved is critical information for improving the internal strategic planning and R&D processes. There is also an expectation regarding the impact on the competitive and co-opetition forces. The observed changes become part of the continuous observation function in the OODA loop and also provides critical input for improving future analyses.

Grading the key results provide a means to discuss and communicate performance against specific targets. If the OKRs are created properly the indicators and graded key results will provide a baseline for where things stand. However, they will not cover the reaction from competitors or potential new entrants or substitutes. This requires watching what they do and holding continuous conversations with key people in the R&D ecosystem. Also, the indicators may significantly lag the actual sentiment of the suppliers and customers. By the time a customer's unhappiness gets reflected in sales numbers or design wins it may be too late. As mentioned earlier, there is nothing better than going to where the work is being done to see for yourself. This means having direct conversations with customers and suppliers. It also means having similar conversations with the people working in the R&D organization who interact with suppliers, complementors and customers. The combination of measured results and direct conversations will build the most complete picture of the state of the change.

Strategic planning for ongoing budgeting discussions

The strategic planning function provides valuable input into ongoing funding discussions within the R&D organization. Companies and their constituent organizations have periodic budget cycles that balance top-down financial constraints with bottom-up requests for more resources. This often requires that the R&D organization review its objectives and rebalance investment. The rebalancing is often driven by tactical considerations of meeting new spending targets or to fill gaps in funding approved programs. However, to be most effective, decisions must still happen within the context of the overall strategy. For example, if there is tension between two separate programs for more funding and a trade-off needs to be made then a good understanding of the relative impact on the overall strategy will help make the right decision.

To be more proactive, strategic planning can provide its evolving views on the strategic importance of each program in the R&D organization. Strategic planning should already have a clear view of the current state and funding levels and can provide this as input into the decision-making process. An example implementation of this would be for the strategic planning function to list all of the programs within the R&D organization and provide their view on which ones should be increased, sustained, reduced, or ended. If there is a recommendation to add a new program then it goes through the regular strategic planning decision process previously discussed. When this input is combined with a similar (more tactical) input provided by each group in the R&D organization then good tension is created between strategic and tactical perspectives. This leads to more fruitful conversations than just entertaining a long list of tactical requests for new resources.

6.4 Summary

To remain vital, the R&D organization must anticipate and respond to that evolving ecosystem. This is true even if the R&D organization is buried deep within a company and delivers only to other internal groups. The process of doing that is what we call strategic planning.

Strategic planning comprehends the competitive and cooperative forces in the ecosystem and determines the best mix of value for the R&D organization and the means to achieve it. This is accomplished by continuously synthesizing new information, creating knowledge, forcing actions, and helping to implement them. As a result, the primary product of strategic planning is not the plan but the continuous loop of observation, orientation, decision and action. Effective strategic planning results in actions that create differentiable and sustainable value for the customer.

Strategic planning differs from tactical planning because it focuses on actions that materially affect the organization's direction that are not easily reversed. Strategic actions may be induced top-down by management or result from autonomous efforts of individuals or groups that do not align with the existing strategy. Managing the strategic planning processes that properly balance induced and autonomous actions and translates decisions into objectives that can be easily communicated to the organization is an important part of the R&D manager's job.

6.5 Key Points

- The strategic planning network spans the entire ecosystem.
- A strategy determines the best mix of value & the means to achieve it.
- A strategy may make the pie bigger for everyone.
- A strategy is built on assumptions. These define the risk.
- Strategic actions materially affect the R&D organization's direction.
- The origins of strategic actions may be induced or autonomous.
- Strategic planning balances induced and autonomous strategic actions
- Strategic planning operates like an OODA loop.
- The R&D manager is responsible for robust strategic planning.
- Observation senses change. It requires conversations.
- Orientation analyzes and synthesizes new information.
- Decisions determine what not to do as well as what to do.
- Action communicates goals, aligns execution and collects reaction.

References

[1] Jackson, E. (2014). Sun Tzu's 31 Best Pieces of Leadership Advice, Forbes Magazine, May 23, 2014. Available from: https://www.forbes.com/sites/ericjackson/2014/05/23/sun-tzus-33-best-pieces-of-leadership-advice/#42eb49995e5e.

[2] Porter, M. E. (1996). What Is Strategy? HBR. November–December. 1996.

[3] Brandenburger, A. and Nalebuff, B. (1996). *Co-opetition: A Revolution Mindset that Combines Competition and Cooperation*. (Doubleday) p. 4.

[4] *Ibid*, p. 69.

[5] Burgelman, R. (2002). *Strategy is Destiny.* (The Free Press) p. 4.

[6] Christensen, C. (2006). *The Innovator's Dilemma.* (Collins).

[7] Burgelman, R. (2002). *Strategy is Destiny.* (The Free Press)

[8] McGrath, B. (2018). Wayne Gretzky and the Mysteries of Athletic Greatness, The New Yorker. November 14.

[9] Polsky, G. (2018). In Search of Greatness, November 2018 (Gravitas Ventures, Director Gabe Polsky).

[10] Boyd, J. R. (2018). A Discourse on Winning and Losing, Air University Press, Curtis E. LeMay Center for Doctrine Development and Education Maxwell AFB, Alabama, 2018 p. 4.

[11] Dyer, J. Gregersen, H. and Christensen, C. (2009). The Innovator's DNA, HBR, December 2009.

[12] Porter, M. E. (2008). The Five Competitive Forces That Shape Strategy, HBR, January 2008.

[13] Christensen, C. (2006). *The Innovator's Dilemma*, (Collins).

[14] Brandenburger, A. and Nalebuff, B. (1996). *Co-opetition: A Revolution Mindset that Combines Competition and Cooperation.* (Doubleday) pp. 16–22.

[15] *Ibid*, p. 17.

[16] *Ibid*, p. 147.

[17] Burgelman, R. (2002). *Strategy is Destiny.* (The Free Press) p. 9.

[18] Bain and Company (2011). RAPID: Bain's Tool to Clarify Decision Accountability. Available from: https://www.bain.com/insights/rapid-tool-to-clarify-decision-accountability/.

[19] Doerr, J. (2018). *Measure What Matters.* (Portfolio/Penguin) p. 121.

[20] *Ibid*, p. 52.

CHAPTER 7

Development

> Diversity in counsel and unity in command.
> — **Cyrus the Great, 600 BC**

Development focuses on the implementation of the engineering work and the coordination required to complete it. For the type of engineering we are considering in this book, the results must satisfy goals related to such things as cost, compatibility, reliability, serviceability, safety and performance and ensure that they are achieved over a range of operating conditions. As a result, development processes and practices involve iterative cycles of planning, modeling, prototyping, analysis, optimization and testing to converge in a final design. The activities are completed inside the R&D organization and with other entities in the R&D system under the direction of the R&D organization. The development function also includes continuous improvement of the whole value stream.

Standards are critical to development. They enable all of the components in the R&D system to act as one. When utilized correctly, standards also focus everyone in the R&D system to attack and improve the same thing. This leads to faster learning and greater innovation. Similarly, delivering in small batches on a regular cadence is an important means to implement rapid learning cycles and increase velocity during development. Being transparent with everyone in the R&D system is critical and having one way to collect, store and communicate all requests, issues and decisions creates alignment and builds trust.

Development velocity is measured from the perspective of the customer. The value stream starts with a customer request and ends when the resulting solution is being used by the customer and feedback has been received. The customer may be the internal strategic initiative, another entity in the same company, or a more traditional paying customer. As a result, the entire engineering flow may involve most of the members of the R&D system.

Continuous improvement of the engineering processes and workflows requires establishing learning cycles between relevant entities of the R&D system. The complexity of the interactions in a large development process means that no single person can understand and improve how the entire R&D system works. As a result, it takes participation from everyone in the system to drive the required improvements.

Not all engineering decisions have the same impact on value, and therefore they need to be prioritized so the results generate the best return. Due to finite resources, saying yes to one thing often requires saying no to something else. Inputs come from many sources and a decision on whether to build, buy, or co-develop what is needed is also important. To prevent paralysis and reduce risk, it is often best to start early development of more than one possible solution and use concurrent engineering to incrementally narrow the choices as more data is obtained.

7.1 Development and Value

Development includes engineering execution and coordination.

Development involves the engineering activities that systematically apply scientific knowledge to design technology and describe its usage. It also includes the coordination needed with members of the R&D system to get the work done. Decisions made during development take into account business, economic, technology, resource constraints and other factors to produce the best possible result. The specification of the design must satisfy goals related to such things as the cost of material, compatibility, reliability, serviceability, safety and performance

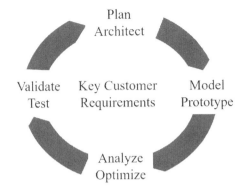

Figure 7.1. A typical engineering cycle during development.

and must ensure that those results are achieved over a range of operating conditions. As a result, development involves iterative cycles of planning, modeling, prototyping, analysis, optimization and testing to converge in a final specification that achieves the objectives as depicted in Figure 7.1. For software engineering, our definition of development also includes coding the software.

7.2 Development Scope

As depicted in Figure 7.2, development relies on relationships with customers and suppliers that can have a big impact on results. The R&D organization's strategic planning process described in Chapter 6 relies on knowledge gained from the development activities to assess the current development pipeline and the capacity of the R&D system. Also, most engineering research will either come from members of the development group or transition to them from somewhere else to be implemented and deployed. Consequently, there is a tight coupling of these entities with development. Development should also continuously evaluate new suppliers, consortia, and institutions while helping the company pursue new customers. Although the figure does not show a direct interaction between the

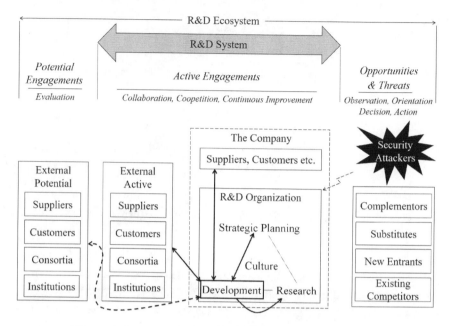

Figure 7.2.   Development and the R&D ecosystem.

development function in the R&D organization and a competitor, there may be indirect interactions via consortia when there is a common cause.

Another way of looking at the entire scope of development is shown in Figure 7.3. This shows development planning in the center and a typical set of interactions. In the figure, there are external interactions with academia, consortia, and industry committees that need to be maintained to influence their direction. There are continuous interactions with customer and supplier working groups needed to coordinate the work at hand. There are, of course, similar interactions with the internal equivalents inside the same company as well as the finance, HR and legal functions. Some are opportunistic or are meant to influence the work of others and some are required to get work done. There are also links to the R&D organization's strategic planning, research and other functions. Establishing an efficient means to manage the network of engagements across

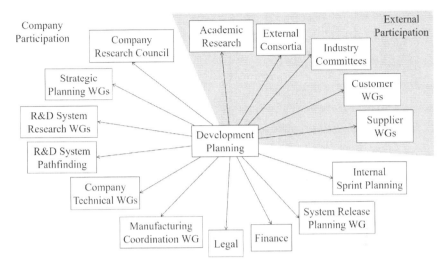

Figure 7.3. **An example of engagements needed for development.**

all of these participants is required to meet the engineering objectives and also continuously improve the R&D system.

7.3 The Importance of Standards

Standards are critical and dynamic.

In 1853 the time of day in the United States was set by towns and cities using local solar time as the reference. In August of that year, in New England, two trains traveling in opposite directions on the same track collided, killing fourteen passengers [1]. The root cause of the accident was that the time on one of the conductor's watch did not match the local standard time and he mistakenly thought he had more time before a track switched. As a result, the New England railroads agreed to set their clocks and watches to the same standard time. However, by 1883, the New York Times reported that there were still fifty-six standards of time used by different railroad companies across the United States [2]. Not only did the risk of accidents remain high but providing information on train schedules became very complicated. For trains that serviced several cities, there were dozens of different arrival and departure times listed for the same train. On November 18, 1883 railroad companies across the United States and Canada became motivated enough to remedy this and established one standard time and the use of time zones for their train tables. On March 19, 1918 the government of the United States followed this by establishing standard time and time zones for the country. Despite the obvious advantages of improving predictability and reducing the risk of accidents, the standard time system was not popular in many local communities who felt they had lost control of an aspect of their identity.

In 2006, the development of the Airbus A380 aircraft was delayed two years at the cost of billions of dollars to the company. A Los Angeles Times report [3] summed it up at the time.

"The installation of wiring, which was expected to take two months for each plane, has grown to nearly six months, mainly because the computer software used to design the wiring system was not compatible with the A380's main design and engineering software, Streiff said. So, when the fuselage assembled in Hamburg came to Toulouse to be mated with other parts, the wiring didn't match."

That initial assessment of the delay proved to be very optimistic since the delay in delivering the first A380 aircraft turned out to be more than two years. To avoid these problems, the development process that has hand-offs between participants across the R&D organization system requires standards.

For engineering workflows, standards encompass such things as the methods, tools, infrastructure, models and collateral needed to develop the technology. Standard processes and workflows mean that everyone can operate with the same assumptions, projects can be shifted between different teams quickly and technology developed by one team can be easily integrated into a bigger system by another team. However, like

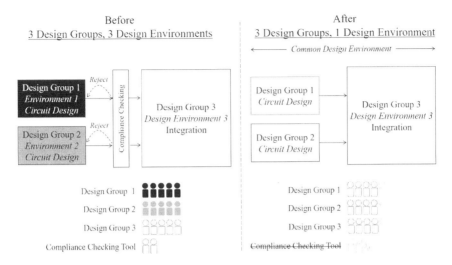

Figure 7.4.   A non-standard and standard development environment.

the move to standard time, moving to standard development processes and workflows can be unpopular since it sacrifices local autonomy for the benefit of the overall system.

Another example of the need for standards is shown in Figure 7.4. It depicts a simple development operation that involves three different engineering groups. The output from design group 1 and design group 2 get integrated into a larger design by design group 3. In the "Before" picture on the left, each group uses a different engineering design environment that is composed of its methods, engineering tools, collateral and infrastructure (called the "environment" in the figure). Differences between the design environments may include separate engineering tools and methods for measuring performance, validating functionality and enabling testing. It is not hard to fathom that if all of the components are developed differently, then the integration will become more complicated and subject to disconnects like the one Airbus experienced. To overcome this, the integration team may decide to staff a team to create and maintain a tool that will conduct an incoming inspection of each component from another group to ensure it meets minimum standards for integration. This identifies potential problems that may prevent integration. The rules are based on past experiences of what created problems and if a problem is found then the component is returned to the sender to be fixed. However, this is not sufficient, nor efficient. Since the tool relies on encoding checks to avoid past issues, it does not catch new ones. Also, as more rules are needed, it eventually leads to a bloated set of rules that prove increasingly difficult to maintain. Eventually, the time spent in dealing with false positives and checking and reworking the collateral can start to far exceed the time spent doing the design of the work in the first place. When that happens, people will find creative reasons to override or wave violations and this can lead to catastrophic results. Although each entity feels like they are being very productive internally, the system's overall velocity may be very low due to the rework and delays that result at the time of the hand-off for integration.

On the right side, we show a more standardized approach whereby the three groups operate with the same design environment. In that case, there is no need for the incoming compliance checks and one does not need to staff that operation. The elimination of the rework associated with a rejection also reduces an important source of waste and one can shift the people to do more value-added work. Also, since all employees use the same design environment, shifting resources between groups is seamless. This rationale for standardization can also be applied for software development, delivering services and other workflows.

Although the example of standard time is a good one to reinforce the value of standards, there is one big difference between standard time and the other two examples. Once the model for standard time is set, it is unlikely to change, however, the standard processes and workflows should be attacked and improved continuously. A major benefit of establishing a standard is that improvement efforts across the R&D system can be focused on the same target and do not get diffused across many different ways of doing the same thing. When this happens, improvement accelerates. This means the standard workflow is not a static document that gathers dust. Instead, it is a snapshot of how the work gets done right now. It serves as a target for others to experiment with and improve. When improvements are completed, they need to become part of the new standard. With this in mind, it is very important for the R&D system not only to standardize and continuously experiment, but it also needs to continuously communicate changes to the standard as it is improved.

Standards enable automation. Automation enforces compliance.

Whether it is the semiconductor, airline, or auto industry the rapid progress in technology has been enabled by making engineers more productive via automating highly repetitive, complex, or error-prone tasks. Setting standard engineering methods and converging onto a standard set of engineering tools, collateral and infrastructure make the automation of the engineering workflows much easier. Conversely, automation enables standards to be uniformly applied by engineers without much effort on their part. Also, if a design methodology mistake is discovered the automation tools can be changed and the fix proliferated to everyone simultaneously. In this way, design automation both benefits from standard engineering methods and enforces them.

As we alluded to in the previous example of the three design groups, design automation can present a problem if it is not properly standardized. The different automation software that Airbus used in 2006 was very productive for each of the teams using it but everything fell apart when all of the parts had to be assembled. This is a problem the authors of this book have seen countless times. Efforts to integrate simple engineering software, models, or hardware that should have taken minutes took days, weeks and even months due to the use of different methods, engineering tools and collateral.

To get the full benefit of standardizing, therefore, one needs a standard design environment like the one shown in

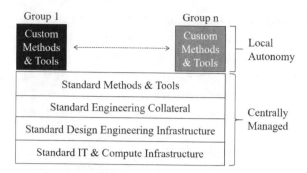

Figure 7.5.   Components of a design automation environment.

Figure 7.5. In that example, the IT and computing infrastructure, engineering infrastructure, design collateral and most methods and tools are centrally managed. The methods and engineering tools that do not need to be standardized can be plugged into the design environment by individual groups.

Learning cycles complement and co-exist with standardization.

As we noted in Chapter 5 and in this chapter, the use of standards does not impede innovation or stifle change. In fact, without standardization, improvement is much slower. Using standard engineering tools, methods, and workflows is also critical when working in the broader R&D system with many interdependencies. Learning cycles are a means to conduct controlled experiments during execution that challenge the current standards and then compare the results with expected outcomes. Changes that meet the expectation can become part of a revised standard. Those that do not can be recorded and then discarded. In this manner, learning cycles continuously attack the standards and everyone is involved in improving them.

Standardizing is hard. Know what to standardize.

Finally, taking a development process full of differences and moving to a common standard is very difficult. There are cultural barriers, technology barriers, training requirements and added delivery risks that are part of the transition. For this reason, the R&D organization needs to know what to standardize. In some cases, like manufacturing, complete standardization may be needed. In other cases, like product development, there may be exceptions to provide incremental value if they do not negatively impact the rest of the system. For example, in Figure 7.4, design group 1 may be responsible for the design of a specialized circuit that requires a unique design methodology or set of engineering tools just for that type of circuit. In that case, there may be some differences between their design environment and other groups. Of course, if this goes unchecked, eventually everything will be claimed to be an exception so it needs to be very clear to everyone across the R&D system what is enforced through a standard and what is not.

7.4 Development Velocity

Development velocity is determined by the customer.

The measurement of development velocity starts when there is a request from a customer and ends when the customer is using the result and has provided feedback. Figure 7.6 depicts an example of this using a rudimentary description of a product development flow in which the engineering organization does not deliver directly to the paying customer. In that case, the customer works directly with a business group responsible for the

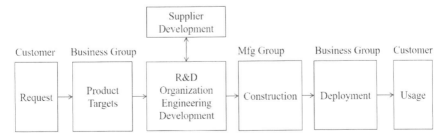

Figure 7.6. The development flow through the R&D system for a customer.

product to describe its requests. The Business group translates those into key product targets for the engineering R&D organization. The R&D organization then works with the supplier to deliver the design to the manufacturing group for construction. The finished product is then deployed by the business group for customer use. In this simple example, even if the R&D organization performs all of its tasks on time, there still may be a delay in deploying the solution if the supplier, business group, or manufacturing group has a problem.

As far as the customer is concerned, they do not care why there may have been a delay or who was at fault. They only see the time it takes between their request and productive usage. It is also important to note that releasing a solution does not mean it is being used. There are many examples of engineering features that were developed and released but sat on the shelf with the customer due to insufficient training, communication, or poor planning. Deployment means that the solution is being used and feedback has been received. So, in summary, the measure of the development velocity starts with the customer request and ends with the usage by the customer no matter how many different entities are involved.

If there is a problem anywhere in the flow shown in Figure 7.6, the leaders in the R&D organization must take responsibility to help fix it. The leaders in the R&D organization cannot just wipe their hands and say they did their job and therefore the delay was not their fault. They need to take the initiative to work with everyone across the product development flow (the R&D system) to fix the problem and continuously improve its operation. For example, in Figure 7.6, the R&D organization can work with the business group and the customer to improve how they capture and communicate the requests so that they are more easily understood by engineering. They can work with the supplier to improve each hand-off so it does not lead to re-work. They can work with the manufacturing group to improve the way the design information is provided so that manufacturing can move faster (and vice versa). They can work with the business group again to improve the way they deploy the technology and train the customer. In other words, the R&D organization needs to work with everyone in the R&D system to increase the development velocity. This requires a culture that values assumed responsibility, continuous improvement and looking holistically at the whole system.

Delivering in small batches on a cadence improves trust.

In Chapter 5, we discussed the benefits of time-boxed development typified by a scrum. Short development sprints enable fast, incremental feedback during development that helps improve the product of engineering. Short sprints also expose problems with the processes and workflows. This is because waste gets exposed more easily when capabilities must be defined, developed, integrated, validated and released in 1 or 2 weeks. Small problems and delays get amplified when time is short. Since each release includes a retrospective and opportunity to improve the process for the next sprint, continuous improvement moves at a rapid rate. Similarly, the actual work capacity of teams can also be understood more quickly and commitments for the following sprints can be adjusted to align with reality. This takes the guesswork out of the equation and reduces the stress on employees who now have some control over the amount of work flowing in.

Adhering to a standard release cadence has the additional benefit of improving predictability and trust with customers and suppliers. If it takes a long time between releases of technology or if a customer is not sure when future releases may occur, they tend to pile all of their requests into the next release and make them all a very high priority. On the other hand, if a rapid release cadence has been established, the customer is more inclined to allocate their requests across the upcoming releases in accordance with their actual priority. To build trust, it is very important in this scenario for the R&D organization to mandate that the release dates will not slip. If it becomes apparent during a development sprint that too much has been committed then some deliverables from that sprint will be moved to the next one. Although that is not great for the customers that may be affected, the good news is that the next one happens very quickly and the delay will be relatively short.

7.5 Continuous Improvement

Visualizing the entire value stream leads to big improvements in velocity.

Doing what we just discussed is not easy. This is because a complete development flow may be complex. One way to visualize the entire flow is by using a value stream map that we discussed in Chapter 4. Figure 7.7 depicts a simple example that highlights how some differences between the entities can create waste. It is similar to the example shown in Figure 7.6 but we eliminate the business unit to make it simpler to draw and add a supplier to differentiate between an internal and external supplier.

In the figure, the R&D system is composed of one customer, one external supplier from another company, one internal supplier from the same company, a manufacturer (also within the

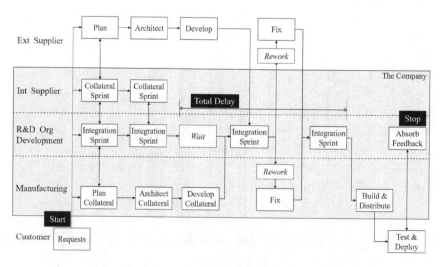

Figure 7.7.   An example development flow with delays.

same company) and the R&D organization. There is a tight relationship between the R&D organization and the internal supplier in this case. They both utilize scrum and have sprints that are synchronized. They also use continuous integration to co-develop, integrate and validate the capabilities as they are completed. In contrast, the external supplier and the manufacturer utilizes a waterfall approach to sequentially plan, architect and develop the solutions. They may make infrequent releases and bundle many new capabilities together in major releases with little interaction during development with the rest of the players.

From the standpoint of measuring velocity, we start the clock when the request is made. In this example, the manufacturer, internal supplier and external supplier will all need to create some collateral for the R&D organization to use to deliver the engineering result to manufacturing who distributes the final technology to the customer. Since the internal supplier is completely synchronized with the R&D organization their work happens together. However, the R&D organization's integration effort will need to wait for the collateral from the external supplier and manufacturer since they are using an unsynchronized process. The additional delay due to this spans the "Wait" box and the extra integration sprints needed to catch up and fix the high likelihood of a problem that requires "Rework". If everything is okay after that then the design information is delivered to the manufacturer who builds the product that is distributed to the customers. Once deployed the users' feedback is provided to the R&D organization and the clock stops. Although each sprint will have its retrospective and improvement cycle that includes the customer and other stakeholders, their effectiveness is reduced because other members of the R&D system have a different development philosophy and release cadence. The result is that velocity is lower than it should be and the learning cycles across the whole R&D system needed to improve are less effective.

In this example, the elapsed time between the customer request and absorbing feedback from the users on the deployed solution determines the time component of the development velocity. The total added delay through the system includes the "wait" time, one "rework' step and two unnecessary integration sprints. For comparison, the ideal flow where all of the participants are using a synchronized scrum with continuous integration is shown in Figure 7.8. As we can see, the alignment to a common development model across the entire R&D system eliminates the delays due to different release cadences, it will also enable a continuous integration model across all entities in the picture. That eliminates the possibility of surprises that lead to re-work and enables everyone to participate in synchronized retrospectives that will generate more useful feedback.

As we stated at the beginning of this section, doing this is not easy. Even for this simple example, the supplier and

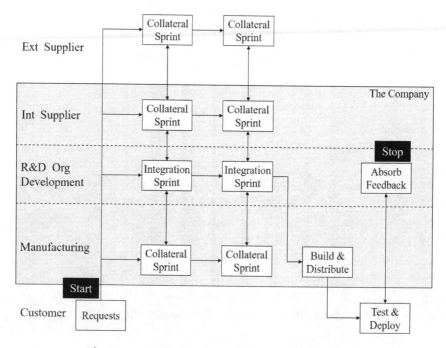

Figure 7.8. An improved development flow.

manufacturing group's resistance to change will likely be very high. However, the improvement possible is also very high, and to keep pace with the competition this type of improvement is needed. As discussed in the more detailed VSM discussion in Chapter 4, it is best to target small incremental steps that will get one from the initial state shown in Figure 7.7 to the state shown in Figure 7.8 over several iterations. Each step should be easily digestible by the participants and the results should be used to plan the next step. This will reduce resistance and make it easier for the leadership of the different entities to commit to the journey.

There are many opportunities for improvement during development.

At the level of abstraction shown in the previous section, some very obvious differences created big opportunities for improvement. There are also many more examples that exist at the level of individuals and small groups of people doing engineering work in the R&D system. To provide a small sample of the challenges and opportunities that exist, we can look at the journey of a request for a small enhancement in a design feature from an external customer through the R&D system.

To keep it simple, we will assume the R&D system is composed of the R&D organization, one external supplier, one internal supplier and one external customer. In this example, a request is made by the customer, the solution is architected by the R&D organization, the specs are delivered to the suppliers who then designs and releases the collateral back to the R&D organization for integration, the R&D organization then releases the solution to the customer who then validates and deploys it to the users within their company.

If we drill down to the level of the workflows it might look something like what is shown in Figure 7.9. The example shows the request coming from the customer on the far right and following the dashed line to the R&D organization where it is turned into a specification that is passed on to the different design groups that need to be involved (one external and

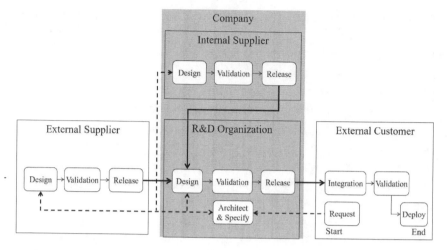

Figure 7.9.   Interaction of the workflows across the R&D system.

internal supplier and the R&D organization). Each entity then uses its own set of workflows to design, validate and release the design for the next consumer in the flow as shown by the solid lines. The two suppliers release the collateral to be used by the R&D organization and the R&D organization completes the design, validates the result and releases the result to the customer. The customer then validates and deploys the solution to the users in the customer's company.

In this case, even if the design steps look the same, the workflows in each entity may be different. As we discussed before, they may use a different set of engineering tools and methods to design and validate the work. The models they use to measure performance, validate functionality and meet other constraints may not agree with the models used by the other entities. It is even possible that the format of the design data is slightly different from the format needed by the receiving engineering tools or models used by groups in the R&D organization. All of this does not matter too much while they are operating in their group but once they hand-off the results to another entity there may be problems with the integration that leads to re-work or quality problems. As we discussed before, this type of analysis can be repeated continuously for the countless number of small, medium and large workflows across the R&D system. Each improvement may not have a big impact in its own right but the accumulation over time of hundreds and possibly thousands of small improvements will have a significant impact.

Applications support is a great way to gain feedback from the customer

As noted, the initial use and feedback from the customer marks the point where the technology has been delivered. The definition of the customer may vary depending on the charter of the R&D organization and who is "paying for" the technology. However, the entire learning cycle does not end until the development teams absorb the feedback. An important resource for gathering feedback from the customer are the people that provide application support. In some cases where the output is delivered to internal customers in the same company, the applications support may be part of the development function. In the cases where the engineering work is delivered to customers outside the company, the applications support may be governed by a separate entity in the company. Regardless, applications support is an important element of the development process. Applications support can provide brutally honest insights into real usage and perceptions of value unfettered by intellectual property or company confidentiality concerns. Employees engaged in applications support often have the best view of the actual current state. Tension between the applications support function and development can be expected and that is ok if and only if it is properly focused on continuous improvement.

7.6 Managing Development Choices

Generative innovation is the backbone of development.

A baseline requirement of development is to retain knowledge and spawn innovation based on its past inventions. An example of this is automotive companies that introduce new models that share innovation from other models. This philosophy was explained by Henry Ford when he discussed the difference between the first automobile that he built 15 years (and several models) before the hugely successful Model T. [4]

> *"The whole design has been refined; the present Ford care, which is the "Model T", has four cylinders and a self-starter — it is in every way a more convenient and an easier riding car. It is simpler than the first car. But almost every point in it may be found also in the first car. The changes have been brought about through experience in the making and not through any change in the basic principle — which I take to be an important fact demonstrating that, given a good idea to start with, it is better to concentrate on perfecting it than hunt around for a new idea"*

When discussing the release of a new version of the MacBook Pro, that included the touch bar Jony Ive of Apple gave a more contemporary description, [5]

> *"There is just no way that you could bring the current MacBook Pro to market if you hadn't gone through the prior products. Each one absolutely required the learning of the previous."*

At a minimum, the R&D organization must do this better than its competition since the competition will also have most of the same information needed as a base for its own innovation. For example, when a technology company releases a new product it is invariably stripped down and analyzed by the competition. Some of the advantages the leading company had due to secrecy up until that point are lost and the playing field gets reset. The only sustainable approach is to move faster than the competition and transform past development innovation into future development innovation faster than everyone else.

The ability to generate future innovation based on past innovation is key to survival and one of the roles of development. However, there will be a point where no additional value can be squeezed from past innovation. So, in addition to enabling value in this manner, the development must strike the right balance between the continued exploitation of the current generation vs. exploring or adopting more dramatic change. The scope of the exploration during development can be managed using concurrent engineering that we will describe shortly.

An ROI framework and concurrent engineering reduce risk.

One of the toughest challenges of development is deciding on the priorities and relative funding for different engineering activities. There are specific requests from customers, directives from strategic planning, local innovation and more that need to be considered and balanced. Not to mention, continuous improvement activities that are going on all of the time within the organization. Each item has an associated uncertainty with the estimated value, development cost, time and risk.

Because the development process has many choices to make, it is important to have a process in place to absorb, synthesize and analyze important information and make decisions that meet tactical and strategic goals. In this process, there are generally more opportunities that may add value then can be done so identifying what to say no to is an important requirement. Also, every new engineering feature added to a technology adds to its technical debt since it will need to be validated in every subsequent release of the technology whether it is used or not. Delivering a new feature that does not get used means time has been spent on something that does not add value and therefore lowers velocity. So, in addition to deciding what to start, the development process must also include decisions about what to end. Keeping in mind that removing an engineering feature will usually take longer than adding one.

The ability to understand what the customer truly needs is important. The real problem or opportunity is likely different from what is being expressed. This requires conversations with the eventual user and other stakeholders. Once the reason is understood, the solution may not require an engineering change and value might be realized in another way. Once the task is understood, the conversations need to continue so that feedback can be obtained and refinements made concurrently with development. Although there are many useful automated systems in place to gather and analyze data from customers and monitor progress, they do not replace the need for direct conversations with customers and suppliers.

The types of development and budgeting can be broken into three parts. The first is the obligation to keep the business running (KTBR) and continue mission-critical engineering activities. This also includes driving continuous improvement. Enabling quick resolution of mission-critical issues for customers and continuously improving development processes and practices is a baseline that should be funded before anything else. After that, there is a balance that needs to be set between work that exploits the current generation of technology to deliver generative or sustaining innovation and work that is exploratory to deliver more disruptive or blue ocean innovation. Strategic considerations influence that balance and we will discuss it more in the next chapter on engineering research.

Several factors guide the prioritization of new development. Most of the factors have a degree of uncertainty associated with them. Although there are a different mathematical models and algorithms that can be help with portfolio and priority management, we will only focus on the important variables. The variables will also depend to some degree on the industry and the charter of the R&D organization. However, no matter what variables or modeling approaches are used, the manager of the R&D organization needs to converge on a framework that everyone can use to guide their decisions. A sampling of variables to consider are below:

- The estimated value (from the customer's perspective).
- The estimation of development time, cost, effort.

- The estimation of risk and uncertainty.
- The degree of adaptability to internal and external changes.
- Dependencies on and contribution to other development.
- The existence of a strong customer advocate.

The estimated value captures the positive benefit of the development in a tangible manner and from the perspective of the customer. The estimation of development time, cost, effort and risk may lower the ROI of the development below that of competing opportunities. If the degree of risk or uncertainty is large for the previous factors, then this may also reduce the predictability of the estimates and make a choice less attractive. Related to this is the durability of the decision relative to changes in the internal and external landscape. If the value associated with the development is highly dependent on the status quo, then changes in the landscape may quickly change the value proposition. Some choices may be dependent on other engineering work or contribute to the value of other technology. When this is the case dropping the development of one thing may have a cascading impact on others. Conversely, developing some technologies may contribute more than their inherent value if they enable value elsewhere. Finally, the existence of an advocate can influence how easily technology will be deployed and used by the customer. Without a clear champion on the receiving side, new technology can often sit unused after it is delivered.

As mentioned at the beginning of this section, the variables all have some degree in uncertainty associated with them that may result in the wrong decision or endless debates that lead to paralysis. Concurrent engineering can be used to reduce the risk of this happening by starting with multiple options, refining the choices as more knowledge is obtained, and then converging on the final answer, at the last responsible moment (when execution can still be completed on time). Figure 7.10 depicts what that looks like. Early in the development process, when there is more uncertainty with each variable, several

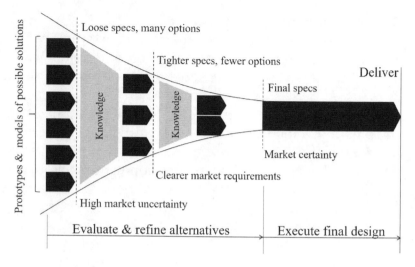

Figure 7.10.   A depiction of concurrent engineering.

options may be valid to keep open. The specs at this point are loose to represent that uncertainty. Early development may be limited to areas that will help reduce uncertainty in key areas. As more information is gained, some options are eliminated and the expectations become tighter. In parallel, customer requirements and the market will become clearer. Ultimately, the development process converges on one answer. Even though there is some early investment in development for options that ultimately do not get chosen, the waste can be reduced by retaining all of the knowledge generated by those experiments so that it is available to inform future engineering discussions.

By using concurrent engineering, the R&D organization does not have to bet the farm on one answer and can permit some dissonance early in the development process. It permits moving forward during the fuzzy front end where there is little concrete data and it can be applied to the development of small components or major systems. It also enables waiting until there is agreement on the solution before more

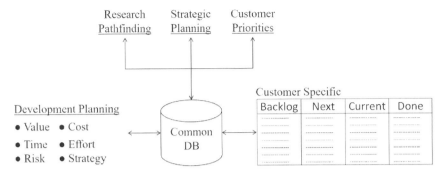

Figure 7.11. A single place for visualizing requests and commitments.

labor-intensive optimization activities start. This is similar to the pathfinding process we describe in Chapter 8 but applied to development. A development process that delivers in batches on a standard cadence, as we described earlier, facilitates concurrent engineering.

Have one way to communicate status and make it transparent.

Having a single way to manage formal development requests, issues and committed actions is critical to prioritizing work. Making that information visible to customers, suppliers and developers eliminates confusion. Transparency in displaying the priorities and committed plans also includes the customer and other stakeholders in managing the queue and builds trust. Figure 7.11 depicts a simple example of a database used to track requests for development from customers, strategic planning and the research process. Ideas or requests for new development are captured in the database and the R&D organization prioritizes them according to the ROI framework we just discussed. This is done with the help of the relevant entities in the R&D system. The initial requests, associated priorities, committed plans and development owners are viewed by everyone using a simple depiction like a Kanban board.

7.7 Build vs. Buy

Have a framework for deciding between build vs. buy.

Another important consideration in the development process is whether to do the engineering within the company or obtain it from a supplier. The decision to build vs. buy is governed by several factors and like the process of prioritization, the manager of the R&D organization needs to ensure that there is a set of criteria for making these assessments. In general, "core" engineering work that provides a competitive advantage for the company should be done by the engineering R&D organization while work that does not provide a competitive advantage should be evaluated to see if it can be obtained from a supplier. Using suppliers has a cost but this is often offset by enabling more engineers within the R&D organization (who have better access to the company's intellectual property) to focus on unique tasks that increase the company's competitive advantage. If a supplier is used, then adherence to industry standards is required to lower switching costs and increase the R&D organization's power in negotiations.

As an example, one can use the levels shown in Table 7.1 to categorize the engineering work and determine if the R&D organization should consider engaging with a supplier. Level 0

Table 7.1. Different levels used to determine engagement.

L	Description of information	Engagement model
0	Core engineering related to a technology that is secret.	Done by R&D organization. Not publicly acknowledged.
1	Core Engineering that enables leadership or unique value.	Done by R&D organization.
2	Engineering based on requirements unique to the company but not critical for leadership.	Done by R&D organization or obtained from suppliers.
3	Engineering work that is common in the industry.	Obtain from suppliers.

covers instances where the work is secret and states that the R&D organization will not even acknowledge to a supplier that the technology exists. Level 1 covers the case where engineering output uniquely provides a competitive advantage to the company. In that case, the work should be done by the R&D organization but can be discussed with suppliers to ensure any relevant supplier technologies are compatible. L2 covers instances where there is no clear competitive advantage but a unique methodology or other requirements makes it difficult to use a supplier. In that case, suppliers can be used, but they may not be interested and work may still have to be done by the R&D organization. L3 covers instances where the engineering work, technology, or collateral is a commodity in the industry and provides no competitive advantage. In that case, the work should be provided by a supplier.

Using this framework, the internal R&D organization's development efforts can be focused on the engineering activities that are secret or provide a competitive advantage for the company. On the other hand, engineering work that is more of a commodity by nature and offers no real reason to do internally can be obtained from a supplier. Using these levels to evaluate the R&D organization's existing portfolio is an important part of the strategic planning loop described in Chapter 6. Of course, a major influence on the final decision is the total cost of using a supplier. It is not listed as a consideration in the list shown in Table 7.1 so that conversation can focus on the core value of doing the work in the R&D organization or obtaining it from a supplier. However, once that is clear, the difference in internal development cost vs. supplier cost may be big enough to motivate keeping the development work within the R&D organization.

7.8 Including Customers in Development

Include lead customers in development and QA.

Some customers may make unique requests that only applies only to them. These may not provide additional value for the rest of the customers but are extremely valuable for a major customer. In many cases, the requests involve capabilities that are meant to differentiate themselves from their competitors and are not meant to be available to anyone else to use. These requests often entail a large and unique development and quality assurance (QA) effort where customer engagement is critical. The R&D organization can do the engineering themselves, but in some cases, it will be better for the customer to do it themselves and hand over the collateral for the R&D organization to integrate. For example, proprietary functions for a System on Chip (SOC) product or a software system may be designed by the customer and integrated by the R&D organization. In some cases, it may be better to enable the customer to work directly in the R&D organization's development flow to integrate the technology themselves.

Whether or not the R&D organization's engineering work delivers the promised value also requires validation and testing. Individual engineering features may operate as specified but the combination may grind to a halt or fail for a customer with an unanticipated usage model, set of interfaces, or different input collateral. When there are many customers, there will likely be a plethora of usage models that create more scenarios than the R&D organization can validate and test alone. As a baseline, the manager of the R&D organization must apply lean thinking and agile principles to ensure the development flow builds quality into each step. However, this will still be a challenge if there is a high volume of scenarios to test. When this happens, the best solution is for the R&D organization to include key customers in testing and validating the technology as it is being developed.

A continuous integration development model enables immediate feedback to all developers on whether their work passes a spectrum of basic tests. These tests can be run concurrently with development as the engineering work is done on components. The same approach can be used to continuously perform system-level testing on the complete design. Although automation and other efficiencies can make continuous integration very productive for the members of the R&D organization, it is better if the methods and infrastructure also enable key customers to run and review the results of their own tests. By doing this, a tremendous pool of resources becomes available for testing customer-specific corner cases during development.

When to build vs. buy, open up customer co-development or enable customer co-validation is driven by business, strategic or tactical reasons. How to safely do that is based on the security culture, policies and procedures in the R&D organization. During development, it is important to protect the intellectual property of everyone involved. This requires a culture that prioritizes the protection of everyone's intellectual property and the means to do so. We will discuss information security in more detail in Chapter 12.

7.9 Summary

Development focuses on the implementation of the engineering work and the coordination required to complete it. The results must satisfy goals related to such things as cost, compatibility, reliability, serviceability, safety and performance and ensure that they are achieved over a range of operating conditions. Development activities are completed inside the R&D organization and with other entities in the R&D system under the direction of the R&D manager. The development function also includes continuous improvement of the whole development flow.

Development velocity is measured from the perspective of the customer. The value stream starts with a request from a customer and ends when the resulting solution is being used by the customer and feedback has been received. Continuous improvement of the engineering processes and workflows requires establishing learning cycles within and between relevant entities of the R&D system.

Not all requests or improvements have the same impact on value and therefore they need to be prioritized so the results generate the best return. A major challenge in development is to decide what should be implemented and what should not. To prevent the management waste of inaction and reduce risk, it is often best to start early development of more than one possible solution with a small resource commitment for each and use concurrent engineering to incrementally narrow the choices as more data becomes available.

Delivering in small batches on a regular cadence is an important means to implement rapid learning cycles and increase velocity during development. This creates focus, improves predictability and leads to better results. Being transparent with everyone else in the R&D system is critical and having one way to collect, store and communicate all requests and decisions is also an important way of maintaining trust and alignment.

Standards are critical to enabling all of the components in the R&D system to act as one. When utilized correctly, standards are dynamic and focuses everyone in the R&D system to attack and improve the same thing. This leads to faster learning and greater innovation.

7.10 Key Points

- Development includes engineering execution and coordination.
- Automation requires standards. Automation enforces compliance.
- Learning Cycles complement and co-exist with standardization.
- Standards are critical. Know what to standardize.
- Development velocity is determined by the customer.
- Visualizing the value stream leads to improvements in velocity.
- Applications support augments feedback from customers.
- Generative innovation is the backbone of development.

- An ROI framework and concurrent engineering reduce risk.
- Have one way to communicate status and make it transparent.
- Have a framework for deciding between build vs. buy.
- Include lead customers in development and QA.

7.11 Case Studies

7.11.1 *Standardization of an Engineering Environment*

Background

The design and validation of a microprocessor is a complex endeavor that involves many engineers. As shown in Figure 7.12, the engineering environment includes specifications of the operating conditions that the finished chip must support. These specifications and other product targets help to define the

Specifications *Frequency, voltage, temperature etc.*	The operating conditions and targets the product must meet.
Methods *Engineering Procedures & Techniques*	How the product will be designed and validated to meet the specifications.
Flows *Automation of Multiple Tasks*	The ordering of the engineering tasks in accordance with the methods.
Tools *Automation of Engineering Tasks*	Specialized engineering software that helps complete each task.
Design Collateral *Standard Design, Validation, Test Components*	Common functions that are implemented in advance and used by all groups.
Manufacturing Collateral *Process Technology Models, Guidelines & Rules*	Manufacturing information required by some tools to complete their task.
Infrastructure *Databases, Data management*	Software that is used to collect, store, process and protect the engineering data

Figure 7.12.   An example of an IC design environment.

engineering methods that will be used to design and validate the chip so it is ready to be manufactured. To increase the productivity of the engineers, a sequence of tasks that make up a method is often automated and called a "flow".

There may be many different flows that span the logic design down to the physical implementation of the transistors. Each flow is typically composed of several Electronic Design Automation (EDA) tools that are designed to complete specific engineering tasks such as logic simulation or circuit simulation. These tools often utilize pre-existing design collateral such as circuit libraries to increase efficiency. Some tools may also require collateral or information related to manufacturing process technology. A common infrastructure is used to manage all of the engineering data utilized by the flows and tools as well as providing other common functions. All of this is called the design environment.

The complete design environment may look similar to what is shown in Figure 7.13. That example spans the design and validation of the logic, circuit and physical design of the custom and synthesizable digital and analog design blocks.

	Logic Design & Verification	Synthesis & Integration	Custom Implementation	Analog Implementation
Flows *Synchronized*	Integration, Verification, Debug...	RTL-Layout Synthesis Timing Analysis. Verification...	Custom Circuit Design Custom Physical Design Verification...	Analog Circuit Design Analog Physical Design Verification...
Tools *Synchronized*	Simulation, Emulation Formal Verification...	Logic synthesis Place & Route Timing Analysis...	Memory Compiler, Custom Place & Route Custom Timing Analysis	Layout Editor Circuit Simulation Co-simulation...
Design Collateral *Unsynchronized*	RTL Macros, Verification IP...	RLS Cell Library for each process technology variant	Custom Cell Library for each process technology variant	Analog Cell Library for each process technology variant
Mfg Collateral *Unsynchronized*	NA	Technology Files & Rules for each process variant		
Infrastructure *Synchronized*	Data Management for Logic Design & Verification	Data Management for Synthesis, Integration, Custom & Analog Implementation		

Figure 7.13.   An example of the contents in a design environment.

The Problem

At Intel in 2013, several microprocessor design groups were operating in parallel. These groups had developed their own design environments. As depicted in Figure 7.14, each design group had autonomy when creating their design environment. As a result. the engineering methods differed between groups. Each group also used a different collection of flows developed by their Design Automation (DA) teams or obtained from the company's central computer aided design (CAD) group. Most of the EDA tools for mainstream microprocessor design groups were common at this time and delivered by the central CAD group but newer microprocessor design groups sometimes made different choices. Similarly, each design group often modified collateral like the standard cell library to insert additional cells customized for themselves. Although the manufacturing

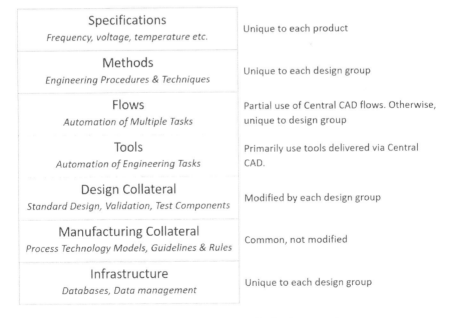

Figure 7.14. Uniqueness of each design environment.

collateral could not be modified, different design groups froze on different versions of a particular process technology generation as they neared completion. Finally, because of the differences in the methods, flows and tools, each design group often maintained their infrastructure. In short, each microprocessor group's design environment was different and managed by its own dedicated DA team. This provided the flexibility to act fast and worked well as long as each microprocessor group operated in their silo.

By 2013 the variety of microprocessors at Intel had increased dramatically to span several new markets. Many of these new markets did not offer high volumes and profit margins and required increasing levels of functional integration with more frequent product refreshes. To be successful in these markets, Intel needed to be able to quickly assemble System on Chip (SOC) style microprocessors with low nonrecurring engineering costs. Re-using the many common design blocks needed in different microprocessor instantiations was an obvious means to enable this. However, as shown in Figure 7.15, the highly siloed operation of Intel's different microprocessor groups made this difficult to achieve. If a design block was initially created using the design environment from Group A it likely used slightly different methods for completing the design. For example, it may have a

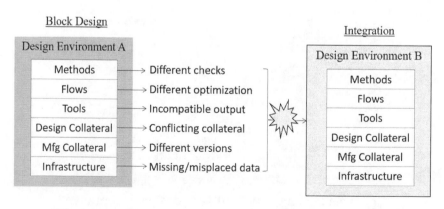

Figure 7.15.   Problems with Integrating IP across groups.

different set of static verification checks or targets for logic verification coverage. The flows may weigh trade-offs differently than other groups and the versions of the EDA tools used may be very old and no longer generating output that is compatible with the newer version of the same tool used by another group. The modifications to the design collateral made to customize them for Group A will not exist in other design groups. Finally, the storage, retention and organization of the design data in Group A may make it very difficult to package and transfer to a group using a different infrastructure. These were just a small sample of the issues that arose.

Actions Taken

Realizing that this was not acceptable, the Intel managers responsible for all of the microprocessor designs decided to standardize onto one common design environment and have that developed and supported by the central CAD organization. This meant converging on a common set of flows, tools, collateral and infrastructure and consolidating them into a common design kit that would be managed and delivered by the central CAD group. In this model, the individual design groups would no longer be responsible for developing their flows, integrating EDA tools and maintaining their infrastructure. Ideally, each microprocessor design environment would be identical and the sharing of design blocks and people across groups would be much easier.

Since the focus was initially on enabling the SOC design style, the emphasis was on the capabilities needed for the synthesis and integration of what was called the unCore. This excluded the CPU core that required custom design and other analog design blocks. As shown in Figure 7.16, the initial design kit, therefore, focused on two of the four columns mentioned before.

Results

To expedite the move to a common design kit, it was also decided to take an existing design environment from a microprocessor

	Logic Design & Verification	Synthesis & Integration
Flows *Synchronized*	Integration, Verification, Debug...	RTL-Layout Synthesis Timing Analysis, Verification...
Tools *Synchronized*	Simulation, Emulation Formal Verification...	Logic synthesis, Place & Route Timing Analysis...
Design Collateral *Unsynchronized*	RTL Macros, Verification IP...	RLS Cell Library for each process technology variant
Mfg Collateral *Unsynchronized*	NA	Technology files & rules for each process variant
Infrastructure *Synchronized*	Data Management for Logic Design & Verification	Data Management for Synthesis, Integration,

Figure 7.16.   Initial design kit content.

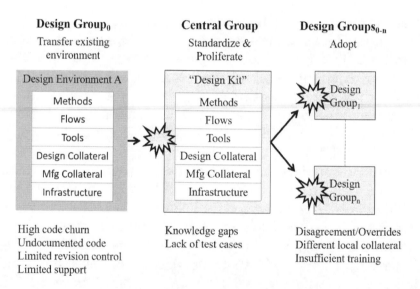

Figure 7.17.   Initial transfer and deployment.

group that was completing the design of a SOC style product. As depicted in Figure 7.17, this design environment would then be standardized by the central CAD group for proliferation to the other microprocessor groups. However, as also shown in Figure 7.17, this ran into some immediate problems. First, the design environment of the microprocessor group in question was still under development itself with a high degree of code churn in its flows. Also, since the microprocessor DA teams did not typically have to proliferate their design environment to anyone else before, there was little documentation, poor or no revision control and hundreds of undocumented scripts and utilities that no one understood. This, combined with the fact that the receiving central CAD group had never constructed a complete design environment themselves and therefore had significant knowledge gaps and few test cases.

The difficulties encountered in the transfer of the initial design environment to the central CAD group was matched by the resistance of the receiving engineers to move to the new kit. Although their management had concluded it was the right thing to do, there was still considerable disagreement with the decision to standardize on the environment that was chosen. As a result, some design groups altered the design kit after it was received to replace some flows and tools with their versions. Also, since local changes in the design collateral were pretty common in each design group, some flows in the source design environment required their custom collateral to be included in the design kit and added to the collateral used by other groups. Most of all, since the design methodology was very different between the design teams, significant training was needed so engineers could properly use the new flows and tools effectively.

Due to the reasons mentioned, the initial release of the design kit was delayed several times. The number of changes that needed to be continuously absorbed meant that the entire design kit was unstable. To remedy this, the central CAD group adopted a sprint model for the development, integration and release of the entire kit. Fundamental to this was the commitment to adhere to a 6-week release cadence for the entire kit with shorter sprints for the development of the constituent flows, tools, collateral and infrastructure. As shown in Figure 7.18, this resulted in a 6-week cycle of continuous development and integration. At the beginning of the six-week cycle changes to the infrastructure and collateral were added to the contour for the next release.

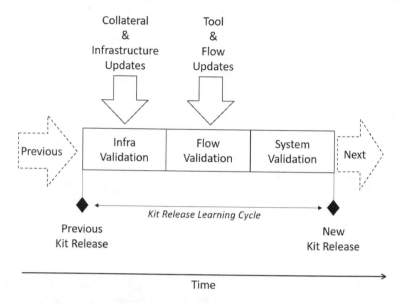

Figure 7.18.   Typical development and integration cycle.

Two weeks later, updates to the tools and flows were added to the release contour and then after two weeks of system testing, the full design kit release was made. In between releases, critical bug fixes were made as a patch to the full kit release. As the design kit matured, it gained customers and the scope increased to include the other content shown in Figure 7.13.

This presented several challenges that included:

(1) Coordinating the increasing volume of collateral.
(2) Managing compatibility while enabling improvement.
(3) Enabling customers who value new features vs. stability.

Questions:

(1) What problems did standardization onto a common design kit try to solve?
(2) What difficulties had to be overcome?
(3) How were these difficulties addressed? Was anything missed?
(4) How should the challenges listed at the end be addressed?

References

[1] Smithsonian's National Museum of American History (1999). *"On Time" Opens at Smithsonian's National Museum of American History*. Available from: http://www.americanhistory.si.edu/news/pressrelease.cfm?key=29&newskey=97 [June 2020].

[2] New York Times Archives (1883). Railway Time Standards; A system by which annoyances will be avoided, New York Times, April 19, 1883. Available from: https://www.nytimes.com/1883/04/19/archives/railway-time-standards-a-system-by-which-annoyances-will-be-avoided.html [June 2020].

[3] Pae, P. (2006). Airbus hit hard by A380 Setback, Los Angelese Times. October 4, 2006. Available from: https://www.latimes.com/archives/la-xpm-2006-oct-04-fi-airbus4-story.html [June 2020].

[4] Ford, H. (2017). *My Life and Work*. (Create Space Independent Publishing) p. 12.

[5] Gugliemo, C. (2016). Jony Ive talks about putting the Apple "touch" on the MacBook Pro, CNet. October 28, 2016, Available from: https://www.cnet.com/special-reports/jony-ive-talks-about-putting-the-apple-touch-on-the-macbook-pro/ [June 2020].

CHAPTER 8

Engineering Research

> Maximizing the value of research is a team game.
> — **Robert J. Aslett**

Although many concepts in this chapter can be applied to other types of research, this chapter specifically addresses engineering research. In our definition, engineering research increases the power of engineering by creating new knowledge and more powerful modeling, analysis and optimization methods to solve design problems. The results can be used to increase the value of current technologies or enable new technologies and, hence, new value. This is very much applied research and not pure research. Applied research also differs from development due to its emphasis on experimentation vs. application.

Most engineers solve problems as part of their job and this often requires some research. As a result, almost everyone in the R&D system conducts some degree of research whether it is in their job title or not. Most of this type of research does not affect others and the focus in this chapter is largely on the more complex and interdependent research that needs to be centrally managed or coordinated. This includes research done within the R&D organization by individuals and groups and research that requires collaboration with other entities in the broader R&D system.

The motivation for research may be induced top-down with specific targets to sustain improvement along a known trajectory, or it may happen autonomously in a way that may disrupt the status quo and open up new opportunities. Ultimately, the value of the research is determined by the customer. The customer may be an external entity or the company itself. With that in mind, research that originates from the company's induced strategic actions is easier to fund since the perceived value is clear and the likelihood of deployment is high. Research that targets disruptive opportunities is often more difficult to fund since the perceived value and likelihood of success are unclear. Establishing the means to encourage and fund an acceptable level of disruptive research is needed to ensure a healthy research pipeline.

The role of the manager of the R&D organization is to create the culture and processes where all of this research thrives and value is realized without delay. To succeed in a competitive environment, the value harvested from all of this research must have a velocity comparable to a start-up. This means ensuring research that is not dependent on others can proceed relatively unencumbered. It also means ensuring more complex research, that has dependencies on others, follows a process that engages the correct people at the right time. This creates a research pipeline of sustaining and disruptive research topics that may come from many sources. It also requires a pathfinding process that samples the research pipeline, ensures relevance and helps those that demonstrate value become implemented and deployed.

Maximizing the value derived from research is a team game. Successful researchers engage with eventual customers and other stakeholders during the research. This often takes the form of iterative prototyping of potential solutions that enable frequent learning cycles and adjustments. Researchers who do this and also help with the implementation and deployment of their research gain credibility with those that will be needed to implement and apply their ideas. They also become better researchers.

8.1 Engineering Research and Value

As we discussed in the introduction to this book, engineering is the process of creating the knowledge needed to build and use technology. Engineering requires meeting goals related to such things as cost, compatibility, safety and performance across different operating conditions. The engineering needed to meet all of the requirements often requires multiple cycles of planning, modeling, analysis and optimization.

Engineering research increases the power of engineering by creating new knowledge and more powerful modeling, analysis and optimization methods to solve design problems. A resulting change that leads to greater value is what we call innovation. Sometimes there is no underlying science that explains the phenomenon engineers must manage. When this is the case, engineering progress is often made using empirical results from experiments and heuristic methods. This may require inventing technology that makes doing research more productive. We will discuss this later when we reference how the success of the Wright brothers preceded much of the theory of aerodynamics.

Research value is determined by the customer who pays.

Engineering research delivers innovation that makes increasing the value of existing technologies possible. It also may deliver innovation that makes new technologies and, hence, new value possible. In these cases, the value of the research is determined by the customer who will use and pay for the technology. However, engineering research may address topics in anticipation of future customers in alignment with the company's vision. Since there is no user yet, the customer is essentially the company itself and it will decide whether or not it will fund the activity. In Chapter 4, we called this potential value. In either case, the value of research is ultimately determined by what these customers will pay. This viewpoint was summarized in a talk given by the former director of IBM Research, James C. McGroddy, in 1993 [1],

> *"For a research institution to be successful over the long run, it must focus on the value it delivers, as perceived by the entity that is its source of funding. And the institution must recognize, and deal with the fact, that the perception of value may change significantly over time.*

> *For a government lab, the funding source might be a fed-eral agency, which in turn represents the values of a constitu-ency like DOD, and ultimately the taxpayers' interest in a strong defense; for a university, it might be the NSF, the students, the state; for me, most of the time, it's the IBM Corporation."*

8.2 Research Scope and Types

Our focus is on engineering research managed by the R&D organization. This includes research within the R&D organization that may be conducted by individuals and development groups for their use and research that is transferred to others for deployment. Research also includes collaborations with other entities such as academic institutions, industry institutes and consortia in the R&D system under the direction of the R&D organization. Therefore, interactions and learning cycles exist between different entities in the R&D system, as shown in Figure 8.1.

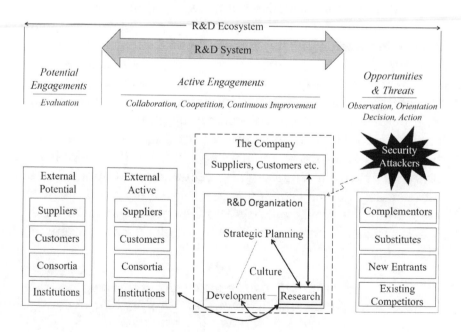

Figure 8.1.   Engineering research scope.

Although we do not show a direct interaction between the research function in the R&D organization and a competitor, there may be indirect interactions via consortia or when there is a common cause that benefits the companies or the industry as a whole. Also, although research is shown as a separate function in the picture of the R&D system, it does not imply that research needs to be a separate organizational unit. Although there may be a need for a dedicated research team, most people across the engineering R&D system do some form of research regardless.

Research is induced and autonomous. It can sustain or disrupt.

The induced and autonomous strategic actions mentioned in Chapter 6 highlighted two paths for innovation in most technology companies. Research that is driven from induced strategic actions has clear objectives aligned with the company's current strategy and processes. This type of research is still very difficult but the timelines and goals are well known and the research can be very structured. This is what Clayton Christiansen called sustaining innovation. On the other hand, research that is driven by autonomous strategic actions may be far less structured and not aligned with the company's current strategy or processes. It often originates as skunkworks below the radar of senior management. As a result, this type of research is difficult to plan and evaluate. What sometimes results from this type of research is what Clayton Christiansen called, disruptive innovation. Although many of the examples given for both sustaining and disruptive innovation refer to technologies or products, the terms also apply to engineering.

We previously discussed the importance of managing these two distinct paths as part of the strategic planning process. In this chapter, we will speak in terms of both the induced and autonomous strategic actions and the resulting sustaining and disruptive innovations and how the R&D organization's research function needs to manage these paths. Before we dig into the details of sustaining and disruptive innovation it is good to keep in mind Clayton Christensen's warning [2],

"It is rare that a technology or product is inherently sustaining or disruptive. And when new technology is developed, disruption theory does not dictate what managers should do. Instead it helps them make a strategic choice between taking a sustaining path and taking a disruptive one."

Engineering research may sustain a required rate of improvement.

As Robert Burgelman [3] noted, induced strategic action of the company is,

"orientated toward gaining and maintaining leadership in the company's core business."

and, as Christensen and Bower noted [4], the resulting technologies,

"tend to maintain a rate of improvement; that is, they give customers something more or better in attributes they already value."

Innovation that sustains technology improvement is critical for the success of the company. Tremendous effort is needed to stay ahead of the competition and deliver the performance improvement that the market expects. Such improvements are often represented as a series of S curves shown in Figure 8.2, where there are hard-earned improvements within each generation until a new architecture or approach is needed to enable the next S curve. The innovation required within the individual S curve may be composed of many small incremental steps or large ones that require a significant investment in resources. It has the added imperative that competitors are also improving with the same targets in mind. The investment required to move from the current generation to the next one (from one "S" curve to the next one) is also significant and comes with additional risk. Technology research that enables innovation along this path also requires engineering research to keep pace.

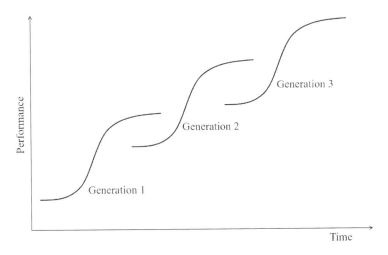

Figure 8.2.   Sustaining innovation along a series of S curves.

Since the rate of improvement needed to obtain leadership on the sustaining path is generally well understood, engineering research can be structured to intercept important milestones. While one generation of engineering solutions is in production, the next one may have several innovations being evaluated in a pathfinding process (we will discuss pathfinding more later) and the one after that may have several other technologies still in research. Figure 8.3 shows a representation of this where one can see that at any one point in time, there is a separate generation in production, pathfinding and research. In the example shown, the research horizon spans two generations. This parallelization of the pipelines is fairly common when the required rate of improvement is understood. The entire pipeline is often aligned to important milestones in the technology or product roadmap. For example, if the release of new product lines may need to be timed with the holiday season, then this process would align the production steps with the optimum date to intercept that target. When this type of pipeline exists for new technology development then the engineering research pipeline will align to it to provide the right engineering solutions at the best points in the pipeline for each

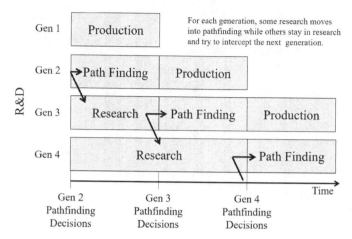

Figure 8.3.   Research and pathfinding aligned to technology generations.

generation. However, it does not mean that research on a topic must start and finish within the short window implied for each generation. Research is continuous and the topics that are deemed ready to evaluate more closely for a transfer into production will move into the next available pathfinding phase. Research that still has potential value but is not ready for pathfinding will stay in research, mature, and attempt to move to pathfinding in the next generation.

This type of research environment is called "complicated" by David Snowden [5] and is characterized by known unknowns. In such a research environment there is high confidence there is an answer but expertise is needed to uncover it. For this type of research, the manager of the R&D system needs to create a culture and processes where the analysis needed by experts can flourish and deliver results on the proper cadence.

Engineering research may disrupt incumbent practices.   .

As Robert Burgelman [6] also noted, autonomous strategic action, *"involves initiatives of individuals or small groups that are out-side the scope of the corporate strategy at the time that they come about."*

If managed properly, autonomous strategic actions can lead to disruptive engineering solutions and technologies. As Clayton Christiansen [7] noted,

""Disruption" describes a process whereby a smaller com-pany with fewer resources is able to successfully challenge established incumbent businesses. Specifically, as incumbents focus on improving their products and services for their most demanding (and usually most profitable) customers, they exceed the needs of some segments and ignore the needs of others. Entrants that prove disruptive begin by success-fully targeting those overlooked segments, gaining a foot-hold by delivering more-suitable functionality — frequently at a lower price. Incumbents, chasing higher profitability in more-demanding segments, tend not to respond vigorously. Entrants then move upmarket, delivering the performance that incumbents' mainstream customers require, while pre-serving the advantages that drove their early success. When mainstream customers start adopting the entrants' offerings in volume, disruption has occurred."

Figure 8.4 is a depiction of the above description whereby a technology originally developed for the low margin economy market improves and ultimately disrupts the incumbents in the mainstream market with an offering that provides good enough performance but at a much lower cost structure. When this happens, it is difficult for the incumbents to react quickly and maintain profitability. Often, the incumbents will surren-der the mainstream market and flea to the high-performance market where they may still have an advantage. However, that is a temporary reprieve since that market will likely to get dis-rupted in time as well.

An example of disruptive technology is low-cost streaming technology that started with low quality and limited usage,

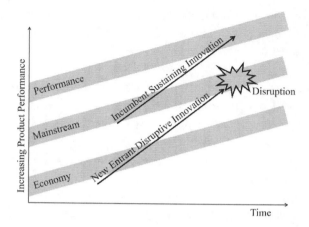

Figure 8.4.   A depiction of sustaining and disruptive innovation.

improved over time, and eventually disrupted video stores and threatened cable TV. Another example described initially by Clayton Christensen was the disk drive industry where manufacturers of more compact disk drives that serviced the growing minicomputer markets disrupted entrenched incumbents targeting the mainframe market. The disrupters, in this case, were similarly disrupted themselves later by companies that provided even more compact disk drives motivated by personal computers.

The idea of disruptive innovation is also applicable to engineering. Individuals or small networks of employees may autonomously explore new engineering theories or practices that do not directly support existing engineering processes. If successful, the efforts could provide a way to develop a new and uniquely valuable technology or dramatically increase the value or existing technology. Such disruptive engineering innovations are often deemed as too risky for mainstream technologies and are usually applied to technology targeting smaller markets first. In the best case, the engineering practices mature and become applicable to mainstream products with less risk. When they do, they bring with them the unique engineering advantage developed for the smaller market segments that disrupt the existing engineering practices of the mainstream.

A disruptive technology or engineering innovation does not magically appear out of the blue. Research and development of the disruptive path may follow the same S curve progression depicted in Figure 8.2. It may follow a similar model for pipelining research, pathfinding and development shown in Figure 8.3 (but addressing a small segment of the company). The challenge is not that disruptive research is any easier or harder, or that it requires a different way of conducting the research. The challenge is that disruptive ideas are not intuitive and often fail. This makes that type of research difficult to identify and justify vs. sustaining research.

Engineering research may enable new markets with no competitors.

In addition to sustaining and disruptive innovation, companies may pursue a strategy to create new markets that start with no competitors. Chapter 6 discussed strategic opportunities to work with complementors and even competitors to change the game and increase the size of the pie for everyone. Chan Kim and Renée Mauborgne [8] coined the term blue ocean to highlight opportunities that open up a new market that is without competition. Engineering may have a large role to play in this if the technologies and products in question require a different way to be designed and be built. Sometimes this may also need to be done in collaboration with industry complementors.

This type of research environment is what David Snowden [9] termed "complex", and is characterized by many unknown, unknowns. In such an environment, the manager of the R&D organization needs to enable a culture and processes that allow experimentation where failure allows important patterns to emerge so the right path can be selected. Also, due to changes that may affect the customers, suppliers and complementors they need to be involved as early as possible to help evaluate, improve and prepare for the potential changes. This requires conversations between the technical leaders across the R&D system during the research stage.

8.3 Research Funding

Sustaining and disruptive research funding should be separate.

Companies that are making a healthy profit have a difficult time investing in disruptive innovation. Similarly, engineering organizations that are successfully enabling mainstream technology have a difficult time justifying the investment in disruptive engineering practices. This is because a vast majority of disruptive ideas do not succeed and are difficult to implement. For every disruptive idea that succeeds, hundreds fail. It is not feasible for an incumbent to invest effort into every single one. The company's current processes and culture also get in the way of those that look promising. Since disruptive innovation often does not address the company's current priorities the standard operational processes that exist to make funding decisions are biased against them.

To solve this problem, companies will sometimes create separate independent business units dedicated to discovering and incubating new business opportunities. Some companies may also form a venture capital arm and invest in technology startups across a wide spectrum of the industry. Although this may operate as a venture capital firm, its goals have less to do with financial results than leveraging the vast entrepreneurial pool of talent in the world to foster disruptive ideas. Intel (Intel Capital), Cisco (Cisco Investments), and Google (Google Ventures) are examples of companies that have venture capital operations for this purpose.

The R&D organization has the same challenges regarding disruptive research. Ultimately, the value of the research is determined by the value of the deployed technology. Research that is sustaining in nature is easier to fund since the perceived value is clear and the likelihood of deployment is high. There are official and well-established budgeting processes that can be followed and clear targets and timelines to judge the progress. Autonomous actions that address disruptive opportunities are more difficult to fund since the perceived value is unclear and the likelihood of deployment is low. These types of research topics do not do well in the mainstream budgeting process. Therefore, establishing different sources of funding and criteria for them is important.

One approach is to create a separate research group with a specific operating budget and staff. If so, the charter should be biased towards disruptive innovation. This is because there will still be significant research along the sustaining path that really should continue to be owned by the entities who are ultimately responsible for developing and deploying the result. As we noted, sustaining innovation follows a known trajectory with specific goals and timelines, and this type of research is best done by the people who are intimately knowledgeable about the current engineering solutions and target technology. The people responsible for this type of research typically are under extreme schedule pressure and cannot justify investment into disruptive ideas with uncertain timelines and a high likelihood of failure. On the other hand, a dedicated research group allows the opportunity to explore disruptive ideas without the pressure of delivering on a rigid timeline or to specific goals. The resulting innovations are still expected to be deployed and the customer is still the ultimate judge of the value of the research but the timeline is more flexible and acceptance of failure is higher. Additionally, since autonomous innovation often starts with an individual or informal network of like-minded engineers, the creation of a dedicated research group mustn't stamp out the bottom-up disruptive ideas coming from others who may not be a part of the research group. This source of innovation needs to be encouraged and nurtured regardless of whether there is a dedicated research group or not. However, the dedicated research group will have the unique capability to provide some oversight and support for those efforts even if they are driven by engineers outside the research group.

When a company creates a venture capital operation, the engineering organization can also take advantage of its existence to propose investment into a variety of engineering start-up companies that can augment its roadmap. This may provide another path for funding disruptive ideas. With a reasonable investment at the right time, the R&D organization will be able to monitor progress and gain early access to important engineering knowledge.

This thinking also applies to academic collaboration. In that case, providing grants, sponsoring internships, or influencing the direction of academic research can provide another low-risk path to explore disruptive ideas. Similarly, participation in industry-academia consortia can provide another path. Although these engagements usually include participants from other companies and even competitors and the research topics are pre-competitive, they still provide a platform for influencing research across the industry that may be beneficial. Often, this will still be advantageous to the company that does the best job translating the useful results of the research into reality.

Allocating a specific percentage of the total R&D organization's budget every year to a dedicated research group and academic or industry collaboration enables enough autonomy to properly explore and disposition ideas with a minimum of overhead. Working with the venture capital arm of the company may also provide a way to invest in engineering start-ups that provide access to a wider pool of talent. By doing this the R&D manager can inject their own judgment on the relative importance of the two paths by setting the budget to match their appetite.

8.4 The Research Pipeline

Research with dependencies should be coordinated in a research pipeline.

As we mentioned, in an engineering R&D system, most employees do some form of research. It is important to distinguish

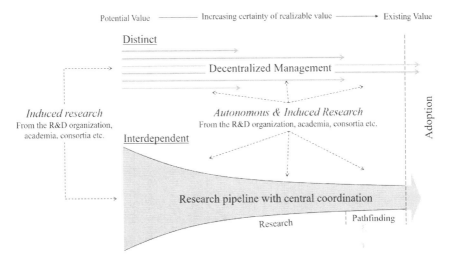

Figure 8.5.   Distinct and interdependent engineering research paths.

between how to manage and coordinate research that is dependent on others and that which is independent. On the one hand, we do not want to burden research that is independent of others with unneeded processes and, on the other hand, we do not want research that is dependent on others to be blocked due to the inability to align all of the necessary stakeholders.

Figure 8.5 depicts two different paths for research. The top half covers what we call *distinct research*. This research is self-contained with no dependencies outside of the development group in which it happens. This type of research does not need central coordination. The bottom half depicts what we call *interdependent research*. This research has at least one of the following characteristics:

(1) If successful it will require others to change how they work.
(2) To be implemented or deployed it will need others to participate.
(3) It is high risk and will benefit from the inputs of other experts.

The sources of research at the top and bottom half of the diagram shown in Figure 8.5 are the same. However, the research captured in the top half of the diagram does not meet any of the criteria and therefore is distinct and largely self-contained. Those research topics are likely induced by the current strategic needs of the R&D system and sustaining in nature with a specific start and end date. Because they have no inter-dependencies the responsible entities conduct the research and then implement and deploy the results on their own with little interaction with the rest of the R&D system. For this type of research, there is no need for central coordination since it will likely just slow things down.

If the research topic meets at least one of the criteria listed, then it will be captured in the research pipeline shown in the bottom half of the picture. These research topics may also be induced by the current strategic needs of the R&D system and be sustaining in nature with specific start and end dates. The research may originate inside the R&D organization or in collaboration with others in the R&D system. Some research may be initiated top-down and start on the far left of the pipeline, others may start from a more mature point and enter the pipeline somewhere in the middle.

Of course, there will also be research going on autonomously across the R&D system that is potentially disruptive. This is shown in the center of the diagram. By its nature, it may start at any point in time. When this type of research starts to mature, it will need to follow one of the two paths indicated depending on how it meets the criteria. On both the top and bottom half of the diagram, most research will start with *potential value* on the left and move to the right as it demonstrates more realizable value. When it reaches a point where it starts to demonstrate *existing value* it enters a pathfinding process that determines if and how it will be deployed for adoption. We will discuss pathfinding in more detail later in this chapter and the case study at the end of the chapter.

There are several additional key points to mention about the research pipeline shown in Figure 8.5.

The research pipeline is linked to strategic planning: The information provided by this pipeline is critical for the strategic planning process we discussed in Chapter 6. It provides very important information about capabilities that are in the research and pathfinding pipeline that may help enable or influence a strategic decision. In many cases, disruptive engineering research topics in the pipeline require a strategic conversation.

All research does not start on the far left: Primarily conceptual Research will likely start there but the research that starts in a more mature state will be classified and placed at the most appropriate point in the picture. For example, autonomous research may start informally by individuals or small groups of people and attain more maturity before becoming visible beyond the originators. In some cases, they move directly into the pathfinding process. The goal is to have an accurate picture of the status of all research and not enforce unneeded steps.

All employees in the R&D organization should have access to the current state: A summary of the status of each item should be visible to employees in the R&D organization. This gives recognition to those that are driving research topics, it encourages the building of informal networks between people who may be working on similar ideas and also encourages others to engage in the process.

Topics can remain in research for a long time: Research that is making progress and still shows promise may continue in the research state for a long time. For example, in the semiconductor industry, the potential for dramatic improvements in semiconductor technology often exists in research for over a decade before they reach the maturity needed to provide real value for high volume production. At the beginning of each generation, the research may be evaluated and fail to meet the criteria for deployment but continue to evolve as a research activity in hopes of intercepting the next generation.

Pathfinding should move fast: Contrary to what was said about research, once a research topic moves into pathfinding, the progress should be fast. When pathfinding starts, there is typically a technology or customer intercept in mind and time constraints become important. Pathfinding is about translating the research's potential value into real value in the shortest amount of time.

When needed, research can be plucked from anywhere in the pipeline: One of the advantages of having a complete picture of the research pipeline is that there will be times where necessity requires a research topic to be plucked from the research and deployed before it otherwise would have been. This may happen to meet a competitive threat. In those cases, the research is augmented with the necessary resources and aggressively moved through pathfinding.

Consumers of the research should be involved early and often: It is very important to engage with the consumers of the research from the very start. These include other members of the R&D system that will need to turn the research into a product and the end customers. This will provide clarity on the research goals and it will make it much easier later to transfer the technology from research to production. In particular, as a research topic moves into pathfinding, there should be a good mix of the original researcher(s), product developers and customers involved in each step.

Research selection/deselection is continuous.

Deciding what research to add or subtract from the research pipeline is a continuous process. Although sustaining research topics that are induced by the corporate strategy have clear intercept milestones and can be pipelined as shown in Figure 8.3, the milestones for disruptive topics driven more autonomously are not predictable. Therefore, the environment needs to be sampled continuously.

Identifying new research topics means casting a wide net across the whole R&D system to engage with industry, academia, industry institutes, government institutes, consortia and others. If we use the semiconductor industry as an example, research may result from engagement in pre-competitive research via industry consortia like the SRC (Semiconductor Research Corporation) and independent research centers like IMEC (Interuniversity Microelectronics Centre). Research topics may also come from the government via agencies like DARPA (Defense Advanced Research Projects Agency). Not to mention, there may be research done by a supplier or customer that may benefit from collaboration with the R&D organization (and vice versa). Providing a complete picture of research across the R&D system is best done by a central entity like a dedicated research or strategic planning group. If there isn't such an entity, then each engineering group within the R&D organization may need to build their own picture of the research system for their domain.

All new research ideas compete for the resources of the R&D system that may already be fully booked on existing projects. So, in addition for determining criteria for selecting what new research projects should be added in the R&D system, there need to be criteria for weeding out projects that no longer provide sufficient potential value or research that is having difficulty demonstrating existing value.

When these research projects are stopped, they are not killed. Instead, the learnings (both positive and negative) should be archived and the work enters a state of hibernation. They are not killed because the environment that made the research impractical today may change later and increase the existing value to the point where the effort needs to be resuscitated. The state of hibernation also encourages the practice of capturing all of the learnings about why it failed. Sometimes knowing what not to do is as important as knowing what to do so even failed research projects provide important knowledge.

The decisions on what research in the pipeline should be stopped to make room for new research or transferred for deployment requires a process with a clear set of criteria. To do this, it is extremely helpful to have a way to categorize research maturity levels with their related expectations and make that transparent to the organization. This can be combined with the research pipeline we just described to provide a visual representation of each research topic's current state. By doing this, all of the people in the R&D system will understand what already is being done (so they can avoid duplication) and how the results are being judged (so they know what to expect with their research). Figure 8.6 is a depiction of a research maturity model for the research pipeline described earlier. It is just an example to clarify the concept. The leadership of the R&D organization should determine the number of levels and associated expectations. The bottom line is that it needs to be clear and transparent.

In the example shown in Figure 8.6, the front end of the pipeline on the far left is meant to capture the topics that are being scoped. This requires limited resources and the barrier to enter this level is low. As a result, there are likely many topics being scoped. To satisfy the expectation of this level, the

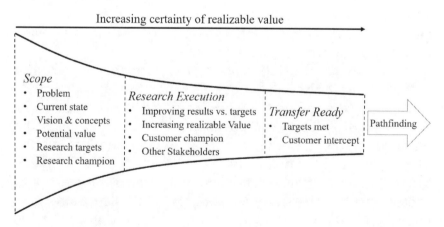

Figure 8.6.   An example of a research maturity model.

researcher must gain consensus on the problem statement, the current state, potential value and research targets. There should also be a good understanding of the vision and main concepts as well as someone representing the end customer or company who will champion the research. If the topic stalls at this level and cannot achieve those expectations, then it will be stopped and put into hibernation. If it does meet all of the expectations, then it will move to the right and into active research. The expectation for that level is that progress will be made against the research targets and these will demonstrate increasing existing value on realistic test cases. Another expectation for this level is that all of the other stakeholders needed to deploy the research are engaged. If the topic stalls at this level and cannot achieve those expectations, then it will be stopped and put into hibernation or returned to the scoping level to be re-defined. If it does meet all of the expectations, it will move to the right and into the level for research that is ready for transfer into production to be deployed. At that level, the expectations are that all of the research targets are met and a potential customer intercept is possible. The research may sit at this level until it is sampled by the pathfinding process that determines if it should be implemented and deployed into production. If the research stalls at this point it may move back into active research for refinement or stop and be put into hibernation.

Induced research topics captured in the pipeline may have specific delivery dates as part of the research targets. In those cases, a roadmap showing when each milestone needs to be met can be constructed and tracked. For autonomous research, those may not be so clear. However, frequent reviews of the status of that type of research are still required. Ideally, a cadence can be established that sets the right pace for each research topic and ensures that pressure is maintained to show progress. Although there may be several resets and the path forward may have several zigs and zags, it should be expected that progress through the pipeline will continue for each research topic. Once a research topic stalls, it is a good time to evaluate whether to stop the research.

The research pipeline provides a transparent view of the status of all interdependent research across the R&D system. It is not meant to be a rigid process that prevents doing the right thing. As we noted before, once the research pipeline is established, it is permissible to identify research in the pipeline that needs to be accelerated for competitive reasons. As we noted earlier, in some cases, the right thing to do may be to extract research before it has completely transitioned to the end of the pipeline and pre-emptively implement and deploy it (skipping over the later pathfinding process). That is the type of strategic discussion that a visual representation of the research pipeline enables.

8.5 Learning Cycles and Rapid Prototyping

Rapid learning cycles in research means rapid prototyping.

Like other functions in the R&D system, research should utilize learning cycles that include rapid prototyping or modeling of new concepts so they can be quickly tested and refined in collaboration with customers and other stakeholders as research progresses. This not only helps refine the research based on continuous feedback but also helps improve the research process by identifying bottlenecks and areas of waste that can be reduced or eliminated. The Look-Ask-Model-Discuss-Act (LAMDA) model discussed in Chapter 5 and shown in Figure 8.7 is an example of a model that can be used to drive innovation and improvement. It emphasizes a thorough understanding of the current state based on direct observation of the work being done and quick iterations of models or prototypes in collaboration with the end customer and other stakeholders.

Increasing the number of learning cycles that can be accomplished in a given time will accelerate learning and refinement. This is so important that creating engineering technologies that speed up the prototyping and modeling iterations often need

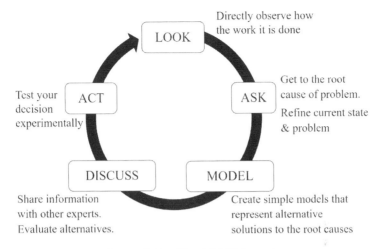

Figure 8.7. The LAMDA cycle.

to be developed along with the research. These could be software programs that automate labor-intensive steps or an engineering infrastructure that facilitates running more experiments in parallel. The Wright brothers are an excellent example of applying this type of innovation to the process of invention. In their pursuit of building the first airplane, they not only captured critical learning by observing what others had already learned, they also created their own technology that enabled them to run experiments at a very low cost and in a much shorter time than others. As shown in Table 8.1, some of the inventions they made before their famous 1903 flight to increase the rate of experimentation included small scale models (kites and gliders) and the world's first wind tunnel.

The Wright brothers' success also highlighted the fact that for engineering, the empirical results sometimes precedes the understanding of the science behind the results. In other words [10],

"Knowledge derived from research does not necessarily or uniformly flow from science to engineering. Engineering progress based on empirical, experimental, and heuristic methods often anticipates underlying scientific principles. Thus, the

Table 8.1.   Wright brothers' chronology.

1899–1902	Conducted extensive literature search and correspondence with researchers on aerodynamics
1900–1902	Built unpowered and easily re-configurable kites and gliders to develop flight control systems.
1901–1902	Invented wind tunnel technology that allowed testing 200 different wing and airfoil models
1900–1905	Conducted 1000 test flights using gliders (and later on powered aircraft) to develop piloting skills.
1903	Developed light-weight engine based on the automobile. Propeller design based on wind tunnel tests.
1903	First flight of self-propelled air aircraft. Refinements continued afterwards into the 1920's.

development of the airplane by the Wright brothers preceded fundamental aerodynamic theories and principles adequate for the design of either airplane wings or propellers. Nevertheless, engineering development techniques, including the use of wind tunnels and flight tests (of gliders), enabled the Wright brothers to design a flyable, controllable machine. Subsequent research, largely in engineering but also in some of the basic sciences, has made possible the tremendous growth in global air transportation over the past century. Engineering research aimed at achieving technical and economic progress of this sort must go well beyond the limited knowledge on which invention or demonstration of technical feasibility of a new device, machine, or system is based. It must produce more in-depth and usually more quantitative information that will allow for continuing improvements in the performance, economics, and range of application of the original invention or technical demonstration."

8.6 Pathfinding

Some research transfer requires a process to align all stakeholders.

Time and effort spent on research that has progressed through the research pipeline and demonstrated existing value but does not have a viable path to be implemented and deployed is unnecessary waste. Ideally, research that has matured to the point where it is capable of contributing value will be transferred into production usage quickly. However, as Jony Ive [11] from Apple points out, there is a difference between the research and exploration phase and the "productization" phase that makes the idea deployable. In this case, he was referring to the MacBook Pro touch bar,

> *"It was a curiously different effort between thinking about the original idea and exploring and experimenting with the original idea, and then working out how [to] make it valuable to a specific product. They are different types of efforts, but we are very much focused on the final product."*

Because of what Jony Ive pointed out, there are often several barriers to adopting and deploying research when it originates in one group but must be implemented or deployed by another. One of these is the resistance encountered by the cognitive bias called "Not Invented Here" or NIH. An example is provided by Arthur C. Clarke [12] who wrote that some of the most important innovations in the world were met with this type of resistance,

> *"The harnessing and taming of electricity, first for communications and then for power, is the event that divides our age from all those that have gone before. I am fond of quoting the remark made by the chief engineer of the British Post Office, when rumors of a certain Yankee invention reached him: "The Americans have need of the Telephone — but we do not. We have plenty of messenger boys.""*

However, the blame for delays in transferring valuable research into production usage is not always the result of NIH or the fault of management. Blame can also be ascribed to the people doing the research if they do not engage early with other stakeholders or help implement the research. In the words of IBM's former Director of Research, James C. McGroddy [13],

"It is often tempting to blame "top management" for such failures to fully capitalize on invention. In my view, this is absolutely the wrong place to point the finger. In any institution, it is the job of the researchers and those who manage researchers (to the degree that they can and should be managed!), to take full responsibility for capturing the value produced by research."

To overcome barriers like NIH, or the lack of engagement between researchers and other stakeholders, a pathfinding process is often needed. The goal of pathfinding is to enable the shortest path to adoption for research with existing value. It is a jointly owned process between those doing research, the consumers and those who will need to implement and deploy the result. It is meant to complete quickly, using short steps to build consensus based on data. After pathfinding is done the innovation is often adopted with the original researcher in tow to help with the implementation and initial application.

As we noted before, the "distinct" research that does not meet our definition of what needs to be tracked in the research pipeline will likely transition without much delay since its scope is contained within the same group and does not affect anyone else. This type of research does not need to go through the pathfinding process. The pathfinding process focuses on topics that have multiple stakeholders and dependencies that will benefit from coordination.

The presence of multiple stakeholders means that there may be disagreements on the following:

(1) The current state and the associated problem or opportunity.
(2) The raw value of the innovation independent of feasibility.
(3) Whether the analysis done represented the real world.
(4) How the innovation should be implemented as a product.
(5) The effort or cost of implementation.

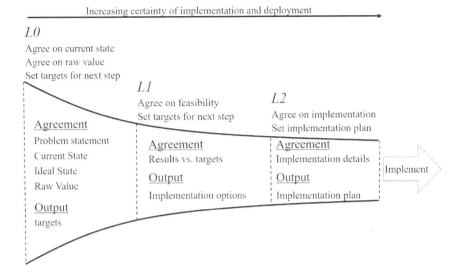

Figure 8.8. **A pathfinding process for interdependent research.**

All of these are valid questions and they need to be addressed systematically to avoid gridlock. For example, without agreement on the problem, current state and the inherent raw value of the research topic any other discussion becomes pointless. If there is no agreement on the raw value, there is no point in debating its implementation. So, to help walk through the questions in the best order the pathfinding process often has multiple levels. Each level addresses a specific question that builds on the answer of the previous one. The decision to transition from one level to the other happens in regularly scheduled meetings. A short description of the three levels (L0, L1, L2) that we have found useful is below and shown in Figure 8.8. Also, a more detailed case study is included at the end of the chapter that goes into detail for each level and how pathfinding was instituted for an R&D organization at Intel in 2005.

L0 Raw Value: The objective of this level is to gain agreement on the problem statement, current state, ideal state and whether the potential value of the solution warrants moving forward. If approved, the output of this level are performance targets that need to be demonstrated on realistic test cases set by the stakeholders to qualify for the next level (L1).

L1 Feasibility: The objective of this level is to gain agreement on the feasibility and cost of integrating the new solution into mainstream engineering solutions or adopting it as a new solution itself. This is done by evaluating the results of the tests stipulated in L0. If approved, the output of this step are options for full implementation that relevant stakeholders must agree with to qualify for the next level (L2).

L2 Initial Integration: The objective of this level is to get agreement on the implementation plan based on an initial integration of the research with existing technology. When this is achieved, the research is formally transferred to a development group for implementation. In many cases, the researchers will continue to help with the implementation for some time.

When first implemented, the pathfinding process can be very painful if the proper work to engage all stakeholders during research has not been done properly. However, over time, perseverance with the model drives changes in behavior that encourages the early engagement that is critical for the success of the research. As we noted, the case study at the end of the chapter provides a more complete example of a pathfinding process.

8.7 Performance Management

Performance management must motivate and reward research.

Employees who are good at conducting research and taking it from concept to existing value have the following traits:

(1) Deep knowledge of the subject matter.
(2) A passion for their mission.
(3) Ability to prototype their ideas.
(4) Excellent communication and collaboration skills.
(5) A willingness to engage all stakeholders.

(6) A willingness to listen to others who disagree.
(7) Clarity of thinking to know when to persist and when to stop.

Not surprisingly, people with these traits are very valuable and in high demand everywhere. Evaluating the performance of people working on long and complex research projects can be difficult. If the performance management, career growth and reward systems are skewed towards short term impact then the best people in the R&D organization will be dissuaded from doing research and the engine of innovation will atrophy. To prevent this, managers in the R&D organization need to include the following when assessing the performance of a researcher:

(1) The quality, relevance and potential value of the research.
(2) The quality and value of research moved into deployment.
(3) The value of industry or academic leadership and influence.
(4) The value of deciding to hibernate stalled research.
(5) The additional value achieved by enabling collaboration.

The bottom line is that if there is an individual that is truly excelling in a research project, they should not have to wait years for it to finally come to production to be recognized. Also, if they decide in the middle that a research direction is not panning out and they cancel it, they should be recognized for making the right decision.

8.8 Summary

Engineering research increases the power of engineering by creating new knowledge and more powerful modeling, analysis and optimization methods to solve design problems. This is very much applied research and differs from development due to its emphasis on experimentation.

The motivation for research may be induced top-down with specific targets to sustain improvement along a known trajectory or it may happen autonomously in a way that may disrupt the status quo and open up new opportunities. Ultimately, the value of the research is determined by the paying customer and this may be an external entity or the company itself.

Research that targets disruptive opportunities is often more difficult to fund since the perceived value and the likelihood of success is unclear. Establishing the means to encourage and fund an acceptable level of disruptive research is needed to ensure a healthy research pipeline.

To succeed in a competitive environment, the value harvested from this research must have a velocity that is comparable to a start-up. This means ensuring research that is not dependent on others can proceed relatively unencumbered. It also means ensuring research that has dependencies on others follows a process that engages the correct people at the right time. This creates a research pipeline of sustaining and disruptive research topics that come from many sources. Coordinating the research pipeline requires a pathfinding process that confirms the relevance and helps overcome barriers that delay transferring research to development.

Maximizing the value derived from research is a team game. Successful researchers engage with eventual customers and other stakeholders during the research. This often takes the form of iterative prototyping the enables frequent learning cycles and adjustments. Researchers who do this gain credibility with those that will be needed to implement and apply their ideas. They also become better researchers.

8.9 Key Points

- Engineering research increases the power engineering.
- The value of research is determined by the customer.
- "Sustaining" research follows a known trajectory.
- "Disruptive" research displaces, incumbent processes/practices.
- Sustaining and disruptive research funding should be separate.

- Rapid learning cycles in research means rapid prototyping.
- Research with dependencies should be centrally coordinated.
- Research transfer requires pathfinding to align all stakeholders.
- Performance management must motivate and reward research.

8.10 Case Studies

8.10.1 *Pathfinding*

In late 1994 the floating-point division (FDIV) bug on Intel's Pentium microprocessor became public knowledge. In 1995 Intel took a pre-tax charge of $475 million against earnings to account for the program to replace defective products. One consequence of this event was an increased interest in applying formal verification concepts to floating-point operations. The initial investment in this area was done by the central Computer Aided Design (CAD) organization at Intel and helped lead to the formation of a dedicated research group called Strategic CAD Labs (SCL) under its direction. The resulting formal verification research on model checking and theorem proving led to breakthroughs in tools and methods that were [14],

> "...applied to a number of design projects, including the Pentium (R) 4 processor. In designing the Pentium 4 formal verification was indispensable, capturing several extremely subtle bugs that eluded simulation. Any of these could have resulted in an FDIV-like recall."

In parallel to research on formal verification, SCL also explored other microprocessor design and validation challenges that spanned architecture, micro-architecture, logic, circuit and physical design and validation. This augmented other applied research done by development groups in the central CAD group. In areas like formal verification, where the user community was very specialized and small, the development and support requirements were minimal. As a result, the researchers in SCL also did much of that work for the engineering tools and methods they created. This complemented small development efforts in other groups.

The Problem

As SCL research expanded into the design and validation of logic, circuits and the physical implementation there were many more potential users and interdependencies with other development groups. In these cases, it was not practical for a small team of researchers to develop and directly support their creations by themselves. They were required to transfer the concepts and early development work to other groups that could complete the implementation and support thousands of engineers. This created some problems familiar to organizations reliant on technology transfer:

(1) Some development groups also conducted their own applied research driven by the immediate needs of their next customer. As a result, SCL research often had difficulty overcoming systematic biases such as "not invented here." Since the adoption of the research was controlled by the development groups, the transfer of promising research was often delayed or never happened.

(2) Partly in response to the above and also due to the pressure to show tangible impact, some SCL researchers would skip by the development teams and work directly with an individual microprocessor design group to implement their research. To gain acceptance, the features were tailored to that group. This increased the competitive tension with the development groups who were trying to standardize microprocessor groups on their solution.

(3) There were also research topics that were disruptive by nature and, if successful, would require a major shift in how

the design and validation of a microprocessor were done. These topics were beyond the ability of even the parent CAD organization to enable. This added another dimension complexity and set of cognitive biases to overcome.

Frustrations grew in the research and development groups and the time and effort it took to move the promising research to development took too long if it happened at all. This friction often bubbled to the top when a research topic was first brought forward for a technology transfer discussion. Often, the research was very far advanced and may have already been in use by one or more design teams. The discussion, therefore, did not just center on the research results but also included a debate on how the solution was built and whether it could be easily integrated into existing systems. In some cases, it was built on its infrastructure, used different data models, or assumed new methods that were not compatible with those supported by the development teams in the central CAD group. This made the technology transfer discussion less about the raw value of the underlying research results and more about opinions on the software languages, data formats and data models the research group utilized. The big bang technology transfer review where everything was considered at once often ended in the rejection of the transfer due to reasons unrelated to the actual value of the underlying idea.

Actions Taken

When one looked at the technology transfer discussions the primary areas of disagreement tended to be in the following areas:

(1) The current state and the associated problem or opportunity.
(2) The raw value of the innovation independent of feasibility.
(3) Whether the analysis done reflected the real world.
(4) How the innovation should be implemented as a product.
(5) The effort or cost of the implementation.

When all of these questions were addressed at the same time, it resulted in gridlock. So, the decision was made to construct a central pathfinding process for highly interdependent research that would break the technology transfer discussions into a series of smaller steps so questions could be addressed earlier and in the right order. Also, the general manager of the central CAD group was made the decision-maker to raise the profile of the research so that researchers would no longer see the need to bypass the development groups in search of recognition.

Selecting Topics for Pathfinding

Research that was self-contained within a development group with no dependencies or impact outside of that group was still managed by those groups and did not have to go through the pathfinding process. Only the research that met one of the criteria below was targeted.

(1) If successful, it will require others to change how they work.
(2) It is complex or high risk and will benefit from more input.
(3) To be implemented it will need others to participate.

Because of the criteria used, the pathfinding process primarily focused on the transfer of SCL research into production and a small number of research efforts that originated in development groups. The pathfinding meeting was called the Roadmap Decision Meeting (RDM) due to its impact on the R&D organization's roadmap.

Roadmap Decision Meeting Framework

The objective of the RDM was to systematically address the usual areas of disagreement in smaller steps and the right order. For example, without agreement on the problem, current state and the inherent raw value of the research topic any other discussion becomes pointless. If there was no agreement that then there was no reason to debate its implementation. So, to help walk through the questions in the best order the pathfinding process was divided into three levels called L0, L1, L2. Each level addressed a specific question that builds on the answer to the previous one. The decision to transition from one level to the next one happens in regularly scheduled path-finding meetings and the following rules were used for each review to reduce waste and encourage offline collaboration with all of the stakeholders in advance:

- Sessions were initially one hour and held every two weeks.
- Three pathfinding topics were dispositioned per meeting.
- Agreement on the problem and current state was required.
- A standard presentation template was used by everyone.
- The GM and relevant staff members had to attend.
- The director of SCL chaired the meeting and set the agenda.
- The GM made decisions if there was no consensus.
- Consumers of the research set targets that needed to be met.

These guidelines provided a clear set of expectations for the participants. The specific template and short duration of each review encouraged researchers to engage early with the stakeholders to get alignment before the review. If there was no agreement between all of the stakeholders on the data being presented then the presenters would be asked to return later after they had worked out their disagreements. The three levels used by the pathfinding process are described below.

L0: Raw Value Discussion

The objective of this level was to get strong agreement in the problem statement, current state and whether the raw value of the proposal warrants moving forward. As we noted, without agreement on the problem and the current state, a discussion on anything else was useless. L0 proposals often ended with a directive to go back and refine the problem statement or current state to reflect reality better. Once this was achieved the ideal state could be presented. The ideal state was the vision, fully realized that described the raw value of the proposal. It was purposely unconstrained by opinions on whether it was feasible. The first question to be answered was whether it (in the ideal state) would solve the problem and create enough value to warrant taking the next step? If the raw value for the ideal state was not considered big enough then the decision to return it to research was simple. If the answer was yes then it is worth taking the next incremental step. By focusing on the problem, current state and raw value first many extraneous debates were avoided and a quick decision could be made to move forward or not.

Agreement on the problem statement, current state and raw value came from the consumers of the result. This included potential customers as well as others in the R&D system that were needed to fully implement the solution later. As a result, representatives from those groups needed to attend the review and give their thumbs up. The purpose of this is to start drawing the recipients and partners into the decision-making process and ensure that the problem and the current state is accurate. It also assigned development groups some responsibility to move the research forward to the next step without having them commit to the complete transfer yet.

The decision to move forward also required agreement on targets for demonstrating existing value on realistic test cases using the existing research prototype. The results would be used in the next (L1) discussion. Since a final production-worthy solution was not available yet, some degree of hand-waving was accepted. Not surprisingly, if the researchers had already been including potential customers and development partners in their research, the discussion of raw value, test cases and targets moved very quickly and that provided positive reinforcement of the right behavior.

L1: Feasibility Discussion

The objective of this level was to gain strong agreement on the feasibility of the technology by evaluating the results of the test cases provided in L0. The focus at this point was on the data and the assumptions made to overcome the fact that the research was not yet production worthy. It was expected that the results were reviewed with the stakeholders in advance to ensure they met the intent and the data was considered accurate. This encouraged conversations between the right people offline to ensure time is not wasted in the official discussion when many more people were there.

Agreement on whether the data proved that there was enough existing value on the test cases provided was based on the feedback from the ultimate consumers. Similar to the previous level this helped ensure the receiving parties had responsibility in the decision-making process and retained a sense of ownership. The review's output was a decision to either move to the next step or return to research for more work. If the decision was to move forward, the review also agreed on the implementation options that needed to be explored. It was expected that the options were already discussed agreed to with the potential development partners.

L2: Initial Integration Discussion

Once the problem, existing value and possible implementations were agreed on, the discussion would turn to how to implement it. Therefore, the objective of this level was to review an initial implementation and get strong agreement in the formal implementation plan. In the L2 review, the implementation plan was discussed and approved by the consumers and development partners. This included the architecture, main functions, integration targets and other collateral that needed to be done to produce an initial production version of the solution. A description of the implementation plans included resource and time estimates. Since the managers who controlled the development resources had been participating in all of the previous pathfinding discussions, getting their commitment at that point became trivial. The output of this level was an approved and implementable plan that was sufficient to enable the productization and deployment of the research in question.

Results

It took over 6 months of RDM meetings before changes in behavior started to be seen. Initial reviews far exceeded the 20 minutes allocated to them or were cut-off as participants struggled to adapt to the set of expectations and the need for pre-work with stakeholders. As a result, there was little improvement at first in the throughput of the research transfer. This was a case of doing the right thing poorly and it took persistence from the GM of the central CAD group and the research director in the face of continuing frustration. However, over time, the richness of the pathfinding conversations grew and researchers saw this an important platform for them to present their research to senior leaders and the GM to gain visibility and valuable feedback. Eventually, all maturing research in SCL was classified as L0, L1, L2. This provided a complete picture of what was moving from research to development. As more topics from groups or individuals from other parts of the central CAD group started to be included it also provided an important vehicle for technical leaders from the rest of the organization to get exposure and gain feedback. Although there were still contentious discussions, most happened at the right stage of the research maturity and reduced the motivation for researchers to work around the existing development groups.

The process was not perfect. Ensuring the right topics went through the pathfinding process required vigilance from the director of research who chaired the RDM. Although the default was that all SCL research would transition to production using this approach, research in the development groups only had to do so if it met the conditions outlined earlier. There were cases of research in development groups that should have been directed through pathfinding but were not. These were similar in scope to the research being done by SCL but did not benefit from the same scrutiny and feedback. In those cases, it was the responsibility of the chair of the RDM to identify those and argue for them to be included.

As the pathfinding process matured and more topics moved through the three different levels, it started to become more difficult to administer reviews such that each topic moved quickly. Due to scheduling conflicts or other disruptions a backlog started to build. To remedy this, some simple decisions were made via email instead of holding a formal meeting. This helped keep everything moving but reduced the quality of discussions that often led to unanticipated insights.

Overall, the implementation of the pathfinding process changed behaviors and made a very positive impact on enabling research transfer from SCL to the development teams. It also had a positive effect on research in the development groups by exposing high-risk topics to greater scrutiny. Finally, it created a path for any individual in the central CAD group who had been working on an idea on their own to have it reviewed by senior leaders and gain incremental feedback and development support. However, its success required persistence by the GM of the central CAD group, their senior staff, and vigilance by the director of research to ensure it was comprehensive, effective and fair. When this level of attention waned so did the effectiveness of the process.

Questions:

(1) What was the problem that the central CAD organization faced?
(2) What was done to reinforce the right behavior from everyone?
(3) How could the increasing volume be handled without slowing down?
(4) How would pathfinding change if there is no central research group?

References

[1] McGroddy, J. C. (1993). Industry, Government, Universities, and Technological Leadership — An Industry Perspective, Honeywell/Sweatt Lecture, University of Minnesota, April 29, 1993.

[2] Christensen, C. Raynor, M. and McDonald, R. (2015). What Is Disruptive Innovation? HBR, December 2015.

[3] Burgelman, R. (2002). *Strategy is Destiny.* (The Free Press) p. 11.

[4] Bower, J. and Christensen, C. (2016). *The Clayton M Christensen Reader,* Chapter 1: "Disruptive Technologies: Catching the Wave" (Harvard Business Review) p. 6.

[5] Snowden, D. Boone, M. (2007). A Leader's Framework for Decision Making HBR, November 2007.

[6] Burgelman, R. (2002). *Strategy is Destiny.* (The Free Press) p. 13.

[7] Christensen, C. Raynor, M. and McDonald, R. (2015). What Is Disruptive Innovation? HBR, December 2015.

[8] Kim C. and Mauborgne, R. (2005). *Blue Ocean Strategy: How to Create Uncontested Market Space and Make the Competition Irrelevant.* (Harvard Business School Press).

[9] Snowden, D. Boone, M. (2007). A Leader's Framework for Decision Making HBR, November 2007.

[10] NASA. *Overview of the Wright Brothers Invention Process.* Available from https://wrigh.nasa.gov/overview.htm [June 2020].

[11] Gugliemo, C. (2016). Jony Ive talks about putting the Apple "touch" on the MacBook Pro, CNet, Oct. 28 2016. Available from: https://www.cnet.com/special-reports/jony-ive-talks-about-putting-the-apple-touch-on-the-macbook-pro/ [May 2020].

[12] Constable, G. et al. (2003). *A Century of Innovation: Twenty Engineering Achievements that Transformed our Lives.* (Joseph Henry Press).

[13] McGroddy, J. C. (1993). Industry, Government, Universities, and Technological Leadership — An Industry Perspective, Honeywell/Sweatt Lecture, University of Minnesota, April 29, 1993.

[14] O'Leary, J. (2004). Formal verification in Intel CPU design, *Proceedings. Second ACM and IEEE International Conference on Formal Methods and Models for Co-Design, pp. 152–.*

CHAPTER 9

The R&D Organization

Great talent is helped or hindered by the organization in which it is asked to work.
— **Gregory Kesler and Amy Kates [1]**

The R&D organization includes the employees and other assets in the direct chain of command of the manager. An organization is needed to handle the complexity of the many interdependent and concurrently running functions and processes required to develop and deploy engineering solutions. We have already discussed the common concepts of culture, diversity, inclusion, value, waste, velocity and learning cycles. We have also detailed the core R&D functions of strategic planning, development and engineering research. In this chapter, we discuss how all of these elements come together.

The R&D organization is much more than an organizational chart. It starts with a strategy and set of values that identify the capabilities needed to deliver value and a competitive advantage. These determine the structure, processes, networks, reward systems and talent required to win. Since the strategy influences the organization's design, a change in the strategy may result in a re-design of the organization. A healthy organization can adjust to enable a change in strategy.

A matrix work model is required to deliver predictability and efficiency. While there are exceptions, most of the work in the R&D organization will need to master working in a matrixed environment whether the organizational structure reflects that or not. Proper governance is needed to create the right balance of power across the R&D organization. Informal intrinsic networks complement the formal hierarchy and enable knowledge sharing, continuous improvement and innovation. Standard policies, practices and processes enable rapid decision making while reward systems and human resource management align the skills and behavior with the strategy. Managing the R&D organization requires different approaches for problems that range from simple to chaotic while inculcating a culture of learning and continuous improvement. This requires the manager to take personal responsibility to instill the right behavior. This cannot be delegated.

9.1 The Role of R&D Organization

The R&D organization governs many interdependent elements.

The R&D organization includes the employees and other assets in the direct chain of command of the manager. An organization is needed to make decisions in a complex environment with many interdependent and concurrently running functions. We have discussed the main elements in Chapters 1–8 and, as depicted in Figure 9.1, they are all connected.

Figure 9.1.   The R&D organization.

At the heart are four common attributes:

- **A common culture (Chapter 2)** that supports the strategy and values of the R&D organization and enables the engineering functions.
- **A diverse and inclusive work environment (Chapter 3)** that encourages the different perspectives needed to create differentiable value.
- **A deep understanding of value, waste and velocity (Chapter 4)** so that everyone contributes to increasing the rate of deployed value.
- **Ubiquitous learning cycles (Chapter 5)** that generate information that is used to continuously improve engineering processes and workflows.

There are three core R&D functions that also run concurrently:

- **Strategic planning (Chapter 6)** that synthesizes and analyzes new information and forces changes that affect the organization's destiny.
- **Development (Chapter 7)** that executes and coordinates the planning, modeling, analysis and optimization needed to implement a solution.
- **Engineering Research (Chapter 8)** that increases the power of engineering to create more value for current and new technologies.

The R&D organization is how these elements come together in a way that speeds decision making and increases velocity.

The R&D organization connects strategy, structure, process, rewards and people.

The organization's design reflects the five interdependent elements given by Jay R. Galbraith's star model [2] shown in Figure 9.2. *The strategy* articulates the direction of the organization

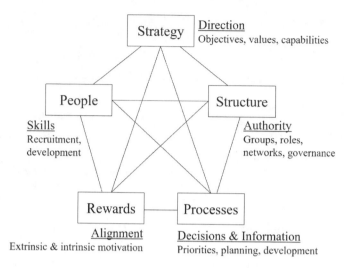

Figure 9.2.   A depiction of the Galbraith 5-star model.

by defining the vision and values as well as the strategic capabilities needed to meet the strategy. *Structure* allocates authority within the organization by combining related strategic capabilities into formal groups, assigning the leaders, establishing intrinsic networks, and setting the rules for governance. *Processes* enable decision making, sharing and planning across the organization and with the rest of the R&D system. *Rewards* motivate the individual employees in a way that reinforces the strategy and values of the organization. *People* ensures individuals with the right skills are hired, developed, and retained and that the overall health of the organization remains high.

A common mistake is to jump to a conversation about an organization's structure before the vision, strategy and values are clear. The structure of the organization and who leads the different groups is probably the last thing that should be decided when creating or changing an organization. Instead, it should follow the sequence depicted in Figure 9.3. In that picture, it all starts with the vision of the organization. As stated in Chapter 4, the vision is an aspirational view of the ideal state that the organization is seeking. It does not tend to change

Figure 9.3.   The sequence for constructing an organization.

over time. In that chapter, we gave an example from Henry Ford's articulation of his decision to manufacture only one model of the automobile (the Model T) in 1909 [3]:

> *"I will build a motor car for the great multitude. It will be large enough for the family but small enough for the individual to run and care for it. It will be constructed of the best materials, by the best men to be hired, after the simplest designs that modern engineering can devise. But it will be so low in price that no man making a good salary will be unable to own one — and enjoy with his family the blessing of hours of pleasure in God's great open spaces."*

The vision helps define the strategy and values. Until there is agreement and alignment between those three elements, it is pointless to move to the next step. As noted in Chapter 2," *culture eats everything for breakfast,"* poor alignment between the vision, the values, and the strategy will lead to failure regardless of what happens next.

The combination of the vision, strategy and values is the foundation that everything else is built. When they are clear, the required strategic capabilities become more obvious. Once the strategic capabilities are defined, they can be grouped and connected via a structure that properly balances power and authority. Similarly, processes and networks can be established to share information and make decisions that span the organization easier. The reward systems, metrics and human resource management then align the motivation, hiring, and development of people with the strategy. Altogether, these elements constitute the design of the R&D organization. We will explore them in more detail in the following section.

9.2 Strategy

As the strategy changes so does the organization.

The strategy comprehends and acts on the internal and external forces in the ecosystem. Since the internal and external forces are dynamic, so is the strategy. Accordingly, the decisions coming out of the continuous loop of strategic planning that we described in Chapter 6 may have a material effect on the R&D organization's structure, processes, reward systems and people as shown in Figure 9.4. As a result, it is important to continuously evaluate the organization's design relative to the changing strategy.

As we discussed earlier, the "rubber band" model described by Robert Burgelman (Figure 9.5) is a good way of visualizing the alignment between the stated strategy (what we say we are doing) and the strategic actions (what we are doing). It also helps compare the basis for competitive advantage (what capabilities we need to win) and the organization's core competencies (what

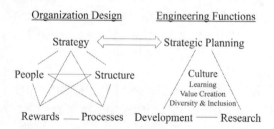

Figure 9.4.   Strategic planning affects the design of the organization.

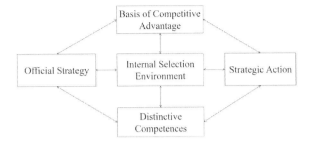

Figure 9.5.   The rubber band model.

we have). The internal selection process shown in the center of the diagram represents the actual culture and processes that drive funding and other decisions.

If the strategy changes, then the other elements in the rubber band model may no longer be aligned. For example, if the strategy is to break into a new market with entrenched incumbents, then the best approach may be to create an organizational capability that targets a niche and can deliver fast. None of the current strategic actions in progress may cover this and the skills needed may not exist. It may also mean changing the organization's structure to temporarily create a vertically integrated group with complete control over all aspects of engineering development. This will allow the group to move fast but may create redundant functions with the rest of the R&D organization. The cost (and lower profits) associated with this may be acceptable in return for success entering the new market but this will require a different set of financial metrics to judge progress. Once a foothold is obtained, the strategy may shift to maximize profit and this may lead to centralizing common functions and eliminating the vertically integrated group.

This type of change may occur more often than one thinks. Often, when a new CEO takes over a company, it is followed by a change in corporate strategy that leads to organizational changes. This can often filter down to the level of the R&D organization. Similarly, this is can also happen when a new R&D

manager takes over an engineering organization. Often, changes that are made to address a new threat or opportunity result in creating one or more autonomous groups followed by a consolidation once a foothold has been achieved or the organization needs to change course again. One can read the annual reports of technology companies and see several instances where the company went through these cycles of change. The case study at the end of the chapter highlights some of the changes in Intel's operating segments published in their annual reports between 2005 and 2018.

9.3 Structure

A matrix is inevitable. A good design properly balances power.

The organization's structure distributes power and authority by grouping capabilities establishing a reporting structure and assigning roles. The grouping and reporting structure reflect the importance placed on such things as target technologies, engineering functions, geographies, or customers. In most large R&D organizations, there are multiple perspectives of equal importance that need to be considered. For example, getting the customer's feedback during engineering and after delivery is important. In that case, the R&D organization may form a group responsible for a major engineering function and a group responsible for customer support. If both of these groups report to the manager of the R&D organization, they will be perceived to have equal influence. Differences of opinion will get resolved at the highest level of the organization with the manager of the R&D organization involved. On the other hand, if the customer support group is embedded in the engineering group, then the engineering group will resolve the conflict in its best interest. Both approaches are valid depending on the aims of the organization but the power of the group responsible for customer support is very different in one vs. the other.

Not surprisingly, care must be taken to establish a structure that properly aligns power and authority with the intent of the strategy. The goal is not to eliminate the tension between groups in the organization. The goal is to create good tension that will expose important perspectives and get to the right answer faster. Good tension occurs when everyone can openly express their perspective, be heard, and then commit to the final decision. Bad tension occurs when different perspectives are not heard, disagreements become personal or when there is no commitment to the final decision. Good tension occurs when the power is properly balanced and the culture is healthy.

Because of the interdependencies in the R&D organization, matrix operation is inevitable. Even if an organization is not explicitly structured as a matrix, it likely needs to perform as one. Figure 9.6 depicts a generic matrix organization where the primary axes may be different combinations of engineering functions, technologies and customers. On the left of the figure, we show the simple example of three engineering groups

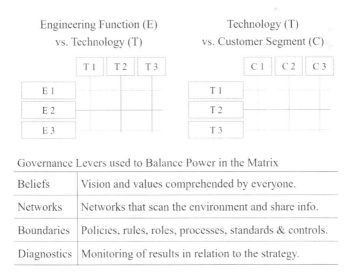

Figure 9.6.   A matrix structure and the levers used to balance power.

that provide solutions to three different technology groups who are then responsible for the delivery of the completed technology to their customers. On the right of the figure, we show an example of three technology groups (that may have their own engineering functions) that each service different customer segments. Below the two examples, we use the framework provided by Gregory Kesler and Amy Kates [4] to highlight the governance levers that are at the manager's disposal to balance the power in the matrix. These are explained in the following.

Beliefs: The first lever is the beliefs inherent in the organization. This is what we call the values or culture of the employees. Because we talked at length about culture in Chapter 2, we will only touch on it here. As we stated, "culture eats everything for breakfast" so its importance cannot be overestimated. Like any other organizational structure, the matrix model has flaws. It can appear complex and it is impossible to map out all possible interactions and provide employees with a rule about what to do in every instance. This can be overcome by setting a clear vision and strategy for the organization and driving the right behavior. A strong culture that is aligned to the vision and strategy of the organization means that people will make the right decisions quickly because everyone understands, "this is what we do here".

Networks: The second lever is networks and specifically intrinsic networks. The management hierarchy of the organization is needed to drive reliability, efficiency and velocity. However, when organizations grow in size the informal networks inherent in a small organization or a startup may vanish. Intrinsic networks complement the organization's hierarchy by mobilizing the organization's leadership to enable a startup-like approach for specific initiatives. Figure 9.7 is a representation of this based on the work of John P. Kotter [5]. The networks are composed of a strong leader and a cross-section of employees from the R&D organization that operates with no hierarchy

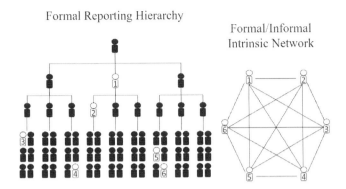

Figure 9.7. The formal reporting hierarchy and the intrinsic network.

to move quickly without any responsibility for keeping the current business running. This allows the traditional management structure to focus on improving reliability, efficiency and velocity while the networks invest in exploring high-risk initiatives.

The Intrinsic networks may be formal or informal. Formal networks have clear objectives from the R&D organization's management and are led by different leaders from within the organization. They may be called working groups or task forces. They may be permanent or come and go as needed. The connection to the rest of the organization is through the people who both report to a group in the organization and are also members of the network. Informal networks form without explicit direction from management. These are often created bottom-up in the organization by employees with a common interest or goal. These networks are important for engaging a cross-section of the right experts from across the R&D organization to assimilate new information, share best-known methods and reinforce standards. They are especially important for enabling the learning cycles and driving continuous improvement since they often explore ideas that lead to bottom-up, autonomous innovation. These networks do not get formed by a management decision and rely on a culture that encourages the creation of informal networks to explore new ideas.

For the most part, the networks we are talking about do not show up on the organization chart but they can play a major role in how the work gets done. A working group composed of architects from different groups in the R&D organization may decide on a standard modeling platform or language that everyone else may need to adopt. A new idea that germinated with the help of an informal network of employees may ultimately get inserted into the formal research and pathfinding process for adoption in a product. Changes in the industry landscape may first get observed using an informal network of industry leaders who meet at industry or academic conferences. All of these networks are an important part of the R&D organization and need to be fostered.

Intrinsic networks also serve other purposes. Intrinsic networks provide an excellent platform for individuals to work with others on important problems with a high degree of autonomy. This is an important ingredient of intrinsic motivation that we will discuss in more detail later. Also, the formal networks provide a great opportunity for leaders from throughout the organization to own important enterprise-level initiatives, develop their skills and gain visibility. Assigning someone deep in the organization to lead an initiative via one of these networks is a great way to evaluate their abilities and decide on how best to continue their development.

Boundaries: Establishing clear boundaries is the third lever. Although Kesler and Kates use the word boundaries, it does not mean that people are prevented from working across organizational boundaries. In this case, it is quite the opposite. The word is used to denote setting policies, rules and guardrails that provide a standard framework for decision making across organizational boundaries. One of the primary complaints about a matrix model is the time it takes to make decisions. This is usually the result of two equally powerful entities being in disagreement with one another on the course of action. In Chapter 6, we discussed the RAPID model for decision making. That model establishes a standard set of expectations for what is needed before any decision can be made. Whether one uses RAPID or another framework, it is important to have one in place to ensure decisions are made expeditiously.

Establishing clear boundaries also involves setting standard development practices that everyone must follow regardless of where they sit in the matrix. We discussed the importance of standards in detail in Chapter 7 and they become very important when dealing in a matrix development model. Standards prevent mistakes when information or other collateral is handed-off between groups in the matrix. It also means having the controls in place to flag deviations from the standard so they can be identified and resolved. By setting standards, the manager of the R&D organization can reduce the power of some groups by centralizing some of their functions while enhancing the power of others by making them the owner of the standard solution.

Diagnostics: The last lever is diagnostic measures. This is the means to determine how the organization performs relative to the strategy and objectives and take corrective action if needed. What is measured can have a big impact on people's performance and choices. Setting goals and metrics that focus on a particular initiative or result will increase its visibility. Connecting the success of different groups in a matrix to a common goal will help drive their priorities in the same direction. Diagnostic measures should also gather feedback from employees on what is working and what is not. It is very educational to sample the organization to understand how employees perceive their work. We will talk more about the reward systems and metrics a little later.

Moving to a matrix model is difficult and takes persistence

Figure 9.8 shows an example of an engineering R&D organization on the left that is based on a business model where the non-recurring engineering cost of development (NRE) is insignificant compared to the total cost of doing business. In this case, there are two product lines (P_1 and P_2) and each has significant profit margins and sales volume. As a result, each product group has their own dedicated functions. Although there will be waste due to redundant functions ($F_{1,n}$ and $F_{2,n}$) in each product group, the proportionally small additional cost is deemed to be acceptable to provide customization and fast turn-around for the customers who are willing to pay for it.

In this example, the industry landscape changes and the company is forced to change its strategy from developing high margin, high volume products to a small number of customers to one that relies on developing low margin, low volume products to a large number of customers denoted by P_1 to P_n shown in the far right of Figure 9.8. In this case, the NRE costs will be a significant percentage of each product's total cost and the previous structure will no longer be justifiable. In the model, on the far right, the common functions are centralized to eliminate redundant work and the product teams (P_1 to P_n) must integrate the standard components into the final products.

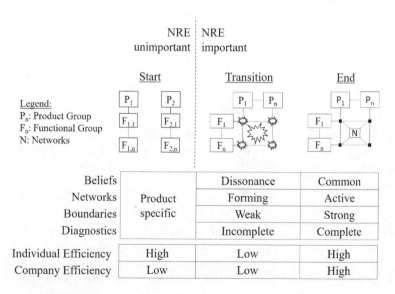

	Start	Transition	End
Beliefs		Dissonance	Common
Networks	Product	Forming	Active
Boundaries	specific	Weak	Strong
Diagnostics		Incomplete	Complete
Individual Efficiency	High	Low	High
Company Efficiency	Low	Low	High

Figure 9.8.   Example transition to a matrix operation.

Making the transition from the model on the left and the matrix model requires a change in the belief system as people shift from valuing customization and working independently to valuing standards and being dependent on others. It requires establishing formal and informal networks to enable sharing information across all of the elements. It requires setting boundaries via standard workflows, policies, processes and decision making that will enable everyone to operate as one team. Finally, it requires the means to diagnose problems with the operation and highlight areas for improvement. As shown in the middle portion of the figure, the period of the transition will most likely be beset by dissonance on the belief system, incomplete networks, boundaries, and diagnostic systems. There is a major chasm to cross from doing the wrong thing well (the far left) to doing the right thing well (the far right). First, you must pass through the state of doing the right thing poorly (the middle) where results suffer, customers grow impatient and employees grumble about their loss of productivity. This is when management often loses their will to follow through on the change and get stuck in the middle. Once the decision is made to move to a matrix model, the organization's leaders will need the fortitude to work through the many changes necessary to make it work.

9.4 Processes

Processes provide means to make decisions and share information.

Processes are the officially recognized way to do such things as make decisions, share information, and resolve conflicts in the R&D organization and across the R&D system. This includes common procedures and policies for such things as, strategic planning, technology transfer, product planning, setting budgets, performance management and more. They exist because many decisions require participants from different groups. They differ from the intrinsic networks we discussed earlier because they have rules of operation and are meant to make the hierarchical structure of the organization operate with predictability and efficiency. No matter how the manager groups capabilities and structures within the R&D organization, there will be decisions needed that involve more than one group. This is especially true when one considers the suppliers, customers and others that are part of the R&D system. When this happens, there is a need for a formal process to make a decision.

Formal processes provide a paradox for managers. The absence of processes to align different groups in the R&D organization or resolve disputes can result in major problems that lead to significant rework or worse. On the other hand, adding more processes or spending more time in a process meeting does not create more value. The time spent in a process meeting is a necessary waste. The ideal state is that everyone shares the same values, understands the strategy and knows what to do without needing to meet. This may be achievable in small teams or new start-ups but as the size and complexity of the system grows formal management processes become unavoidable. Nevertheless, the goal should be to reduce the number and duration to as close to zero as possible without sacrificing the organization's output. Fortunately, advances in such things as information systems, communication and collaboration technology can significantly reduce this form of necessary waste. As we discussed before, continuous improvement and learning cycles also apply to process meetings. They should also be subject to retrospectives. The manager can also reduce the need for formal processes by clearly communicating the strategic imperatives, making the R&D organization values part of the culture and utilizing a clear decision framework.

Processes are needed to make decisions and share information across the whole R&D system as well as the R&D organization. As a result, some processes involve members from other suppliers, customers, consortia or institutions. Since the R&D organization's manager directs the efforts of the R&D system,

Table 9.1.   An example of common processes in an R&D organization.

Aligned with company	Internal to R&D organization	Shared with R&D system
Financial	*Planning*	*Customer*
Annual Plan	Strategic Planning	Release Planning
Quarterly Update	Resource Allocation	Request Management
		Issue Management
Human Resources	*Research*	*Supplier*
Organizational Health	Research Pipeline	Release Planning
Performance	Pathfinding	Request Management
Hiring		Issue Management
HR Policies		License Management
Legal	*Development*	*Institutions & Consortia*
Agreements	Standards	Industry Standards
Licenses	Development ROI	Technology Transfer
Publications	Buy vs. Build	Knowledge Transfer

the owners of those processes will likely be leaders from the R&D organization. Table 9.1 highlights some of the processes that are common for a large R&D organization. Some processes must align with the company processes (e.g., for finance, HR and legal). Some processes are restricted to members of the R&D organization (e.g., strategic planning, research pipeline, pathfinding, development, etc.). Some processes include customers (e.g., release planning), suppliers (e.g., issue management) and consortia/institutions (e.g., technology transfer).

Most of the processes shown in Table 9.1 relate to each other in some fashion and often will have members from different groups in the R&D organization. As depicted in Figure 9.9, the processes that are internal to the R&D organization that address continuous improvement (Chapters 2–5), strategic planning (Chapter 6), development (Chapter 7) and research (Chapter 8) are continuous and tightly connected. Together

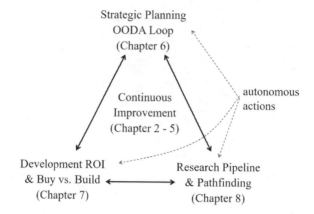

Figure 9.9.   Key R&D processes already discussed.

they form a set of concurrent processes that determine how quickly the R&D organization can react to major strategic changes in the ecosystem. The R&D organization's culture should enable the vast majority of tactical decisions to be made autonomously, at the lowest point of the organization without the need for a process.

As noted in Chapter 6 and Chapter 8 autonomous actions may be happening within the organization that starts outside of any of these processes. They may be the result of informal intrinsic networks that address new or disruptive ideas outside the scope of the current strategy. The proper R&D culture encourages this as long as there are ways to fold them into the appropriate process once they mature require additional help.

9.5 Rewards

Extrinsic motivation is needed but intrinsic motivation is best.

Motivating employees comes in two forms. Extrinsic motivation includes such things as additional compensation, tangible awards, recognition and promotion. Intrinsic motivation comes from the employee's sense of satisfaction.

Rewards that address extrinsic motivation are necessary. They are used to align employees to the strategy and motivate high levels of performance. This is done by tying rewards to accomplishing objectives derived from the strategic actions. The rewards come in different forms since employees are individuals and they react to rewards differently. The simplest level is personal recognition from a superior. For some people, being pulled aside by someone higher in the management hierarchy and being directly thanked for their work can have a big impact. It means even more if the manager can speak to the details of how their work contributes to achieving the strategy. A short, informal chat or telephone call from a senior executive doing the same can help reinforce the message that their work is having an impact. For others, recognition in a public gathering with a monetary reward and token of appreciation is highly valued. These types of awards can scale from peer to peer recognition or those handed out by team leaders in a small team meeting to corporate level awards that reward impact across the company. For most people, a big source of extrinsic motivation comes from their performance reviews, promotions and compensation. To some people, this is the bottom line for understanding how they are valued and no amount of personal or public recognition can overcome a disappointing performance review and a smaller than expected increase in compensation.

The impact that the culture has on the organization's success is so great that effort should also be taken to seek out and reward those that are doing the right thing, the right way even if the results are not there yet. If one of the organization's values is risk-taking then those that took an informed risk did everything right but failed to achieve the result should be recognized for demonstrating that value. This should also be done at the appropriate level to achieve the right effect.

Once there is an equitable process for delivering extrinsic rewards, intrinsic rewards become a powerful means to motivate additional effort and make employees happy. Intrinsic motivation is why some people go home after work and contribute to open-source software or write a chess program to compete against others. They are not doing it for the money. As Daniel Pink [6] observed, they are doing it because it gives them a sense of purpose (doing something that matters), permits autonomy (being able to direct oneself) and a sense of mastery (continually improving at something that matters). The R&D manager enables intrinsic motivation by clearly articulating the strategy and objectives in a way that each employee can see how their work contributes to the goals, delegating decision making to the lowest feasible point in the organization's hierarchy, and enabling career growth for individual employees.

With this in mind, the employee's motivation can be strongly affected by management behavior, especially if it is the direct manager. Machiavelli [7] famously wrote that it is better to be feared than loved. Although it is highly unlikely that any manager would defend this extreme viewpoint today, the reality is that when a manager loses the faith of the people in their organization they sometimes revert to fear and punishment as the only means left to motivate their employees. Fear and punishment might temporarily motivate employees to work long hours to achieve a major milestone but it will ultimately come at a severe price. The high performers will seek employment elsewhere since, unlike the era when Machiavelli lived, people have more options.

Metrics are often defined to try and provide a quantitative way to evaluate an organization's progress vs. a goal. They are also used to justify rewards to employees. Expectations for such things as quality, productivity, turn-around-time and others can be translated into metrics that continuously measure how the organization is doing. When done properly, these metrics can be an excellent tool for driving continuous improvement. However, when the metrics are used to judge individuals and are tied to their performance reviews it can create unintended consequences. Even when the metrics seemingly align with the strategy, the employee may spend more time trying to change how the metric is measured or find workarounds to manipulate the results than actually trying to meet the spirit of the metric.

More often than not, these types of metrics are not properly aligned to the strategy or are beyond the control of the individuals who are being judged. In these cases, they divert effort from the real problems and can de-motivate employees who feel they are being unfairly judged.

The bottom line is that establishing goals and associated metrics should primarily be used to shine a light on the assumptions embedded in the strategy and expose problems so they can be fixed before it is too late. If employees feel that the metrics are used to expose problems so they can be fixed and not for punishment, they will participate in setting aggressive and meaningful goals. If they feel that metrics will only be used to judge their performance than they will fixate on achieving the letter of the law even if it is counterproductive to achieving the strategy.

9.6 People

People are the most important part of the R&D organization.

It is a cliché to state it, but people are truly the most important part of the R&D organization. They not only get the work done but they are also the champions of innovative ideas and dedicate their blood, sweat and tears to turn those ideas into value for the R&D organization. Finding, hiring and retaining a diverse set of high performers with the right skills is the most important ingredient to continuing success. Human resource management (HRM) is the function that establishes the policies and manages the processes that attract, develop and retain the people.

The competition for the type of people that the R&D organization needs is intense. HRM works with others in the R&D organization to establish the talent pipeline. This should go far beyond collecting resumes. It means establishing relationships with schools to promote their investment in the fields of study that interest the R&D organization. It means creating innovative approaches to foster and find talented diverse candidates that are missed using traditional approaches. It means promoting the R&D organization on campuses, job fairs, social media and other forums. Each hiring event also constitutes a learning cycle. HRM should hold retrospectives after each event to determine what needs to be changed the next time to improve results.

In addition to finding and hiring new people, HRM must also drive continuous performance management of the current employees based on meritocracy. Some people will exceed the expectations set for their job and others will not. For the people that do not meet expectations, there needs to be a means to intercept issues early and help them improve. In most cases, problems are a result of poor training, poor direction, a bad environment the wrong role, or something else beyond the control of the employee. It is important to get to the root cause and correct what is wrong with the system because the individual performance of others is probably being affected as well. If these problems are identified early, they can be resolved quickly and everyone benefits. Similarly, there are going to be people that exceed the expectations for their job and warrant a bigger role. It is also important to identify these cases early and recognize the employee as well. All of this should not wait for an annual review process. The R&D organization and direct managers need to have the tools and training to make this a continuous and lightweight process.

An important part of HRM's performance management process is a robust career development path for individual contributors and managers. Some employees are most effective as individual contributors and have no aspirations to join the ranks of management. These are usually employees with deep technical expertise or specialized skills. These people should not become managers. However, they need a clear career growth path that mirrors the management path. In each path, there need to be clear expectations for each pay grade or title and a process for evaluating employees against those expectations.

With this foundation in place, the HRM function should also build a curriculum to grow the skills of individual contributors and managers. This should extend from a new hire to the R&D organization manager and their staff. As part of this process, there should be regular retrospectives with the employees on the value of the development opportunities so that they can be continuously improved.

The HRM performs other critical functions as well. They amplify the voice of the employees by conducting surveys and forums to collect feedback on the overall organizational health of the R&D organization. The HRM function should identify areas of strength and weakness across the R&D organization and provide feedback to the employee population and especially senior management on the status. HRM should then identify the areas for improvement and work with the leaders to implement necessary actions. This becomes an organizational health learning cycle that should be repeated at a specific cadence to continuously improve the health of the R&D organization.

9.7 A Practical Example

Let's look at how everything we just discussed works in practice using an example R&D organization. Figures 9.10 and 9.11 show two extremes, the primary elements are:

- Functional groups (F_1, F_2, F_3,)
- Customer-facing groups (C_1, C_2, C_3,)
- Site managers (S_1, S_2, S_3,)
- Legal, finance and HRM representatives

The functional groups provide engineering solutions to the customer-facing groups who then deliver them to the customer and provide support. The site managers are responsible for geographic-specific regulations, policies and local employee development and communications. The HRM, finance and legal

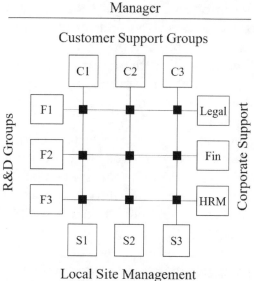

Figure 9.10.   A depiction of a highly matrixed organization.

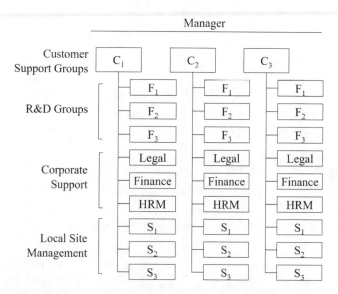

Figure 9.11.   A depiction of a siloed organization.

representatives provide specialized support and align the organization with corporate policies and practices.

Figure 9.10 is a depiction of a highly matrixed implementation of the organization that has been designed to maximize the sharing of the engineering work to reduce cost. The functional groups work with each customer group to deliver what they needed for the customers they support. Because each functional group does work for every customer, they can spot similarities and produce some standard solutions that apply to them all. This reduces their costs. However, they also need to prioritize their work with each customer and there are likely to be disagreements. From the perspective of the customer, they may feel that work is taking too long or feel that they are not getting the right level of priority. The site managers oversee the interests of all of the employees at the local site regardless of what group they are in. They are the R&D organization's leadership at the site and provide a single point of contact for all local employees. The legal, finance and HRM functions are shared and utilized by all of the groups in the R&D organization as well. From the standpoint of the reporting structure, they all start with equal positional power since they all have equal access to the R&D manager and are involved in decisions as part of the manager's staff.

Figure 9.11 is a structure that consolidates all of the other elements under a customer group. It is designed to enable moving fast for each customer. In this model, each customer group has its dedicated functional groups, site managers, legal, finance and HRM resources. The work and decisions happen in each of the customer group silos without interaction with other customer groups. This allows each group to move fast to satisfy the needs of their customer but it creates duplication of effort in the other customer groups across the larger R&D organization. In this model, the customer groups have very big positional power since they are the only ones reporting to the R&D organization's manager. It is fast and customers will like the

Table 9.2. Example governance and process elements in the matrix.

Governance	*Beliefs*: Vision, Strategy, Values *Internal Networks*: Development WG, Research WG *External Networks*: Customer, Supplier, Academia *Boundaries*: Policies, Practices, Standard Methods *Diagnostics*: Velocity, Quality, Efficiency
Processes	*Continuous Improvement*: Workflow Retrospectives *Planning*: Strategic Plan, Roadmap, Development *Research*: Research Pipeline, Pathfinding Decisions *System*: Supplier Management, Customer Management *People*: Recruitment, Hiring, Performance *Finance*: Budget Allocation, Employee Expenses

attention they get but it is expensive and becomes more so as additional customers are won.

Although one may encounter both extremes as well as variations in the middle, we will explore the matrix structure in Figure 9.10 in more detail.

Table 9.2 lists important elements of the matrix model spanning the governance and processes that may exist for a matrixed organization. Governance starts with the manager's leadership team internalizing the vision, strategy and values for the organization and communicating it to the rest of the organization. Together, the R&D manager and the leadership team create the intrinsic networks that connect employees with common skills, functions and ideas to establish standards, share best-known methods and enable autonomous innovation. They also set clear boundaries by setting policies, practices and standard engineering methods and tools. To track progress, the leadership also establishes the means to measure velocity, quality, efficiency and the organization's health. Knowing the governance model helps determine the processes that will be needed. In this case, there are processes for such things as continuous improvement, planning, research, and more.

There is a lot to digest in Table 9.2 and the manager of the R&D organization cannot own all of this themself. The responsibilities must be distributed across the organization using the R&D manager's staff and other leaders. An example of how this can be accomplished is described and shown in Figure 9.12. In this case, we assume that the R&D organization has a legal, finance and human resource manager (HRM) reporting to the manager of the R&D organization, as well as one research group, one development group and one customer-facing group.

The R&D manager is given responsibility for governance. The manager may assign the administrative work to an assistant or small operations group but they are held responsible for directing governance. This means working with the leaders in the organization to align on the beliefs (vision, strategy and values), creating networks with leaders, establishing boundaries, and setting the right diagnostic measures to track progress. In this

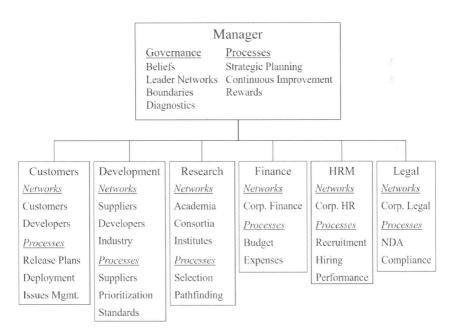

Figure 9.12.   An example of roles in a simple organization structure.

example, the manager also leads the strategic planning process, the awards process and the overall continuous improvement process for the organization.

The development group manager is responsible for the industry standards and supplier networks and the internal developer networks that allow developers in different teams to interact. The development group is also responsible for supplier management processes.

The research group manager is responsible for academia, government and consortia networks. They are also responsible for the research pipeline and pathfinding processes.

The customer group manager is responsible for customer networks and the release planning, and customer management processes.

The legal representative is responsible for the non-disclosure agreement (NDA) and intellectual property protection processes.

The finance representative is responsible for the budget allocation and expense reporting processes.

The human resources representative is responsible for the recruitment, hiring and employee performance evaluation processes.

Figure 9.12 is just an example for illustrative purposes. The distribution of responsibilities will vary depending on the strategy, grouping of capabilities etc. It will also depend on how imbalances in positional power may need to be mitigated by providing ownership of processes.

9.8 Governance

The R&D manager must have strong leadership and management skills. We will discuss leadership in detail in Chapter 13.

The manager must not delegate the governance of the organization.

In Chapter 2, we described the culture as the muscle memory of the R&D organization that guides the behavior of each employee such that when faced with ambiguous or incomplete information, they instinctively know what to do. We also gave some examples showing where the stated values or beliefs of an organization did not match the actual culture and lead to disastrous consequences. By now, we hope we have walked through enough examples to convince everyone that a large R&D organization and the associated R&D system is complex to organize and manage. As a result, the organization's culture must align with the stated values and strategy. When there is strong alignment, many deficiencies that exist in the structure, networks and processes can be overcome by employees using common sense. When there is not strong alignment, no structure or set of processes will overcome that basic flaw. Similarly, the organization's strategy, structure, networks, boundaries and diagnostic measurements establish the framework that the rest of the organization uses to conduct their work. They are all strongly linked and require constant care and attention from the manager of the R&D system. Although many people will help, the manager of the R&D organization ultimately is responsible for governance. Owning that function highlights to the rest of the organization that the manager takes all of the components very seriously and goes a long way to creating a culture that aligns with the strategy and values.

The manager is responsible for continuous improvement.

Chapter 4 discussed the importance of everyone understanding what constitutes value so that they can eliminate waste and continuously increase velocity. In Chapter 5 we highlighted the importance of learning cycles and observed that continuous improvement works when everyone in the R&D organization views themselves as scientists who run experiments to improve their workflows and processes. Big sources of waste are generally easy to identify. However, the total waste in a system is dominated by the many small wastes that individually do not have a big impact but accumulate to create a big problem. Identifying and removing these sources of waste requires everyone in the organization to participate. Extrinsic motivation in the form of awards and recognition for employees that show results may work temporarily but to make it self-sustaining, employees need the intrinsic motivation to commit their time and effort. This means providing clear top-down direction on the importance of continuous improvement and how each employee's contribution supports the organization's strategy. It means providing the employees the skills and tools to see and eliminate waste and the autonomy to take action on their own when needed. Finally, it requires that the employees see and experience how their effort are improving the performance of the organization and making their own work more productive.

All of this requires the manager of the R&D organization to make continuous improvement part of the organization's culture and create the necessary networks, processes and rewards. As a result, the manager of the R&D organization needs to assume overall ownership of the continuous improvement process for the organization. In our experience, organizations that do not have their manager personally engaged in driving continuous improvement invariably fail to make sustainable progress. On the other hand, managers who understand that eliminating waste in workflows and other processes directly leads to greater innovation and velocity get personally involved. They make it their top priority and set the example for others. When this happens, organizations make it part of their culture and it becomes self-perpetuating.

Different problems require different management approaches.

We discussed earlier that the R&D organization's structure and processes provide systematic means to manage standard work. We also noted that intrinsic networks complement this by mobilizing the leadership and talent everywhere in the organization to act on initiatives that require more of a start-up mentality. Similarly, the management of the R&D organization needs to comprehend the wide range of problems that are encountered every day and address each one properly. A good way of looking at this is captured by David J. Snowden and Mary E. Boon [8] who wrote.

> *"Good leadership requires openness to change on an individual level. Truly adept leaders will know not only how to identify the context they're working in at any given time but also how to change their behavior and their decisions to match that context. They also prepare their organization to understand the different contexts and the conditions for transition between them. Many leaders lead effectively — though usually in only one or two domains (not in all of them) and few, if any, prepare their organizations for diverse contexts."*

Snowden and Boone defined the elements of a decision framework depicted in Figure 9.13 and described in the following. We focus on the different domains and how they fit with the other concepts in this chapter. Starting at the lower right of the figure and moving counter-clockwise we have the following domains.

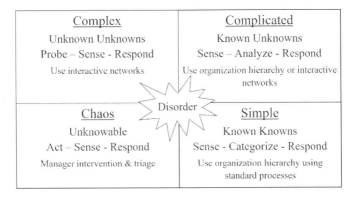

Figure 9.13. A depiction of the Cynefin framework.

Simple

In this domain, problems are understood by everyone. Work is generally in the form of standard flows, policies and practices. Decisions and continuous improvement of the standards are delegated to the lowest possible point in the organization and when issues occur, they are fixed without escalation. Although standards exist, they need to be challenged and continuously improved to avoid complacency and falling behind. The decisions needed are handled by the existing organization structure since there are clear owners for the workflows.

Complicated

In this domain, the solution to the problems are not immediately clear but they can be discovered by experts in the organization after some analysis. Decisions related to sustaining innovation or induced strategic actions often fall into this category since there is a path to an answer. In most cases, a group in the existing organization structure may already have the charter for the area in question so the problem can be assigned to them. In some cases, an intrinsic network may be needed to pull in other experts. Conflicting expert opinions on the right course of action can happen and delay decisions so a good decision process like RAPID is needed. If insufficient data exists to make a decision, the problem space may be "complex," and the manager should treat it as described in the following.

Complex

In this domain, a single answer is difficult to zero in on since the environment may be new and there is little expertise within the organization. The correct answer may only become obvious through experimentation to gain knowledge and experience. In this case, failure is an acceptable option if it leads to learning. Once the data is clear, decisions can be made. Since there is not a good match with an existing group within the R&D organization an intrinsic network will need to be formed to gather the right leaders and experts from across the organization to complete the experiments. This requires a clear direction on the objectives and support of the manager. However, dictating a specific process to follow and deadlines may prove counter-productive. In the R&D organization, this is where the manager often needs to spend most of their time.

Chaos

This represents a class of problems that do not have a solution. Either a catastrophic event has already occurred, there is no time to gather data or there is so much flux that no analysis or experiments can be defined to answer the questions. In this case, the manager needs to respond quickly, prevent the situation from deteriorating further and try and transform the problem into one that experiments can be run (e.g., turn it into a complex problem). This will create knowledge and prevent problems from recurring. This is a problem that the manager needs own since it needs quick action to triage the problem without much opportunity for discussion.

Disorder

In the center of the 2 × 2 matrix in Figure 9.13 is disorder. This indicates that there is no agreement on which domain the problem should reside. In this case, it may be necessary to break the problem into separate components that people can agree on and then reassemble the problem statement later when there is more knowledge. The manager will need to take the initiative or else the problem will fester.

An important thing to note in the figure is that moving clockwise from the lower left (from chaos to simple) indicates an increase in knowledge. This is the direction most problems should take since it moves solutions to a place where they can ultimately become part of a standard workflow or process and managed thereafter by existing groups in the organization. Similarly, moving counterclockwise from the lower right (from simple to chaos) can also occur if standard processes become stale and do not improve to reflect changes in the environment. This is one reason why standards need to be challenged and continuous improvement should never stop. There is also the possibility that if standard work processes and policies become very out of date, they may lead to a catastrophe that quickly moves the problem into the domain of chaos.

The manager must create time for exploration.

One of the biggest challenges for the R&D organization and its manager is achieving the right balance of exploitation vs. exploration. On the one hand, the R&D organization is responsible for keeping the business running (KTBR) and delivering on objectives dictated by the existing strategy. That type of work exploits the existing engineering solutions and squeezes out as much new value as possible via incremental innovation. In most cases, there is significant innovation and hard work needed to meet those commitments, and failure to meet them harms the company. On the other hand, the R&D organization needs everyone in the organization to continuously improve the workflows to increase velocity. It also needs to be able to reliably dip into the pool of leaders and experts to create the intrinsic networks needed to solve complex problems and enable autonomous strategic actions. These are actions that require exploration and are vital for continued success.

Some gimmicks can be used to try and set aside employee time for the exploration we discussed. For example, some organizations encourage employees to block time off on their calendars for exploration activities. However, if the same organization bases employees' annual performance reviews and other success metrics on meeting aggressive schedules and predetermined targets, this will not work. Generally, the only approach that will work is for all project planning and related management decisions include slack to create the time for employees to use. For organizations that use a scrum development model this is easy to accomplish. Since the scrum team's capacity becomes apparent after a few cycles future planning can include stories in each sprint that are dedicated to exploration or the team may decide to use every third or fourth scrum for exploration. Either way, there is a concrete way to build exploration into the planning. For organizations not using scrum, there needs to be a similar way to build in the slack needed. The management of the R&D organization needs to role model this by supporting the model and not forcing teams to commit above their capacity. With this in mind, an effective R&D manager needs to be ambidextrous and properly balance exploitation vs. exploration.

9.9 Summary

The R&D organization includes the employees and other assets in the direct chain of command of the manager. The R&D organization is much more than an organizational chart. It starts with a strategy and set of values that identify the capabilities needed to deliver value and a competitive advantage. These determine the structure, processes, networks, reward systems and talent required to win. Since the strategy influences the design of the organization, a change in the strategy may result in a re-design of the organization.

While there are exceptions, the R&D manager will need to master working in a matrixed environment, whether the organizational structure reflects that or not. Proper governance is needed to create the right balance of power across the matrix to enable good tension. Formal and informal intrinsic networks are used to complement the formal hierarchy and they are critical for agility and enabling autonomous innovation. Formal processes enable decision making and sharing across the organization, while reward systems and human resource management align the skills and behavior with the strategy.

Managing the R&D organization requires different management approaches that depend on the type of problem being addressed. In the R&D organization, a single approach to solving every problem will not work. Finally, a culture of continuous improvement by everyone in the R&D system is the foundation for winning. Inculcating that culture in the R&D organization requires the manager to take personal responsibility to instill that behavior. This cannot be delegated.

9.10 Key Points

- The organization links strategy, structure, process, rewards, people.
- As the strategy changes so does the organization's design.
- A matrix is inevitable. A good design balances the power.
- Processes provide means to decide and share information.
- Extrinsic motivation is needed but intrinsic motivation is best.
- The manager must not delegate the organization's governance.
- The manager is responsible for continuous improvement.
- Different problems require different approaches.
- The manager must create time for exploration.

9.11 Case Studies

9.11.1 *Intel Operating Segments 2005–2018*

Figure 9.14 shows the major operating segments described in the Intel annual reports between 2005 and 2018 that span the tenure of two CEOs (Paul Otellini and Brian Krzanich) [9]. Each operating segment often correlated to a business group reporting to the CEO.

The 2005 Intel annual report highlighted a change in strategy to reflect a focus on platforms vs. individual chips. This was accompanied by a major reorganization and the creation of new operating segments and associated business groups related to those platforms.

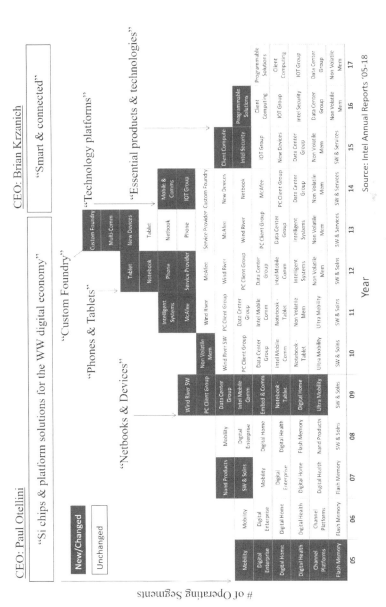

Figure 9.14.   Intel operating segments for 2005 to 2018.

Source: Intel Annual Reports '05–18

> *"Today, we are reinventing Intel once again, to focus on the growth opportunities presented through platforms — advanced solutions that integrate Intel® microprocessors and other technologies, such as complementary chipsets and communications chips, all optimized to work together."*

The 2009 Intel annual report noted its response to the growing popularity of low power devices and notebook computers by adjusting the strategy and adding a focus on devices. This was followed by another major reorganization of the business units and operating segments.

> *"Driven by the Intel Atom processor, the spectrum of products based on Intel® architecture is expanding beyond PCs and servers to include handhelds, consumer electronics devices, and hundreds of embedded applications."*

The 2012 Intel annual report noted, "we entered the markets for smartphones and tablets". This resulted in the formation of new operating segments to address those markets. 2012 also marked the point where Intel tentatively began providing foundry services.

The 2015 Intel annual report summarized a shift towards enabling a "smart and connected" world via the development of "technology platforms". This resulted in some consolidation and the creation of new operating segments focused on the Internet of Things (IOT).

> *"Our business model is evolving. The data center and Internet of Things businesses are the primary growth engines for Intel and memory and field programmable gate arrays (FPGAs) can accelerate these opportunities — forming and fueling a virtuous cycle of growth."*

In 2017 Intel annual report made the changes more concrete by exiting some more businesses and consolidating into fewer operating segments:

"Intel has been undergoing one of the most significant trans-formations in corporate history. We are evolving from a PC-centric to a data-centric company, delivering products that play critical roles in processing, storing, analyzing, and shar-ing data to enable amazing new experiences and competitive advantages."

As one can see from this small sample, each change in the corporate strategy resulted in new operating segments and asso-ciated business units (the black boxes in the figure) that allowed a focused effort in that area. Over time these units were consoli-dated with others or abandoned as the strategy matured or changed once again. Although it is not highlighted in Figure 9.14, some of these transitions were also accompanied by acqui-sitions of other companies that provided key ingredients.

Questions:

(1) What events motivated the creation of new operating segments?
(2) How was Intel's strategy changing over this period?
(3) What do you think happened to the operating segments that were no longer reported in the annual report?
(4) How is the change in Intel corporate values discussed in the case study in Chapter 2 related to the change in structure?

References

[1] Kessler, G. and Kates, A. (2011). *Leading Organization Design*. (Jossey-Bass, San Francisco CA.)
[2] Galbraith, J. R. *The Star Model*, Available from: http://www.jaygal-braith.com/images/pdfs/StarModel.pdf [May 2020].
[3] Ford, H. (2017). *My Life and Work*. (Create Space Independent Publishing Platform).
[4] Kessler, G. and Kates, A. (2011) *Leading Organization Design*. (Jossey-Bass, San Francisco CA.)
[5] Kotter, J. P. (2014). *Accelerate — Building Strategic Agility for a Fast-moving World*, (HBR Press, Boston MA.)

[6] Pink, D. (2009). *Drive: The Surprising Truth About What Motivates Us*. (Riverhead Books, New York.)

[7] Machiavelli, N. (1988). *The Great Books of the Western World*, 23, ed. Robert Maynard Hutchins, Chapter 1 *"The Prince,"* (Encyclopedia Britannica) p. 24.

[8] Snowden, D. and Boone, M. (2007). A Leader's Framework for Decision Making, HBR, November 2007.

[9] Intel (2005–2018). *Annual Reports*, Available from: https://www.intel.com/ [May 2020].

CHAPTER 10

The R&D System

> It is management's job to direct the efforts of all components
> towards the aim of the system. [1]
> — W. Edwards Deming

The R&D system is the specific collection of entities, assembled and directed by the manager of the R&D organization to work together and deliver the engineering product or service. This includes the R&D organization, along with its suppliers, customers, institutes, consortia, and others. Although these entities do not report to them, the manager of the R&D organization directs the co-development, collaboration, coopetition and continuous improvement across the R&D system for the benefit of the R&D organization.

Working together to improve the velocity of every entity in the R&D system benefits the R&D organization. Doing so means enabling the safe exchange of ideas and information. This requires investing in the culture, infrastructure and processes needed to collaborate. As a result, collaboration is not free and collaboration for collaboration's sake is not the objective. Collaboration needs to be disciplined and focus on areas where the objectives are clear and the return on investment is high.

When collaboration is justified, the type of collaboration and the associated effort and expectations will vary depending on the purpose. Some require hierarchical management with select participants from other entities in the R&D system to execute critical work for the R&D organization. Others rely on flat networks that are open to everyone in the R&D ecosystem to sense important changes, and gain knowledge.

10.1 The R&D System Scope

The R&D system exists to meet the R&D organization's objectives.

As shown in Figure 10.1, the R&D System includes the R&D organization along with its associated suppliers, customers, institutes, consortia, and others. The manager of the R&D organization is responsible for assembling the components and directing the co-development, collaboration, coopetition, and continuous improvement needed to meet the objectives of the R&D organization. This includes working with entities in the same company as well as those in other companies.

Directing the R&D System requires evaluating the R&D ecosystem.

The R&D ecosystem shown in Figure 10.2 is the collection of entities that affect each other even if they do not work together directly. These relationships are constantly evolving. Directing the R&D system requires observing and evaluating the whole R&D ecosystem to understand and act on the threats posed by competitors, new entrants (that may become

Figure 10.1.   The R&D system.

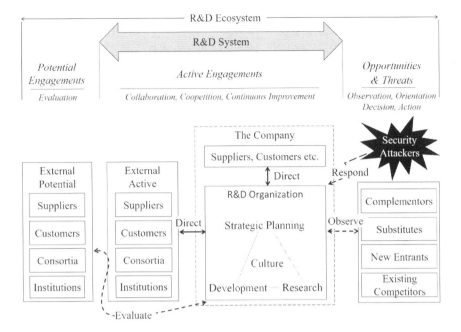

Figure 10.2.   The R&D ecosystem.

competitors), substitutes (that may displace the need for the R&D organization's solutions) and security attackers (whose intent is malicious). It also means evaluating new opportunities presented by complementors as well as potential suppliers, customers, and partners. Generally speaking, there is little or no direct working relationship between the R&D organization and these entities and this is why they are not considered part of the R&D system. However, there may be indirect engagement with some of them via a consortium, industry standards bodies, industry forums, etc. or via other coopetition ventures.

The R&D system has interdependent functions, workflows and networks.

Another view of a simple R&D system is shown in Figure 10.3. It depicts one supplier and one customer and shows the

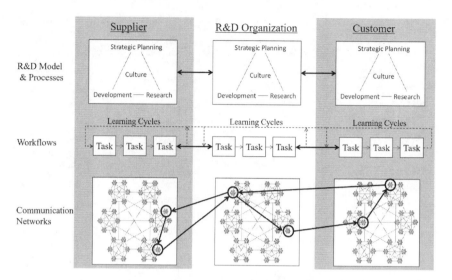

Figure 10.3.   The interdependent R&D models, workflows and networks.

connections between entities at the level of processes work-flows and communication networks. In the top horizontal slice of the figure, there are connections between the different strategic planning, research and development processes in each entity's R&D model. For example, the R&D organization's strategic planning function may engage with the supplier and customer as part of its planning cycle. There may be joint strategies that need to be managed. There may also be instances where the research pipeline in the R&D organization includes research from the supplier and interaction with the customer.

Similarly, during development, there is a lot of interaction between the R&D organization and the supplier and customer. Each has its development process that needs to be synchronized. At the heart of each entity's R&D model is the culture. Differences in the culture may have a large impact on how well the different entities operate together. In some cases, the differences in culture can be bridged with the right coordination.

In other cases, the differences may be more problematic and need to be changed to create better alignment.

The middle horizontal slice of the figure represents an abstraction of the workflows that may connect the supplier, R&D organization and customer. They represent how different activities are completed and connect. Because of the iterative nature of engineering, these workflows can become very complex across multiple entities.

In the bottom horizontal slice are networks of people that communicate with one another. This can be done via a formal exchange of information or informally through conversations that happen as part of a collaboration or other interaction. In many cases, the original targets of the conversation pass on the information to others in the same group or outside. This is very useful when trying to gather the right information from the proper experts but can be a challenge when the information needs to be protected.

When we refer to the interdependencies across the R&D system, we are discussing the concurrent and interconnected processes, workflows and networks. Detailed examples of those three views were given in Chapter 7, so we will not repeat them. Each requires coordination and benefit from standards.

Standards are critical to enabling working in a complex system.

We have mentioned the importance of standards in the development chapter and in enabling continuous improvement in general. It is even more important when we consider the complexity of working in the R&D system. This is not a surprise; whole industries have recognized the value of standards. An excellent example of this was the horizontalization of the personal computing industry summarized by Intel's CEO at the time, Andy Grove in 1999 [2].

> *"Before the PC started, the industry back in the early '80s consisted of largely vertically integrated companies. Companies like IBM, Digital, Bull, Nixdorf, a number of international computer companies, were all completely integrated. They build their own chips, they build their own software, they build*

their own computers, they sold their own computers, and then they competed one block against another.

The introduction of the personal computer brought a completely different economic structure, business structure, and competitive structure to this industry. It horizontalized the industry. By and large, a handful of processors got adopted and got used by all the companies, a handful of pieces of software got used industrywide through all this industry, and on and on and on.

The industry became horizontal also in terms of competitiveness. People who resided in each of these horizontal bars competed with each other very ferociously, and all of that gave an enormously productive, enormously competitive offering, much more so than the vertical industry did before."

Similarly, the R&D organization needs to have standards to make it easier to work with suppliers and make it easier to replace one supplier with another. As we discussed in Chapter 6, this increases the power of the R&D organization by lowering the switching cost, which, in turn, drives greater competition between the suppliers for the R&D organization's business. Not to mention the increased velocity that may result from being able to easily plug and play different supplier solutions into the engineering system as needed to provide differentiable value. As a result, influencing industry institutions or consortia to converge on common standards for suppliers is important.

10.2 The Role of the Manager

The R&D system requires the R&D organization's manager to direct it.

The entities in the R&D system see tangible and intangible benefits from participating in the R&D system. Suppliers are compensated for their contributions, but they are also motivated to do well to build loyalty and a reliable revenue stream, get feedback to improve their solutions, discover ways to lower their costs, get early access to their customers' engineering advances, etc. Customers are motivated because they want to have a say in the definition and testing of the solution to ensure it meets their needs. They may also want to co-develop engineering features that help differentiate themselves from their competition. Academic institutions are motivated to engage with industry partners to obtain practical test cases and results that increase their knowledge and develop their students. Consortia and other institutions are naturally interested in engagements as part of their charter and may rely on funding from the company. So, although the reasons may be different, the motivation to engage is strong.

Several forces may cause all of these engagements to falter as well. For most suppliers, there is not a lot of profit margin and the pressure to meet their market segment share goals, protect their intellectual property (IP) and strike the right balance across all of their customers may conflict with the interests of the R&D organization. They may also have difficulty working in a system that also includes their competitors. Customers also have their strategies and objectives to meet and IP to protect. They most likely have alternative suppliers and may pit them against the R&D organization to gain leverage. They may request custom work that does not align with the strategy of the R&D organization. Similarly, other institutions and consortia may have their own strategies, objectives and IP that may not align with the needs of the R&D organization. Of course, there are also barriers associated with human nature that may need to be overcome. Confirmation bias, not invented here, stereotyping, and groupthink are just some of the cognitive biases that may negatively affect innovation and decision making in such a system.

Because the R&D organization is ultimately held responsible for the results, the R&D manager must direct the efforts of the R&D system. This not only includes the operational aspects of keeping the whole R&D system running. It also means building the relationships, culture and technology that will enable everyone to improve together and deliver mutually beneficial results. For practical purposes, much of that work may be delegated to leaders within the R&D organization but the buck stops with the R&D manager.

10.3 The Shared Objective of the R&D System

The goal is to increase the velocity and profit for everyone in the system.

In Chapter 4, we defined velocity as the rate of deployed value and wrote that engineering tasks create value if they contribute to the company's vision in a way that the customer will pay. Many other entities in the R&D system have activities they perform to enable the engineering tasks of the R&D organization. In each case, they also want to contribute to their own company's vision in a way that they will get paid. Although the R&D system exists to meet the objectives of the R&D organization this is accomplished best if it is done in a way that the velocity and profitability of all the entities benefits.

Figure 10.4 shows a high-level depiction of an analysis of a simple value stream that includes a supplier, the R&D

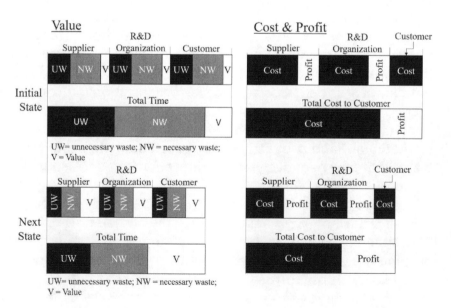

Figure 10.4.   Depiction of value, cost and profit for all parties.

organization and a customer. In the example, we assume that the customer must do some integration work to add additional value once it receives the solution from the R&D organization. We also assume that the supplier and R&D organization are interested in making a profit on the work they complete.

The total time it takes to complete all activities that span the value stream is shown on the left. This includes time spent on unnecessary waste, necessary waste and value. The sum of those equals the total time through the value stream. The initial state shows what it might look like before any improvements are made. The next state is shown below and depicts what it might look like after everyone worked together to reduce waste in the supplier, R&D organization and customer workflows and increase the overall value and velocity across the entire value stream. If only one member increased their velocity, then the total impact is less. Worse, if one member made changes that improved their velocity but reduced the velocity of another member then the overall improvement will be reduced further. In this simple example, everyone should work together to reduce waste in each other's processes and workflows.

On the right side, the figure also depicts a hypothetical breakout of the sum of the costs and profits across the value stream. The total determines the cost to the customer. The initial state shows what it might look like before any improvements are made. The next state shows what it might look like after each member has worked together to find ways to reduce costs in the supplier, R&D organization and customer workflows. Each sees a reduction in their own cost and may they take some of that to improve their profit margin while still providing their output at a price reduction to the next one in the chain. The individual improvements sum to an overall improvement in cost across and profit margin across the entire value stream. The result is that the profits for the supplier and R&D organization increase while the total cost to the customer goes down. Similar to the example with value, if the R&D organization only decreased its costs then the total impact is less and if one entity made changes that reduced their costs but increased the cost of another member in the value stream, then their savings may not be realized by the customer. In this simple picture, it is also better for everyone to work together to reduce costs in each other's processes and workflows. Similar examples can be shown to justify the need to work together to improve each entity's quality, productivity, efficiency, etc.

As an example of how this has been accomplished in practice, we can look at the automotive industry. Supplier management plays an important role in each company's success and there has been a wealth of research on the topic. Although much of it relates to manufacturing, it provides important lessons for engineering as well. For example, a study by Aoki and Lennerfors in 2013 [3] concluded what many others have also concluded:

> *"increasingly the locus of competition is between supply chains rather than between individual companies."* and
> *"Having conducted interviews and gathered data during 39 visits to auto plants and 192 visits to parts makers in Japan and overseas, and analyzed two decades' worth of auto-manufacturing data, we believe that Toyota's current supply-chain system represents one of the company's greatest advantages."*

Toyota's approach continues to evolve, but it is based on Taiichi Ohno's philosophy [4],

> *"The achievement of business performance by the parent company through bullying suppliers is totally alien to the spirit of the Toyota Production System." The key word in that statement is "parent," which signals a long-term relationship that involves trust and mutual well-being."*

In the previously referred to analysis from Aoki and Lennerfors, this was summarized as,

> *"Instead of abandoning suppliers when others offer lower prices, it provides support for operational improvements, organizing "study groups" and dispatching engineers to help vendors improve efficiency and quality and bring prices down."*

In some cases, as a reward, Toyota will allow the supplier to retain the cost savings for themselves temporarily before requiring them to set a lower price. In this way, they provide a bigger incentive for the supplier to engage.

> *"As part of its product-development process, Toyota provides physical spaces that facilitate cooperation with and among suppliers. Vendors may be invited to a meeting known as an obeya — literally, a big room — where they work with Toyota representatives from several departments, including design, engineering, production, quality, and purchasing."*

This is done while evaluating and utilizing other suppliers, so no one takes the relationship for granted.

> *"Instead of buying exclusively from companies with which it has long-term relationships, Toyota also sources from the global market, including from mega suppliers whose stream-lined operations allow them to offer very low prices. This gives it flexible sourcing and keeps costs down.*
>
> *When setting target prices for long-term suppliers, Toyota looks at the prices offered by multiple global companies, another boon in containing costs.*

They also expect the supplier to have a bigger role in the development of the final product,

> *"Instead of buying individual parts, the automaker demands that suppliers provide integrated systems of components. This helps it develop high-quality products while reducing costs and development time.*
>
> *Toyota encourages suppliers to enhance their ability to provide these integrated systems and to become involved in product development at the planning stage."*

One way of capturing what Toyota and some other auto manufacturers do is provided by Liker and Choi's Supplier-Partnering Hierarchy [4] shown in Table 10.1. It summarizes the steps they feel are important in creating deep Supplier

Table 10.1. Supplier-partnering hierarchy.

1　Understand how your suppliers work 　　Learn about the supplier's businesses 　　Go to see how suppliers work 　　Respect their capabilities 　　Commit to co-prosperity
2　Turn supplier rivalry into opportunity 　　Source each component from two or three vendors 　　Create compatible production philosophies and systems 　　Set up joint ventures with existing suppliers to transfer 　　　knowledge and keep control
3　Supervise your suppliers 　　Send monthly report cards to core suppliers 　　Provide immediate and constant feedback 　　Get senior managers involved in solving problems
4　Develop suppliers' technical capabilities 　　Build suppliers' problem-solving skills 　　Develop a common lexicon 　　Hone core suppliers' innovation capabilities
5　Share information intensively but selectively 　　Set specific times, places, and agendas for meetings 　　Use rigid formats for sharing information 　　Insist on accurate data collection 　　Share information in a structured fashion
6　Conduct joint improvement activities 　　Exchange best practices with suppliers 　　Initiate Kaizen projects as suppliers' facilities 　　Set up supplier study groups

relationships. They are meant to proceed sequentially from top to bottom. It is a good framework for the R&D organization as well.

10.4 Collaboration and the R&D System

Optimizing the entire R&D system relies on enabling collaboration.

Since the functions, workflows and networks cross multiple entities, the ability to collaborate effectively becomes an important requirement. To do it properly requires an investment above and beyond funding the internal development activities. This was noted in a study of successful global collaboration by Alan MacCormack and Theodore Forbath [5] who described four areas of investment common in successful firms:

> *"People. Successful firms alter their recruitment, training, evaluation, and reward systems to focus on "soft" skills such as communication so that managers can better learn to motivate and coordinate team members who are outside the firm and, sometimes, in vastly different cultures. Many of these companies also help to train partners — for example, by inviting them to internal development programs so that future teams learn together what it takes to collaborate.*
>
> *Processes. Leading firms use a learning-driven approach to designing collaborative processes. German electronics giant Siemens recruited several university teams around the globe to test different strategies for managing distributed teams. Among the lessons learned: Teams from different cultures have different strengths and working methods, which must be matched to their assigned tasks; and formal requirements are no substitute for frequent, high-bandwidth communications, which are critical for resolving unanticipated problems.*
>
> *Platforms. It's important to create an infrastructure — a set of tools and standards for sharing data — that allows dispersed teams to work together seamlessly. Failure to do so puts a project at risk. In 2006, Airbus revealed that its flagship A380 aircraft would be delayed two years, at a cost of billions of dollars, because partners' use of different versions of design software resulted in 300 miles of wiring and 40,000 connectors that did not fit together.*
>
> *Program. Successful firms manage collaboration as a coherent program, not a series of stand-alone efforts. Many companies achieve this by, in effect, designating a "chief collaboration officer," who oversees all partnered efforts and focuses on building the firm's overall collaborative capabilities."*

For an engineering organization, we can be more specific.

People: Among other things, this requires an investment in employee training to identify and mitigate the cognitive biases that often grind collaboration to a halt. It also requires training and systems that reinforce the correct privacy and security mindset to avoid cross-contamination of participants' intellectual property. Finally, it requires an investment in training, systems and role modeling that enables a culture of continuous improvement.

Processes: Investing in the continuous improvement of the collaborative processes themselves is required. This includes informal interactive networks and more formal hierarchical networks and processes. When deciding on a process, the bias still should always be towards keeping it simple and instead focusing efforts on building the right culture. In this way, most problems can be avoided altogether or resolved quickly without a lengthy process.

Engineering Platform: Investing in the infrastructure that enables collaboration across engineering entities is essential. The most direct way to ensure the seamless operation is to also have 100% alignment on the engineering tools, methods and collateral that are used. However, it is not required if there is a clear understanding of what requires alignment and what does not. Depending on the scope and complexity of the collaboration effort, investing in the collaborative infrastructure means investing in capabilities such as:

- A common database that participants use to share engineering data.
- Product life cycle management to track specs, test cases and results.
- A common database for capturing and reconciling requests and issues.
- Common engineering tools, collateral and methods that must align.

Governance: In the R&D system, there may be a very large number of collaborative programs underway at any one time. Because of their diversity, having a single person to oversee all of these efforts is not advisable. However, designating a dedicated person to manage the development of the collaborative culture, processes and platforms is a good way to maintain management focus on the investment required to succeed. This person should have a separate budget to enable their work. Within each major collaboration domain, additional investment in collaboration governance is still required. This includes setting the specific objectives, calculating the expected premium and managing the work.

Collaboration needs to be disciplined and doesn't always make sense.

Enabling the ability to collaborate across the R&D system is necessary. However, that does not mean that everything must then become a collaboration effort. Like everything else, collaboration requires a clear set of objectives, measures of success and should provide a positive return on the investment. For example, when discussing the results of a study on internal collaboration, Morten Hansen [6] wrote,

> *"We ultimately determined that experienced teams typically didn't learn as much from their peers as they thought they did. And whatever marginal knowledge they did gain was often outweighed by the time taken away from their work on the proposal.*
>
> *The problem here wasn't collaboration per se; our statistical analysis found that novice teams at the firm actually benefited from exchanging ideas with their peers. Rather, the problem was determining when it makes sense and, crucially, when it doesn't. Too often a business leader asks, how can we get people to collaborate more? That's the wrong question. It should be, Will collaboration on this project create or destroy value? In fact, to collaborate well is to know when not to do it."*

So, although the ability to collaborate is essential, each potential engagement still needs to be evaluated on its merits. Spinning out multiple ineffective collaborations with a negative ROI will end up siphoning critical resources from more productive efforts. To resolve this, one can apply Morten Hansen's equation to calculate the collaboration premium:

projected return – opportunity cost – collaboration cost
= collaboration premium

The projected return is the estimated value of the proposed collaboration. The opportunity cost is the estimated value of the most attractive project not undertaken due to the decision to invest in the collaboration. The collaboration cost is the additional costs that result from the collaboration.

It is not always easy to calculate the collaboration cost and the projected return is often overestimated in the beginning. However, the exercise of walking through this calculation for each collaboration proposal helps create a necessary conversation on whether the proposed collaboration should go forward or not. Also, it is important to note that as the R&D system gets better at the act of collaboration, then the collaboration costs start to reduce and the number of collaboration activities that will produce a positive ROI will grow.

The types of collaboration vary depending on the strategy.

All collaboration is not the same. So, in addition to making sure the collaboration effort provides a positive premium, it is important to structure the effort properly. At one end of the spectrum, there may be a collaboration open to anyone and any idea and operates as an informal network. At the other end of the spectrum may be a collaboration that is limited to a select group of experts, focused on a critical problem and relies on the work of multiple organizational units. Some may have the autonomy to make their own decisions while others have that determined by the R&D organization. Gary P. Pisano and Roberto Verganti [7] proposed a framework for categorizing the types of collaboration. Their framework is based on two questions:

> *"Given your strategy, how open or closed should your firm's network of collaborators be? And who should decide which problems the network will tackle and which solutions will be adopted?"*

Although they refer to firms, the same thinking applies to an engineering organization. Their 2×2 matrix groups collaboration into four types shown in Table 10.2: "a closed and hierarchical network (an *elite circle*), an open and hierarchical network (an *innovation mall*), an open and flat network (an *innovation community*), and a closed and flat network (a *consortium*)." In their framework, a collaboration that begins in one quadrant may change and move to a different quadrant if the strategy changes. For example, Apple started the development of iPhone applications using collaboration similar to the elite circle model, but it later moved to the innovation

Table 10.2. Pisano and Verganti collaboration types.

		Governance	
		Hierarchical	Flat
Participation	Open	**Innovation Mall** A place where a company can post a problem, anyone can propose solutions, and the company chooses what it likes best.	**Innovation Community** A network where anybody can propose problems, offer solutions and decide which solutions to use.
	Closed	**Elite Circle** A select group of participants chosen by a company that also defines the problem and picks the solutions.	**Consortium** A private group of participants that jointly select problems, decide how to conduct work, and chooses solutions.

Table 10.3. Collaboration types for the R&D organization.

		Governance	
		Hierarchical	Flat
Participation — R&D Eco-system		Open & Directed Problems posted for which anyone can propose solutions. The R&D organization chooses what it takes forward.	Open & Autonomous A network where anybody can propose problems, offer solutions and decide which solutions to use.
Participation — R&D System		Restricted & Directed A select group of participants chosen by the R&D organization that also defines the problem and picks the solutions.	Restricted & Autonomous A private network that jointly selects problems, decide how to conduct work, and chooses solutions.

community model to enable everyone to participate in and rapidly grow the number.

The translation of this into the equivalent 2 × 2 matrix for the engineering R&D organization is shown in Table 10.3. In our case, closed participation means membership is limited to the entities in the R&D system since those entities have the necessary agreements in place to enable tight collaboration on proprietary topics. Open participation means there is open participation from anyone on the rest of the R&D ecosystem who are not yet part of the R&D system. They may be potential customers, suppliers and also competitors, new entrants and substitutes. In those cases, there is no pre-existing framework that enables direct collaboration on sensitive intellectual property and the topics reflect that.

Hierarchical governance refers to efforts that are directly managed by the R&D organization with specific objectives and associated staffing from the participants. Flat governance looks like the intrinsic networks we discussed in Chapter 9. They are not tied to the existing reporting structure.

Table 10.4. Cost and expected benefits of each collaboration type.

		Governance	
		Hierarchical	Flat
Participation	R&D Ecosystem	Low collaboration cost Speculative potential value *Identifies new talent & partners*	Low collaboration cost Speculative potential value *Exposes unknown unknowns*
	R&D System	High collaboration cost High existing value *Executes strategic actions* *Improves workflows and processes*	Medium collaboration cost High potential value *Executes autonomous innovation* *Improves workflows and processes*

Each collaboration type has an associated collaboration cost and expected benefit that can be used to estimate the collaboration premium and therefore determine if the particular effort is warranted. As shown in Table 10.4, the lower left quadrant (hierarchical governance with participation limited to members of the R&D system) has the highest collaboration cost since those efforts require an investment in the common collaboration culture, processes, engineering platform and governance we described earlier. These collaborations tend to be mission-critical, require the sharing of engineering data and detailed co-development work. A collaboration effort that falls into that category is also expected to bring large value. The topics tend to originate from the induced strategic actions or top-down improvement efforts of the organization.

The lower right quadrant (flat governance with participation limited to members of the R&D system) has a lower cost since the networks are informal. These networks may form on their own and others may be initiated by management, but their operation is focused on ideation and experimentation.

This is also the primary means by which continuous improvement of the workflows happens. The realizable value is more speculative but the lower collaboration costs justify the investment in many cases.

The upper right quadrant (flat governance with participation open to everyone in the R&D ecosystem) has a low collaboration cost. In most cases, it is just a matter of connecting experts from the R&D organization into industry think tanks, conference forums, industry committees, or helping employees create their networks. It is for the benefit of the employee development but also can be used to better understand the R&D ecosystem and emerging threats and opportunities posed by entities that the R&D organization does not normally interact with. The primary risk with this type of collaboration is the accidental leakage of sensitive company information or contamination by accidentally hearing sensitive information from other participants. Investment is therefore needed to train employees on how to avoid those situations.

The upper left quadrant (hierarchical governance with participation open to everyone in the R&D ecosystem) also has a low collaboration cost. It may include providing a specific challenge to academia, suppliers, or the general community that is based on an important problem defined by the R&D organization. The R&D organization must provide the framework for this and evaluate the responses. This type of collaboration provides a means for the R&D organization to identify and acquire individual talent or evaluate potential new suppliers.

Strategic planning utilizes all collaboration types.

The strategic planning OODA loop described in Chapter 6 looks across the entire R&D ecosystem to observe the current state and then analyze and synthesize that information into a proposed set of actions. As shown in Table 10.5, all of the collaboration types are relevant to the strategic planning OODA loop. Strategic deep dives with experts from the R&D system to sense and analyze

Table 10.5. Strategic planning collaboration types.

		Governance	
		Hierarchical	Flat
Participation	R&D Ecosystem	Open Challenges *Address known unknowns* *Sense-analyze*	Open Conversations *Expose unknown unknowns* *Probe-Sense*
	R&D System	Restricted Deep Dives *Address known unknowns* *Sense-analyze*	Restricted Conversations *Expose unknown unknowns* *Probe-Sense*

the known unknowns will occupy the lower left quadrant. On the other hand, informal networks used to uncover unknown unknowns occupy the lower right quadrant. This is an important means of identifying potentially disruptive events that the analysis of the induced strategic actions may not cover. Probing and sensing of the rest of the R&D ecosystem to discover opportunities or threats occupy the upper right quadrant. This can be done by enabling the participation of experts from the R&D organization in industry forums, conferences, think tanks, etc. that may prove disruptive. Engagement with the rest of the R&D ecosystem on specific topics based on the R&D organization's strategic direction occupies the upper left quadrant. Although the topics in this quadrant may not be secret the results of the work will be the property of the R&D organization.

In general, the cost of collaboration in strategic planning is relatively low, while the potential penalty of missing an important inflection point in the R&D ecosystem can have a large negative impact. So, although it is very difficult in the case of strategic planning to estimate the projected return of each collaboration, the investment, in general, is low relative to the potential cost of not doing it. As a result, there may be a large number of strategic topics to track and disposition. The means to do that was discussed in detail in Chapter 6. As with all of the 2 × 2 matrices, we will look at, the most expensive type of collaboration is in the lower-left quadrant since it does rely to some extent on creating the common culture, as well as the collaboration platform, process and governance needed to work effectively together. However, since there is no development activity in strategic planning, it can be fairly light-weight.

A healthy research pipeline utilizes all collaboration types.

As shown in Table 10.6, all of the collaboration types are also relevant to engineering research. Collaborative research that targets the induced strategic actions of the company or R&D organization occupies the lower left quadrant. The research is directed by the R&D organization and will occur with experts drawn from the R&D system and the use of a common collaboration framework is important. Meanwhile, more autonomous engineering research will result from informal networks and

Table 10.6. Research collaboration types.

		Governance	
		Hierarchical	Flat
Participation	R&D Ecosystem	Open Challenges *Address known unknowns* *Analyze-Transfer*	Open Conversations *Expose unknown unknowns* *Probe-Sense*
	R&D System	Induced Action *Address known unknowns* *Analyze-Prototype*	Autonomous Action *Address unknown unknowns* *Analyze-Prototype*

occupy the lower right quadrant. Similarly, engaging with the rest of the R&D ecosystem on other pre-competitive research topics open to everyone occupies the upper right quadrant. This can be achieved through participation in such things as consortia. The R&D organization has limited control over the topics there but, as with strategic planning, it is a very good way to identify potential threats and opportunities. Engagement with the rest of the R&D ecosystem on specific research controlled by R&D organization occupies the upper left quadrant. These may be research topics opened up to all of academia or the industry for proposals. This requires some investment by the R&D organization to communicate the proposal, select the participants, analyze and absorb the results. The results of the work are not shared with the rest of the R&D ecosystem and become the property of the R&D organization.

Overall, the cost of collaboration during research is slightly greater than that of strategic planning since investment is needed in some cases to build prototypes. So, as was discussed in Chapter 8, there needs to be a continuous process of evaluating the potential and realizable value of research as it matures. This will help re-calculate the collaboration premium as more knowledge is gained. The most expensive type of research collaboration is in the lower-left quadrant since it requires oversight and the use of the common collaboration framework.

Pathfinding primarily utilizes hierarchical collaboration.

As shown in Table 10.7, research that enters the pathfinding process and that is dependent on collaboration will typically require the collaboration framework we described earlier to move quickly. At that point of maturity, pathfinding of a particular solution only includes members of the R&D system. Therefore, most collaboration during pathfinding falls into the lower-left quadrant. However, during pathfinding, other experts can attend reviews to provide their inputs and this may spawn offline discussions on alternative best-known methods (BKMs) or solutions. Those types of discussions happen in

Table 10.7. Pathfinding collaboration types.

		Governance	
		Hierarchical	Flat
Participation	R&D Ecosystem	Supplier Evaluations *Analyze-Transfer*	NA
	R&D System	Research Selection *Analyze-Transfer*	BKM Sharing *Sense-Share*

informal networks characterized by the lower right quadrant. During pathfinding, there may be a need to evaluate new suppliers as part of the effort to make the research deployable. These efforts fall into the upper left quadrant. For pathfinding, no work falls into the upper right quadrant.

The cost of collaboration during pathfinding can be significant for complex topics since the output of the process is a solution that is ready to implement. These costs can be greatly reduced through the use of a common collaboration culture and related processes, platforms and governance. So, to a large extent, the collaboration cost becomes dependent on how well that framework operates. Organizations without such a framework may experience prohibitive costs that will prevent effective pathfinding altogether and them at a disadvantage vs. their competition who have such a framework.

Development uses hierarchical and flat collaboration.

Development involves narrowing down the engineering options, conducting the actual engineering work and continuously

Table 10.8. Development and collaboration types.

Governance		
	Hierarchical	Flat
R&D Ecosystem — Supplier Evaluations *Analyze-Transfer*		NA
R&D System — Engineering Execution *Analyze-Execute*		Continuous Improvement *Sense-Analyze-Execute*

(Participation axis label on left)

improving the workflows to increase velocity. As shown in Table 10.8, collaboration on narrowing the choices and conducting the engineering work falls into the lower left quadrant since the engineering work has specific targets and requires the collaboration framework to enable all of the participants to work productively. Continuous improvement of the workflows themselves is happening in parallel with development and this is largely driven bottom-up in the R&D system with everyone, every day involved in experiments to evaluate and implement incremental improvements. This type of work is done in networks formed by employees in the R&D system as needed and therefore resides in the lower right quadrant. Similar to the pathfinding example, there are also informal or formal networks of employees who discuss and share engineering and technology BKMs. These efforts also fall into the lower right quadrant. During development, there may be evaluations of new suppliers as well. If so, the collaboration will fall into the upper left quadrant until it is decided to take the next step and more closely engage. At that point, the effort will move down to the lower-left quadrant. Due to the nature of development tasks, it is unlikely that engagements with entities outside the R&D system that relies on open and unstructured interaction will occur and the upper right quadrant is unoccupied. By combining the hierarchical governance and collaboration framework for selecting and conducting the actual engineering work but allowing BKM sharing and workflow improvement using networks of people within the R&D system a good balance can be achieved between the oversight and agility needed to succeed.

Like pathfinding, the cost of engineering collaboration during development can be significant for complex topics since the output of the process is a solution that is ready to send to the customer. Similar to the other cases, the collaboration cost of development depends on how well that collaboration framework operates. Organizations without such a framework may experience prohibitive collaboration costs, opportunity costs and delays that reduce their velocity and put them at a major disadvantage. Moving forward with a collaboration that has a negative premium means that other worthy projects may not get funded so it is important to make the right decision.

As mentioned, another important collaboration that occurs during development targets continuous improvement of the workflows and occupies the lower right quadrant. These are low-cost networks of employees in the lower right quadrant that engage in the continuous improvement of the workflows. They are composed of autonomous activities that occur during development and at all levels of the system. These informal networks are critical to increasing the velocity of the R&D system. Calculating the collaboration premium for these types of networks is difficult since they should permeate the entire system and generate hundreds (probably thousands) of incremental improvements over time. In this case, the collaboration cost should just be considered the cost of doing business. Ideally, continuous improvement also addresses deficiencies in the collaboration culture, processes, platform and governance so that the overall cost of collaboration also declines over time.

Targeted adjustments to cultures in the R&D system can benefit everyone.

As we discussed in Chapter 2, and referred to in this chapter, there is a case to be made for selectively optimizing cultures across the entire R&D system to create the best collaborative framework. The goal is not to converge on one common culture for the entire R&D system but to see where there is a large return in investment for all parties and then work together to drive the change. The direct control that the manager of the R&D organization has on other entities in the R&D system is limited. Even though the case study we explored in Chapter 2 highlighted a change in culture in two companies that provided benefits for both, it also demonstrated that it is not easy to implement. It starts with getting the leadership of the different organizations to agree with the value of making the change and then is dependent on each one driving a cultural shift inside their organizations. This is tough to achieve inside one organization, let alone across a number of them. One example we are familiar with it took several years to drive before seeing a positive return. However, the return on the investment was very large.

10.5 Summary

The R&D system is the specific collection of entities, assembled and directed by the manager of the R&D organization to work together and deliver the engineering product or service. This includes the R&D organization, along with its suppliers, customers, institutes, consortia, and others. Although these entities do not report to them, the manager of the R&D organization directs the co-development, collaboration, coopetition and continuous improvement across the R&D system for the benefit of the R&D organization.

Working together to improve the velocity of every entity in the R&D system benefits the R&D organization. Doing so means enabling the safe exchange of ideas and information. This requires investing in the culture, infrastructure and processes needed to collaborate. As a result, collaboration is not free and collaboration for collaboration's sake is not the objective. Collaboration needs to be disciplined and focus on areas where the objectives are clear and the return on investment is high.

When collaboration is justified, the type of collaboration and the associated effort and expectations will vary depending on the purpose. Some require hierarchical management with select participants from other entities in the R&D system to execute critical work for the R&D organization. Others rely on flat networks that are open to everyone in the R&D ecosystem to probe the environment, sense important changes, and gain knowledge.

10.6 Key Points

- The R&D system exists to meet the R&D organization's objectives.
- Directing the R&D System requires evaluating the R&D ecosystem.
- The R&D system has connected functions, workflows and networks.
- Standards are critical to enabling working in a complex system.
- The R&D system requires the R&D manager to direct it.
- The goal is to increase velocity and profit for everyone in the system.
- Optimizing the entire R&D system relies on enabling collaboration.
- Collaboration needs to be disciplined and doesn't always make sense.
- The types of collaboration vary depending on the strategy.
- Targeted adjustments to cultures can benefit everyone.

10.7 Case Studies

10.7.1 *Background: Three Eras in a CAD R&D System*

A computer is composed of one or more central processor units (CPU) that receive data and then process and perform actions on it according to a specific set of instructions defined by the CPU. The CPU is often referred to as the brain of the computer. Early in the history of computing, a CPU was constructed using circuit boards that connected a variety of small integrated circuits. Improvements in semiconductor manufacturing processes eventually enabled all of the functions on the circuit boards to be placed on one integrated circuit called a microprocessor. The first example was the Intel 4004 microprocessor announced in 1971 that contained about 2,300 transistors [8]. Improvements in manufacturing process technology continued to reduce the dimensions of the transistors and interconnections on the integrated circuit so that more functions with greater sophistication and higher performance could be placed on a chip of the same size. Forty-six years later, Intel's 28-core Xeon Platinum 8180 microprocessor had approximately 8 billion transistors (about a 3.5 million-fold increase).

The Instruction Set Architecture (ISA) defines the CPU and has a major impact on the performance of the CPU. A computer operating system (OS) is tuned to the ISA of the microprocessor. Once an OS becomes popular, software developers tend to write their applications to work first with that OS. Since a licensed version of the Microsoft OS and Intel's microprocessor were both used on the early IBM PC a virtuous cycle of growth was created that established a huge legacy of software applications for the Intel architecture that, in turn, increased its attractiveness.

As microprocessors matured, they integrated multiple CPU Cores, graphics processing, I/O management, video display, WiFi and other functions in the same integrated circuit chip. This increased level of integration made it possible to shrink computer systems or platforms. Even greater levels of integration enabled devices like smartphones and the microprocessors designed for those applications became known as a System on a Chip (SOC).

A new microprocessor generation generally occurred every two years. To keep pace with this and the growing complexity of microprocessors, new engineering tools and methods were developed to model, simulate and optimize the design and increase engineering productivity. This was called Computer Aided Design (CAD) or Electronic Design Automation (EDA). CAD tools were also applied to help product engineers test the finished product and to design and validate the larger computer systems that contained the microprocessor.

These case studies explore how Intel's central Computer Aided Design (CAD) group and related R&D system changed with Intel's growth and evolving microprocessor strategy. Although the central CAD group supported other product segments at Intel during this period, these case studies concentrate on the interaction with the microprocessor groups. To avoid confusion, we refer to Intel's central CAD group as "central CAD" (its actual name changed over 30 years) and we refer to comparable external suppliers as "EDA vendors".

Overview

1988 marked the year that Intel completed the design of the 486 microprocessor (It was introduced in 1989). Intel's 1988 annual report [9] noted that it was now a

> *"microcomputer company, providing the component, board, and system level building blocks from which its original equipment manufacturers (OEM) customers can build their end products."*

When the 486 DX microprocessor was introduced in 1989 it contained 1.2 million transistors. This was over 4X the number of its predecessor (the 80386) and subsequent generations of microprocessor continued to grow in size and complexity as described by Moore's Law. As some of the key industry innovators at the time later wrote [10],

> *"Intel's microprocessor design teams had to come up with ways to keep pace with the size and scope of every new project. This incredible growth rate could not be achieved by hiring an exponentially-growing number of design engineers. It was fulfilled by adopting new design methodologies and by introducing innovative design automation software at every processor generation."*

The use of design automation software (Computer Aided Design) and its importance was also highlighted in Intel's 1987 annual report [11],

> *"Thanks to new methods and new computer-aided design technology, we're able to create chips faster, less expensively and with better quality — even though the complexity of those designs continues to rise rapidly..."*

The integrated device manufacturing (IDM) model that tightly linked Intel's internal microprocessor design, manufacturing process technology and CAD enabled Intel to lead the industry. As a critical part of the IDM model, Intel's central CAD organization delivered revolutionary design methods and design automation that were tuned for microprocessor design. Although this model led to high non-recurring engineering (NRE) costs the resulting sales volumes and margins that leadership brought far outweighed the additional expense.

By the end of 2012, a lot had changed. In 1994 the floating-point division (FDIV) bug on Intel's Pentium microprocessor elevated the importance of validation and verification. In 1998, Intel responded to increased competition by segmenting its microprocessor products. In 1999, Intel acquired several communication companies and referred to itself as a computing and communications company. In 2005, it proclaimed itself to be a platform company. In 2009, it added consumer electronics devices to its portfolio. This ultimately led to officially entering the markets for smartphones and tablets.

Through all the corporate strategic changes, the central CAD and the microprocessor groups and their relationships to each other evolved from one that was optimized for a small variety of high volume, high margin custom microprocessor design to one that was meant to deliver a large variety of microprocessors across a growing number of market segments.

The period from 1988 to 2012 is divided into three eras that coincide with the tenure of each Intel CEO. They are described below and shown in Figure 10.5. Each CEO drove well-publicized changes in strategy that influenced the organization and work of the microprocessor engineering and central CAD groups. Each era is a separate case study but viewed together they provide a picture of how the work models evolved and, like a pendulum swung between centralization to decentralization and back again. The different depictions shown in the case studies are abstract and could apply to any company going through similar transformations. They also do not reflect every change. Within each era, there may be multiple incremental changes needed to get to the latest era shown. For brevity, the most representative example for each era is shown.

Era 1: Rapid Growth and Rapid Invention (circa 1988–1997): Characterized by one microprocessor market segment, rapid invention and the synchronized release of a suite of centrally developed CAD tools.

Era 2: Competition and Product Autonomy (circa 1998–2004): Characterized by competition that led to new microprocessor segments and design group autonomy to select EDA industry or central CAD tools.

Era 3: New devices and Platform Scope (circa 2005–2012): Characterized by the advent of smartphones, tablets and other devices, a move to System on Chip (SOC) designs and the increasing scope of central CAD.

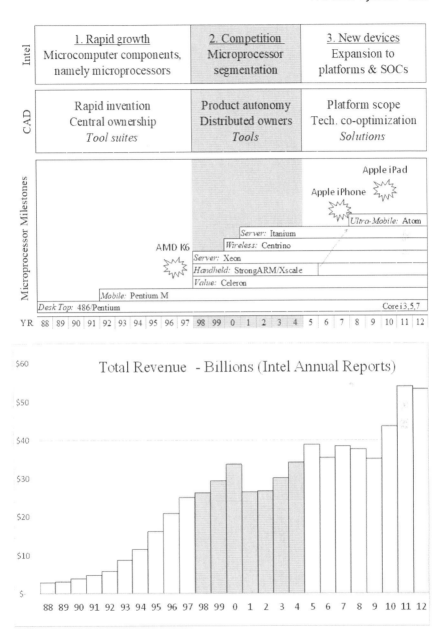

Figure 10.5. Four eras of central CAD and microprocessor interaction.

10.7.2 *Era 1: Rapid Growth (1988–1997)*

Intel Business

In this period, Intel characterized itself in the following way [12],

> *"Intel operates predominantly in one industry segment. The Company designs, develops, manufactures and markets micro-computer components and related products at various levels of integration."*

Although it also produced EPROM (Erasable Programmable Read Only Memory), Flash memory, and micro-controllers, Intel's primary business was microprocessors. In support of the microprocessor business, Intel also produced microprocessor peripherals, motherboards and systems. Notable microprocessor products in this era were the derivatives of the original 386 microprocessor (386 SX, 386 SL) and the next generation of 486 (486 DX, 486 SX, 486 DX2, 486 DX4) and Pentium, Pentium Pro, Pentium II and Pentium MMX products. The focus on microprocessors was exemplified by the CEO of the company (Andy Grove) whose mantra, "Pentium is Job 1" focused the company through the 1990s.

This era was marked by an extreme focus on a single microprocessor segment targeting desktop computing and delivering increasingly more powerful processors to improve the PC experience. In the 10 years from 1988 to 1997, Intel's total revenue grew from $2.9 to $25 Billion with a 42% compound annual growth rate of its stock price. By 1997, revenue growth was driven by the Pentium II microprocessor product line. [13]

Microprocessor Design

The Microprocessor Products Group (MPG) developed the lead microprocessor. The lead microprocessor was followed by a process technology "compaction" of the design on the next generation manufacturing process to reduce the size of the chip and hence the manufacturing cost. The compaction was completed by a design team in the manufacturing organization and this compacted design was the basis for subsequent derivatives designed by MPG. The number of transistors, process technology and architecture complexity increased with each generation and increasing the number of chips that could successfully be manufactured per silicon wafer was also important to the bottom line. With a growing reliance on newly hired college graduates to fill its ranks MPG depended on CAD to increase engineering productivity, eliminate human error and enforce reliability and manufacturability rules.

The lead microprocessor design team was located in one building on the same site and the compaction design team was located in one building at a different site. The fact that all members of each team were located in their own building meant that there was a strong common culture with many opportunities for informal interactive networks to form. Similarly, internal standards were easy to communicate and enforce. However, because the lead design and compaction were done by design teams in different locations, in different organizations, they often implemented the designs using different design philosophies and methods. This created redundant work and inefficiency when viewed at the Intel level but the associated NRE cost was a very small fraction of the recurring manufacturing costs and minuscule compared to the revenue generated by the microprocessor products.

Central CAD

The organization responsible for CAD was part of the MPG organization. The era started with almost all CAD software (referred to as CAD tools) being developed internally by Intel's central CAD group. Significant inventions were made in collaboration with a small number of universities like the University of California at Berkeley [14]. These were often the first of their kind in the industry and gave Intel a critical competitive

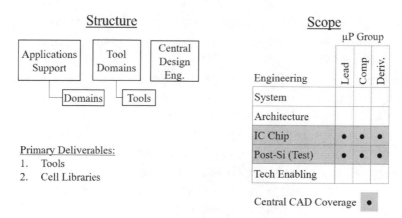

Figure 10.6.   Central CAD circa 1988.

advantage. The first centrally managed corporate standard cell library was developed and delivered by the central CAD group around 1988 and new methods to improve such things as design re-use were pursued in a central design engineering group. At this time Intel also started to develop motherboards and systems and did its own manufacturing, testing and assembly. However, central CAD did not yet develop CAD tools for architects or to those who were developing the computer systems or for those enabling the development of the process technology. As shown in Figure 10.6, the primary focus of central CAD in this era was on enabling the correct implementation of the integrated circuit once the architecture was defined as well as a small number of tools used by product engineers who did the testing of the finished product. This meant enabling the design, verification of the logical description of the design, the corresponding circuits and the physical layout of the transistors and interconnect. In the case of test CAD, it meant translating test vectors from the pre-silicon simulation into input used by tester equipment.

As the number of CAD tools increased, they became more interdependent and problems were encountered with their unsynchronized development and release. When one group in

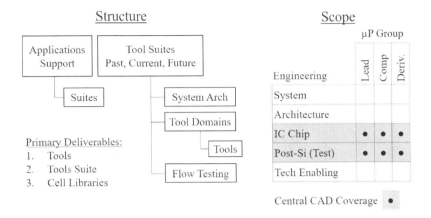

Figure 10.7. Central CAD circa 1997.

Central CAD released a new version of their tool it often broke compatibility with tools downstream in the design flow. As a result, central CAD started delivering a complete suite of tools that were validated (flow tested) together to ensure compatibility. The organization also shifted from an organizational structure based on tool development silos to an organization structure based on tool suites shown in Figure 10.7.

The development of a new suite of tools targeting the next generation of usage was done in parallel. This generational model increased the predictability of the existing set of tools and enabled major innovation between tool suite generations. However, it came at the cost of dedicating a significant part of the organization to building a completely new suite of tools in parallel to supporting the existing one. As the generational model matured, three suites needed to be staffed simultaneously. The previous suite (that was being end-of-lifed), the current suite and the next generation suite all required investment and competed for resources. The generational model also meant that the first design team to transition from one tool suite generation to the next went through a very large change.

In 1988, the EDA tool industry was still very young, especially with respect to microprocessor design. As a result, EDA vendor tools were rarely adopted by central CAD or the micro-processor design teams. In many cases, there was no equivalent solution to what central CAD provided or could create and there was a very strong culture within the central CAD group to build everything in-house. Nevertheless, by the end of this era, some EDA vendor tools were getting traction in the industry and a small number were adopted by the central CAD group. The problems associated with migrating a design team to a completely new tool suite also motivated some microprocessor design groups to start advocating for greater adoption of EDA vendor tools.

The R&D System

As shown in Figure 10.8, the interaction between central CAD and the microprocessor design teams was straight forward. Central CAD worked with the lead design team and the com-paction design team to define what new tools or enhancements would be needed and worked with them to test and deploy them. The derivative products then used these tools for their designs. In this era, the invention of new tools was still common and there was a standard product proposal (POP) process for proposing a new tool and, if that was accepted, a process for converging on the product requirements description (PRD).

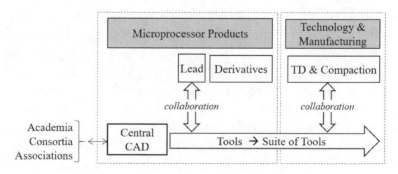

Figure 10.8.   R&D system in the first era.

In general, central CAD did not have any major suppliers other than the manufacturing technology group that supplied manufacturing process collateral needed for several tools. Central CAD also relied on deep collaboration with a small but important group of leaders in academia who were at the forefront of CAD innovation. Several critical new tools resulted from these collaborations that led to dramatic improvements in engineering productivity.

By automating many steps in the design methodology using push-button design flows, the central CAD tools restricted the freedom of engineers to deviate from accepted practice. This served the lead microprocessor teams well since they were tuned for their specific needs and reduced the risk of human error from the many college graduates being hired to fill the needs of the rapidly growing company.

As the number of products began to grow, MPG began an initiative that encouraged design groups to share their design methods and help central CAD define tools and tool suites that met the needs of a growing number of microprocessor design teams. This took the form of joint engineering teams composed of engineers who were meant to converge on the technical recommendations and joint management teams that resolved issues and made final decisions.

Finally, the rapid pace of CAD technology invention ran concurrently with advances in process technology and the microarchitecture of the microprocessor. The degree of change in each area was very large for each microprocessor generation and the concurrent nature of the development meant that many things came together as engineers were doing their design work. Since the engineering and CAD community was small, the interactive networks were strong the learning cycles were very short. This created a great environment for rapid invention needed to increase engineering productivity. Also, simultaneously being on the cutting edge of the process technology, architecture design and CAD also meant there were many problems when new CAD tools were initially rolled out, the tight collaboration between central CAD, engineering and the process technology development teams proved essential to maintain trust and manage the chaos at those times.

Era 1 Questions:

(1) In your own words, state the problem that central CAD faced.
(2) Describe the value central CAD provided Intel at this time.
(3) Why did central CAD change and deliver a suite of tools?
(4) Will central CAD's generational model scale as Intel grows?
(5) What choice did the arrival of EDA vendors pose to the manager of the central CAD group?

10.7.3 *Era 2: Customer Autonomy (1998–2004)*

Intel Business

Craig Barrett became Intel CEO in 1998 but the first era ended with the advent of the AMD K6 microprocessor that was introduced in 1997. The AMD K6 was compatible with Pentium based PCs but was offered at a much lower price. This created a low cost (sub $1,000) PC market that caught Intel by surprise as expressed in its 1998 Annual report [15]:

> *"With hindsight, it's clear that we were caught off guard by the increase in demand for low-cost PCs. We were late in recognizing the emergence of this value PC market segment — and the competition took advantage ... In response to the evolving computing marketplace, it was clear that we had to drive our business in a new way. We developed a broad game plan that would enable us to participate in every level of the newly segmented computing market. We revamped our microprocessor lineup with new products created specifically for each computing segment"*

This resulted in three product lines, the Celeron microprocessor (entry-level PC buyers), Pentium II microprocessor (performance desktop and entry-level servers and workstations), Pentium II Xeon microprocessor (for mid- and high-range servers and workstations). As Intel stated,

"Our segmentation strategy is designed to allow us to participate profitably in various segments of the computing market and to pursue new growth opportunities in the high-end server and workstation market segments."

During this period, Intel also introduced the Centrino brand for wireless/notebook products based on modified Pentium and Celeron microarchitectures. Intel also introduced a new ultra-low-power XScale architecture (for hand-held and embedded applications). This was based on the StrongARM architecture acquired from DEC in 1997. Similarly, the new Itanium architecture (for enterprise servers) was introduced.

In 1999 Intel spent $6 billion acquiring several communications companies and began to describe itself as a company that [16],

"designs, develops, manufactures and markets computing and communications products at various levels of integration"

Over this period, Intel's total revenue grew from $26 to $34 Billion. In 2004, 85% of Intel's revenue came from microprocessors and related chipsets and motherboards and 15% from the communications group. [17]

Microprocessor Design

With a new CEO and a focus on segmentation, the Intel Architecture Group (IAG) was formed to oversee the broadening product lines. IAG included the Business Platform Group, the Consumer Products Group, the Mobile/Handheld Products Group, and the Enterprise Server Group. The MPG organization still existed in parallel to these business units and focused on the development of the microprocessors. In 2000, the MPG organization was merged with IAG. The lead microprocessor designs were still done by MPG and the initial compaction was still done by a design team in the manufacturing organization.

Segmentation was meant to enable greater flexibility to customize designs for different markets. For example, a new microprocessor design group was formed in MPG to address the low-cost computing segment. This required microprocessors with more transistors packed into a smaller area (greater logic transistor density) to reduce die sizes and therefore manufacturing costs. Similarly, new design methods were needed to reduce the non-recurring expense (NRE) of the design effort. In that market, trading off raw performance for development effort and die size was a viable option. In both of those respects, Intel lagged its competition and needed a focused effort to catch-up and then gain leadership. In the beginning, many of the new microprocessor designs targeting the new segments were heavily based on the current microprocessor micro-architectures but, by the end, customizations were made to address them.

In order to catch-up, the new microprocessor design groups evaluated new design methods with Intel's CAD organization. However, facing more intense pressure, they also demanded more control over the choice and integration of CAD tools. The nascent EDA industry was maturing and design groups were soon provided the freedom to select and integrate EDA vendor tools themselves. This spurred the creation of design automation teams in the microprocessor design groups who worked with Intel's CAD organization and external EDA suppliers to integrate the tools of their choice.

Central CAD

During this era, Intel's CAD organization stopped releasing a complete suite of tools and moved the responsibility for the

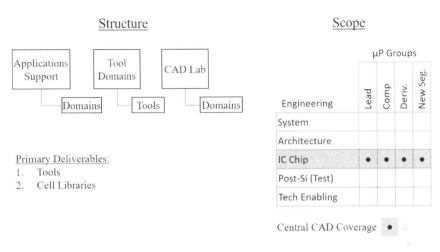

Figure 10.9. Central CAD in the second era.

integration and flow testing to the microprocessor design groups. Also, the floating-point division (FDIV) bug on Intel's Pentium microprocessor that became public knowledge in the previous era (in 1994) generated interest in applying formal verification concepts to floating-point operations. The work in this area was done by central CAD and this contributed to the formation of a dedicated research group called Strategic CAD Labs (SCL). The result was that by the end of 2000, the central CAD organization was once again organized around specific CAD tool domains with the addition of the Strategic CAD Labs (SCL) as shown in Figure 10.9.

Although there were some cases where central CAD worked with an external EDA supplier to utilize their tools, they essentially saw the EDA vendors as a threat and competed with them to achieve designs wins at Intel to increase the share of the internal market serviced by central CAD tools. The move back to becoming a tool provider instead of delivering a suite of tools also meant that there was no pressure to drive internal development standards or standardize on common capabilities such as user interfaces, data models and software languages across the different tool domains. Also, several initiatives were started to create new internal tools that would dislodge EDA vendor tools that had been selected by a microprocessor design group. In many cases, the advantage offered by the competing central CAD tool was marginal.

Although central CAD moved away from delivering a completely validated suite of tools and flows it still provided subsystems that needed to operate together and the data models and infrastructure that often required the use of their tools (and vice versa). As a result, the microprocessor design groups already using central CAD tools did not deviate too much from those offerings. The exceptions were new microprocessors or other design groups that were just forming or that were acquired from other companies. They had more flexibility or initially came with their own CAD organization and preferred set of EDA vendor tools. During this era, Intel also began to expand its microprocessor engineering groups into other countries. As a result, the central CAD group also began operating in the same countries to provide the local support needed.

The culture of design wins created a tense dynamic between the tool domain, application support and the SCL groups. The tool domain groups felt that the CAD lab's work with design organizations made it harder to get design wins and the applications support group sometimes supported the design groups move to an alternative EDA vendor tool.

The R&D System

As shown in Figure 10.10, the major change in the work model for this era was central CAD becoming a tool provider in competition with the EDA vendors and the resulting growth in the number of separate EDA supplier evaluations conducted by the design groups. These were not coordinated and often led to different microprocessor design teams selecting CAD tools from different EDA suppliers to do the same function. Within each microprocessor design group, the incremental

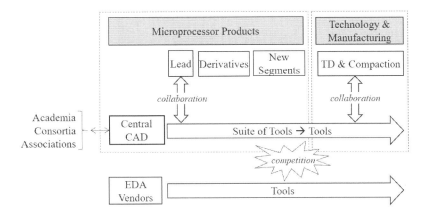

Figure 10.10.   The R&D system in the second era.

costs seemed manageable but integrated across all of Intel the spending on EDA vendor tools started to grow at a rapid rate. Also, the EDA suppliers soon began to get overwhelmed supporting redundant evaluations occurring across the company. The high demand for new features and support coming from different Intel design teams used to operating closely with an internal CAD organization put even more strain on EDA vendors.

The relative inefficiencies related to the decentralization of EDA tool decisions, and the resulting costs associated with a lack of standardization, were a trade-off that enabled the autonomy for some design groups to act fast to address new markets. Formal management and technical networks still existed between Intel's CAD and the design team employees despite the move away from delivering a complete tool suite. Although by the end of this era, the formal networks that were in place previously to enable the delivery of a complete suite of tools had dissipated, the technical working groups remained very active in each specific area.

Despite the changes, major innovations for microprocessor design still resulted from the close collaboration between the lead and compaction microprocessor design teams and the central CAD group. This was because the design methodology for each new generation still changed significantly to meet aggressive design goals. This required big changes in the tools for each generation and needed a very close working relationship.

Design challenges in the new microprocessor segments also required close collaboration. For example, the development of a new design methodology targeting microprocessors for low-cost PCs was a co-development effort between central CAD and the design groups.

Era 2 Questions:

(1) What benefits did Intel get with more autonomy to design teams?
(2) What cost did Intel pay with more autonomy to design teams?
(3) Why did central CAD see EDA suppliers as competition?
(4) What advantages did central CAD have vs. EDA supplies?
(5) What advantages did EDA supplies have vs. central CAD?

10.7.4 *Era 3: New Devices and Platform Scope (2005–2012)*

Intel Business

On May 18, 2005, Paul Otellini became CEO. In the 2005 annual report, he wrote [18],

> *"Today, we are reinventing Intel once again, to focus on the growth opportunities presented through platforms — advanced solutions that integrate Intel® microprocessors and other technologies, such as complementary chipsets and communications chips, all optimized to work together."*

This resulted in 6 new business units: the Mobility Group (notebooks & handheld devices), the Digital Enterprise

Group (business & consumer desktop PCs), the Digital Home Group (emerging digital home market), the Flash Memory Group (storage in cell phones and other devices), the Digital Health Group (healthcare research, diagnostics & productivity) and the Channel Platforms Group (for developing and selling Intel products to meet the unique needs of local markets). To achieve the platform vision and enable competing in new highly segmented markets Intel started to speak in terms of delivering Systems on a Chips as part of the platform (SOCs).

In 2006, there were 5 microprocessor brands: Itanium, Intel Xeon, Intel Core, Pentium, and Intel Celeron. In June 2007, the first Apple iPhone was introduced. Apple had met with Intel early in the development of the product to see if Intel would be willing to provide the ultra-low-power microprocessor needed for its unnamed product but Intel declined to pursue the business. Apple then chose a design based on the competing ARM architecture. Independently, Intel introduced its low power microprocessor called Atom that targeted handheld devices in June 2008.

Intel recognized the growing importance of the handheld and consumer devices market in its 2009 annual report,

> *"Driven by the Intel Atom processor, the spectrum of products based on Intel® architecture is expanding beyond PCs and servers to include handhelds, consumer electronics devices, and hundreds of embedded applications."*

That year, Intel's business units were restructured again under a newly reconstituted Intel Architecture Group (IAG) where it established a separate business unit to focus exclusively on handhelds and devices called the Ultra-Mobility Group. At the same time, Intel also recognized the importance of graphics to the user experience and the growth possibilities in the server market by creating the Visual Computing Group and on the Data Center Group respectively. To remain focused on its core business it combined the mobile and desktop products under the same business unit call the PC Client Group. Finally, as part of the same reorganization, Intel also created the Embedded Communications Group and retained the Digital Home Group.

In April 2010, the first Apple iPad was introduced and energized the tablet market segment. It used an ARM-based microprocessor and Intel soon found itself behind in both the smartphone and tablet markets. In August 2010, Intel announced it was acquiring Infineon's wireless business for $1.4 billion [19]. This was done with the eye on the smartphone market and enabling the integration of Infineon's wireless modem technology with Intel's Atom microprocessor on a SOC.

This era was marked by improvements in Intel's competitive position in the desktop and mobile computing segments enabled by an aggressive Tick-Tock development model, the adoption of a re-configurable "Core" micro-architecture (used in Core i3, i5, i7 and Xeon products), the integration of HD graphics on the same die as the microprocessor and clear manufacturing technology leadership. However, at the same time, Intel was having trouble gaining traction in the new and exploding smartphone and tablet markets that were dominated by companies with more cost-efficient and quicker development models based on a SOC design methodology that leveraged an industry ecosystem that was growing around the combination of ARM (architecture), foundry services (standard circuit libraries and fabrication), and EDA vendors (tools tuned and validated with the rest of the ecosystem).

Over this period, Intel's total revenue grew from $34 to $53 Billion. 84% of that came from the client and server microprocessor products. [20].

Microprocessor Design

Around 2005, Intel adopted a "tick-tock" development model for its microprocessors. In this model, it introduced a microprocessor with significant microarchitectural changes one year (the tock) and then followed it up one year later with a version with minor microarchitectural changes on the next manufacturing process technology to reduce cost and improve performance (the tick). This cycle repeated to create an annual tick-tock release cadence. It provided customers with a predictable roadmap of new microarchitectural changes and performance improvements. With the implementation of this model, the compaction microprocessor design that was previously designed by the technology manufacturing group became the tick and was now the responsibility of a microprocessor design group in the PC Client business unit.

The Central Processing Unit (CPU) is the highly tuned heart of the microprocessor and has a big influence on the overall performance. Over time, microprocessors evolved from primarily being a CPU to being a system composed of the CPU, memory and related functions that used to be located in other chips. In this era, Intel started to use a common CPU "core" design for its family of microprocessors in the desktop, mobile and server segments. In general, the microprocessor group responsible for the tock also created the common core CPU that was re-used in the design of the other microprocessors. The products based on the use of the common cores became the Intel Core i3, i5 and i7 branded products.

Also, the new visual computing business unit evolved into a design organization that developed the graphic execution units that were ultimately integrated directly into the microprocessor itself. Until then, computers utilized separate graphics processing chips. By integrating them on the same die as the microprocessor Intel enabled basic graphics functionality in a smaller small form factor for use in mobile computers.

In parallel, a design engineering organization was formed to create a repository of commonly used microprocessor functions known as Intellectual Property (IP) blocks. These common IP blocks were meant to be re-used in different microprocessor designs and would ultimately expand to cover different functions that could be integrated quickly with a microprocessor. As a consequence of these changes, the microprocessor designs included a common CPU core, a set of graphics execution units and some special design blocks imported from different design groups. The microprocessor design was also no longer completely designed by one group and started to resemble what was called a system on a chip (SOC).

The Atom microprocessors also resembled a SOC but were based on a separate microarchitecture and CPU core and typically integrated many more additional functions on the same integrated circuit. The design organization defined its own design methods and selected the CAD tools they felt they needed. Some of them came from Intel's central CAD group but several did not. To facilitate the unique needs of the handheld and devices market served by the Atom microprocessors, the manufacturing organization began to provide a manufacturing process technology tuned for a System on a Chip (SOC) style of design. However, the perceived importance of maximizing the CPU core performance meant that many process technology decisions for SOC designs were still dictated by the needs of the CPU core.

Central CAD

This era saw four significant changes in strategy for central CAD. First, shortly before this era began, the central CAD organization moved to the Technology and Manufacturing Group (TMG) and no longer reported into a microprocessor design group. TMG was responsible for the research and development of the semiconductor manufacturing process technology as well as the manufacturing and packaging of the chips. The move to TMG was part of a strategy to focus more effort on the co-optimization of the internal CAD tools with the manufacturing process during its development. The objective was to do a better job translating Intel's manufacturing technology leadership into microprocessor product leadership. As part of this move, two small CAD groups in TMG were merged with the central CAD group and the central CAD group became more integrated into the process technology development activities.

Second, the central CAD organization began to focus on delivering solutions in addition to tools. A solution incorporated

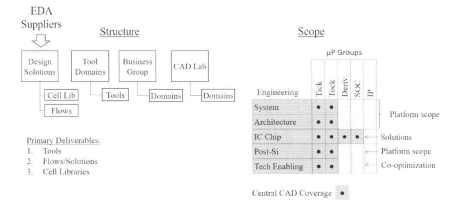

Figure 10.11. Central CAD in the third era.

an existing EDA vendor tool in an automated design flow that was specific to Intel. By this time the EDA industry had matured to the point where much of central CAD's tools had a counterpart offered by one or more EDA vendors. The objective of this strategy was to use EDA vendor tools when feasible to free up central CAD resources to focus on developing CAD tools that EDA vendors did not supply or that provided Intel a clear competitive advantage. Responsibility for delivering solutions and tools was divided into different groups within the central CAD group as shown in Figure 10.11. This enabled freer collaboration with EDA vendors without creating a conflict of interest with redundant internal tool development. It also created an internal advocate for using EDA vendor solutions to balance the bias towards internal development.

Third, in alignment with Intel's new focus on platforms, central CAD expanded its scope to more explicitly address architecture and system-level design and validation along with post-Si manufacturing test, debug and analysis. The merger of the CAD group from TMG also meant the central CAD groups scope now included the development of some of the CAD tools needed for process technology development. These were areas that either involved company-specific technology, were secretive, or relied on unique Intel methodology that EDA vendors did not see much value investing in. This complemented the strategic shift to solutions by opening up areas for applying CAD that could greatly benefit Intel and where the current list of EDA vendors had little to offer.

Fourth, central CAD became more active driving design tool convergence and entered into a primary supplier agreement with one EDA vendor. Due to the freedom design teams had to select EDA tools from either central CAD or the EDA industry directly, the spending across the company on EDA industry tool licenses was growing at an unsustainable pace. To flatten the curve, central CAD worked with the microprocessor design teams to drive convergence into a smaller variety of EDA tools and suppliers that they could then include in central CAD solutions. To facilitate this, central CAD entered into an agreement with one EDA supplier to become the primary vendor and centralized the management of the EDA tool licensing budget. This enabled greater control and predictability of spending and also opened up opportunities for collaboration and co-development. As a result of this approach, Intel design groups could reduce their costs if they used the central CAD solutions that already included EDA vendor tools.

As a consequence of these strategic shifts, the culture of competition that had existed between central CAD and the EDA industry began to subside and strategic planning began to differentiate between what value can be delivered via EDA vendor engagement vs. that which needs to be done internally to provide a distinct competitive advantage.

The R&D System

As shown in Figure 10.12, this era saw the R&D system expand to include EDA vendors as true suppliers to central CAD. It took several years but, eventually, fewer EDA vendor tools were delivered directly to the mainstream microprocessor design

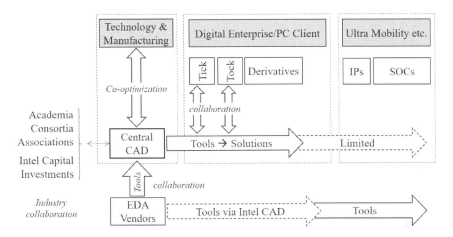

Figure 10.12. The R&D system in era 3.

teams and more were delivered via central CAD from the new solutions groups. On the other hand, the new SOC design and IP groups in the other related business groups at Intel had their own dedicated design automation teams that worked directly with the EDA vendors in support of a different design methodology.

The new tick, tock microprocessor design groups remained the primary customers of central CAD and collaborated closely on defining new capabilities for each generation. The microprocessor design groups still completed the final integration, validation and first-line support of the complete design environment they used but considerable effort was made to replace their design flows using the EDA vendor tools and design flows now developed by central CAD.

The move to solutions and the primary supplier model by central CAD necessitated greater collaboration and coordination with the different EDA suppliers, greater involvement in IEEE and other standards committees, and consortia like the Semiconductor Research Corporation (SRC). In parallel, central CAD also became more active in proposing and enabling Intel

Capital's investment in start-up companies in the EDA industry. This was part of Intel's broader approach to foster and then harvest the creative energy in the industry as a whole.

The move to the Technology and Manufacturing Group enabled much tighter collaboration between central CAD and the technologists defining the next-generation manufacturing process. This enabled earlier exploration and testing of CAD tools and methods that had strong dependencies on the process technology. The manufacturing technology leadership Intel held at this time was its crown jewel. It had enabled the company to weather tough competitive threats. As a result, any leakage of process technology information could be very damaging to the company and data was tightly held. Gaining the trust of the key technologists for collaboration was very difficult. Moving the central CAD group to TMG helped reduce those barriers.

Although the company was starting to move to a SOC style of design to enable new markets in handhelds and devices, the vast majority of the revenue still came from the traditional desktop, mobile and server microprocessor product lines. central CAD's focus remained on those microprocessor groups. Engagements with the graphics engineering organization, the Atom design teams and IP organizations generally focused on areas where existing central CAD solutions would provide additional value over what the design teams had already had incorporated themselves. For the most part, the design automation teams in those design groups still operated independently of central CAD and worked directly with the EDA vendors.

Progress was made on converging onto fewer CAD tools during this time but progress on the converging on the design methods was very slow. By the end of 2012, as more and more design sharing started to happen, these differences resulted in significant problems with the integration of other internal IP. Even between microprocessor groups using the same tools from central CAD, the design methods differed significantly enough that validating that an EDA vendor tool worked for one design team did not mean it worked for another. This resulted in complex software implementations of the design flow that enabled multiple ways to use the same EDA tool depending on the design group that was using it.

Era 3 Questions:

(1) How did Intel respond to the changes in its environment?
(2) How did the design of microprocessors start to change?
(3) How did central CAD's relationship with EDA suppliers change?
(4) What were the benefits of central CAD's new model?
(5) What are the benefits of the ultra-mobility group's autonomy?
(6) What are the risks of allowing the above autonomy?

References

[1] Deming, W. E. (1994). *The New Economy, For Industry, Government, Education,* 2nd Edition. (MIT, Center for Advanced Educational Services) p. 50.
[2] Grove, A. S. (1999). *Intel Keynote Transcript, Los Angeles Times 3rd Annual Investment Strategies Conference.* Available from: https://www.intel.com/pressroom/archive/speeches/cn052499.htm [May 2020].
[3] Aoki, K. and Lennerfors, T. (2013). The New, Improved Keiretsu, *Harvard Business Review*, September 2013.
[4] Liker, J. and Choi, T. (2004). Building Deep Supplier Relationships, *Harvard Business Review*, December 2004.
[5] MacCormack, A. and Forbath, T. (2008). Learning the Fine Art of Global Collaboration, *Harvard Business Review*, January 2008.
[6] Hansen, M. (2009). When Internal Collaboration Is Bad for Your Company, *Harvard Business Review*, April 2009.
[7] Pisano, G. and Verganti, R. (2008). Which Kind of Collaboration Is Right for You?, *Harvard Business Review*, December 2008.
[8] Intel, *The Story of the Intel® 4004,* Available from: https://www.intel.co.uk/content/www/uk/en/history/museum-story-of-intel-4004.html [May 2020].

[9] Intel (1988). *Intel Annual Report 1988*, Available from: https://www.intel.com/content/www/us/en/history/history-1988-annual-report.html [May 2020].

[10] Gelsinger, P. Kirkpatrick, D. Kolodny A. and Singer. G (2012). Coping with the Complexity of Microprocessor Design at Intel — A CAD History. Available from: https://www.researchgate.net/publication/268005718_Coping_with_the_Complexity_of_Microprocessor_Design_at_Intel_-_A_CAD_History [May 2020].

[11] Intel (1987). *Intel Annual Report 1987*. Available from: https://www.intel.com/content/www/us/en/history/history-1987-annual-report.html [May 2020].

[12] Intel (1997) *Intel Annual Report 1997*. Available from:https://www.chiphistory.org/chc_upload/content/pdf/19/1506018723/1506018723.pdf [May 2020].

[13] *Ibid.*

[14] Gelsinger, P. Kirkpatrick, D. Kolodny A. and Singer. G (2012). Coping with the Complexity of Microprocessor Design at Intel — A CAD History. Available from: https://www.researchgate.net/publication/268005718_Coping_with_the_Complexity_of_Microprocessor_Design_at_Intel_-_A_CAD_History [May 2020].

[15] Intel (1998). *Intel Annual Report 1998*. Available from: file:///C:/Users/15037/OneDrive/Documents/HTM%20Book/Reference%20Docs/Intel/history-1998-annual-report.pdf [May 2020].

[16] Intel (2002). *Intel Annual Report 2002*, p. 74. Available from: https://www.intel.com/content/www/us/en/history/history-2002-annual-report.html [May 2020].

[17] Intel (2004). *Intel Annual Report 2004.* Available from: https://www.intel.com/content/www/us/en/history/history-2004-annual-report.html

[18] Intel (2005). *Intel Annual Report 2005*. p. 3. Available from: https://www.intel.com/content/www/us/en/history/history-2005-annual-report.html

[19] Shah, A. (2010). Update: Intel to Acquire Infineon's Wireless Division, Computerworld, August 30, 2010. Available from: https://www.computerworld.com/article/2515085/update--intel-to-acquire-infineon-s-wireless-division.html [May 2020].

[20] Intel (2012). *Intel Annual Report 2012*. Available from:https://www.intc.com/investor-relations/financials-and-filings/annual-reports-and-proxy/default.aspx [May 2020].

CHAPTER 11

Engagement with Academia

> Industry and academia have different goals.
> — **John M. Acken**

For our book, the R&D organization's engagement with academia spans activities that include: small one time projects, education and hiring of new college graduates, continuous informal connections at conferences, major joint research projects, joint research goals within consortia (such as the SRC), through long term interaction based upon national funding organizations (such as the NSF). There is a great opportunity for both the R&D organization and a university to benefit if they engage. However, for the engagement to be beneficial, all parties must understand the objectives of the other parties. As described in previous chapters, the R&D organization's success is determined by delivering value to the customers. Academic success is determined by delivering educated students and by research that discovers truth.

11.1 The R&D Organization and University Goals

The R&D organization's goal is to deliver value to the customer.

Ultimately, corporations must deliver products that customers are willing to buy. How well the corporation identifies and delivers features that satisfy the paying customers determines corporate success. The value described previously in this book is the R&D organization's contribution to corporate success. Specifically, for the R&D organization, an engineering task has value if it contributes to the vision of the company in a way that customers will pay. In other words, the value of the research and development effort in the R&D organization is measured by how much it contributes to the corporation's goal of delivering products to customers that they are willing to buy.

The fast learning cycles described in this book are the mechanism for the R&D organization to gain the knowledge needed to identify and refine the features that the customer desires. Learning cycles are also used to gain the knowledge needed to improve the engineering tools, workflows and processes used to develop those features. The R&D organization applies that knowledge to solve the engineering problems associated with providing technology features and capabilities that add value. From an engineering perspective, knowledge is valuable in as much as it is applied to solving practical technical problems. One way to do this is to embed that knowledge in an engineering solution (such as software) that is part of the final technology. However, the knowledge needed by an R&D organization also includes processes for applying knowledge to building and improving the technology. Thus, an engineering R&D organization will be creating knowledge when it identifies technology development process improvements and will be applying the created knowledge when the R&D organization implements those process improvements within its workflows. A key reminder is that the R&D organization's goal is to perform the engineering activities that create value and that contribute to the company's goal of satisfying paying customers.

One goal for academia is to educate students.

Education of students is a primary goal for universities (and academia in general). Educating students is a mechanism for dispersing knowledge. As mentioned above, from an engineering perspective, knowledge is valuable in as much as it is applied to solving practical technical problems. Academia contributes to the pool of people capable of applying knowledge to problem-solving. The knowledge imparted to students includes basic science and math concepts, problem-solving process ideas, the ability to search for existing solutions, and the ability to create the knowledge for solutions when needed. Restated, academia prepares students to become productive engineers by providing them with the basic facts, training them on the tools to apply the basic knowledge, and the ability to become life-long learners so they can solve new technical problems as they arise in the future. Teaching the ability to become a lifelong learner requires assigning learning tasks beyond simply reading the textbook, listening to the lectures and running lab experiments. For example, assigning the reading and reporting on published technical papers. The education effort disseminates knowledge via the students. The success of the education effort is measured by the output of students capable of contributing engineering knowledge to solving technical problems.

Another goal of academia is to discover basic principles and processes.

Academic research strives to discover knowledge. The knowledge discovered or created can be basic theoretical concepts or new practical applications of basic concepts to technical problems. The goal is the discovery of knowledge for basic principles and processes. This does not depend upon the utility of the created knowledge, nor does it preclude finding utility for the created knowledge. Research can be viewed as fundamental research or applied research. The R&D organization uses applied research to create new knowledge and more powerful modeling, analysis and optimization methods to solve design problems. Some university research projects do this also. There can be engagement between a company's applied research effort and a university's fundamental research effort, but both parties must be clear about the different objectives for the parties. The easiest engagement to coordinate is applied research that fits the university's existing expertise and that is not in the critical path for the company's delivery schedule. Many scientific and mathematical discoveries were not applied to practical implementations for years or even decades. Academic research is a search for technical truth. Academic research creates researchers.

Both the academic goals of educating students and performing research are advancing the frontiers of knowledge independent of economic value. There may be economic value in the education of students, but the education content is not measured by the economic value. Similarly, the academic research may lead to economic value, but the real goal is advancing the frontiers of knowledge.

11.2 Conditions for Engagement

Different objectives do not preclude cooperation.

Although the R&D organization and universities have different primary objectives, cooperation on the tasks to achieve those objectives can be beneficial to both parties. Specifically, the successful completion of a project or task can meet independent goals. As a simple example, a university researcher may discover and publish a new technical principle that the R&D organization uses to improve a product.

The legal environment is different in academia than for industry.

The R&D organization must protect its created knowledge (as intellectual property) from unauthorized disclosure to give its products a competitive advantage. A university must expose and share its created knowledge to both get the credit and contribute to the general knowledge of the field. One view of this difference is that an R&D organization uses policies and procedures that have strict legal requirements to protect the ownership and use of its intellectual property. Whereas a university must use policies and procedures that meet the legal requirements of attribution to the creators of the knowledge while dissemination and expanding the use of that knowledge.

Both parties must understand the other parties' objectives.

For engagement between an R&D organization and a university to be successful, each must understand the other's objectives and environment for cooperation to occur. As described in previous chapters, the R&D organization's success is determined by delivering value to the customers. Academic success is determined by delivering educated students and by research that discovers truth. Successful engagement also requires an understanding of the basic cultural differences between industry and academia. The culture in a university is to share and discuss everything. Learning from discussing and sharing knowledge is seen as a benefit in and of itself. Within the R&D organization activities need to be evaluated relative to creating value. While some activities involve discussions and sharing knowledge, the activities must contribute to the R&D organization's definition of value. As we discussed in Chapter 4, this means the activity contributes to the company's vision in a way that the customer will pay.

Mutually beneficial projects must be identified.

Many R&D organizations and universities make statements that they need more engagement between industry and academia. However, without clearly defined objectives and projects, the participants may be very disappointed with the results. An example is hiring new college graduates. The university has a goal of educating students. The R&D organization has a goal of hiring qualified students. The university creates the curriculum and standards for the degrees it grants. If the university ignores the needs of the industry it will fail to teach students the necessary technical skills to get hired and perform well once hired. This does not mean only teaching courses or topics that an R&D organization wants. There are basic, general, and advances specific technical topics that students need and should be the bulk of the curriculum. One project that can facilitate this engagement is a university advisory board made up of industry representatives. This helps members from industry communicate their expectations when hiring a new college graduate.

Selfish objectives must be avoided.

For projects to be mutually beneficial and to ensure cooperation, self-serving attitudes and practices must be avoided. One such self-serving attitude is to view a company's R&D organizations as a source of free money. Yes, the R&D organization may provide money and resources for a particular project or task, but it properly expects a return on its investment. An example of this mistake may occur when a university professor takes a sabbatical to work for a period at a company's R&D organization. While this is a common practice, and with proper implementation can benefit both parties, in some cases, the professor may view the company as a source of money for an extended paid vacation. On the other hand, a mistake can occur when the R&D organization provides grant contracts to support students. Once again, this is a common practice and can benefit both parties, however, it is a mistake for an R&D organization to misuse this channel to get cheap labor. The university expects the students to benefit intellectually from the project, and as one professor described it: "I am not providing my students as coding monkeys".

11.3 Providing Value in Joint Projects

The R&D organization evaluates projects based on potential value.

The R&D organization must evaluate each project based upon whether it adds value. Sometimes a joint project increases the ability of the R&D organization to create value by reducing waste in its workflows, processes and practices. Sometimes a joint project directly adds value to a specific technology as measured by the incremental increase in what the customer is willing to pay.

Academia evaluates projects based on potential for teaching/ discovery.

Not every project will be successful. University research seeks answers to new questions, teaches student practices, answers old questions and discovers new questions. When a university selects projects and tasks in which to participate it must select promising topics that contribute to one of its two goals:

(1) Research for knowledge discovery, and
(2) Teaching to prepare students.

Measuring the merits of a joint project is different for both parties.

The R&D organization measures the merits of a project based on whether it contributes to the vision of the company in a way that the customer will pay. Academia measures the merits of a project based on the education of students and the discovery of new ideas. A contribution to the education of students can be in the form of transferring knowledge to the students, improving the university methods and processes for educating the students or increasing the capacity for the students to learn on their own. The contribution to the discovery of ideas can be determined by how much the task either enables that or how well the task facilitates the dissemination of university generated knowledge.

There are contrasts in benefits for the R&D organization and academia.

The same result can be viewed as a benefit to academia and a downside to industry, and vice versa. For example, the R&D organization does not benefit from training employees who promptly leave to go to work at a competitor. Whereas, a university does benefit from training students who go to work at competing universities. Conversely, the R&D organization will benefit from finding new knowledge that gives them a competitive edge and keeping it a secret. Whereas a university will not benefit from finding new knowledge and keeping it a secret. In the R&D organization's case, the secret adds value to the product because no one outside knows it. In the university case, the very value of the knowledge it creates is the dissemination of that knowledge through either teaching it to the students or publishing it is journal articles.

The R&D organization can benefit from joint investigations with academia on very speculative or basic research.

As we noted, the R&D organization must justify joint investigations based on its determination of value, associated costs and resulting return on investment. University research must justify any joint investigation based on the existing research goals. These two justifications overlap when there is a basic or fundamental scientific or technical principle that is an open question (hence, it has an unclear probability of a result to be found) and a potential game-changer for the R&D organization. This is an ideal case for university research but a risky one for the R&D organization. As a result, this type of engagement is often supported by the R&D organization (because of the potential upside), but the work is done by the university (because of its interest in searching for, creating, and dissemination knowledge).

Academic research benefits from a connection to real-world applications.

There are many possibilities when it comes to research topics. A professor must identify which research topics to pursue and which topics to skip. The criteria vary widely among professors. One example of the criteria to decide is the potential application for a research topic. When faced with a new research proposal topic, funding agencies will ask "So what?" If the answer can be put in the context of an industry problem in general, or even more pressing, a particular opportunity with a particular company then it often appears more attractive.

11.4 Engagement Examples

Examples of successful cooperation.

As mentioned before, one way the university benefits from engagement with the R&D organization is by introducing students to real-world problems. Conversely, the R&D organization benefits from academic engagement by getting access to students and injecting example problems into the students' education. The clearest example of this case working for the benefit of both is internship programs. The university and R&D organization teach the students to be productive employees and the student is motivated by seeing that school knowledge is useful. Another example is Ph.D. student employees. In this case, research is guided by company needs and skills are acquired that the company can utilize. Visiting scholars, guest lecturers, and adjunct professors are other examples. Industry experts also provide guidance to universities for course work and what students should be learning. Another byproduct of successful cooperation is the connections created that may result in hiring students, not to mention the enthusiasm generated in the students for basic concepts that have practical benefits to the industry.

Examples to avoid.

There are several examples of tasks and projects to avoid when considering candidates for engagement between the R&D organization and a university. A company should never allow mission-critical tasks to depend upon a university project. The R&D organization's schedules and product delivery requirements do not match a university schedule, time frames, and knowledge delivery mechanisms do not match. Another example to avoid as a candidate for engagement between universities and companies are mundane projects. The R&D organization can contract or internalize repetitive and mundane tasks but the university needs tasks to educate students and create knowledge. This is important when there is a visiting scholar, either from the university to the company or from the company to the university. Without a clear project or task area, and a responsible mentor at the host institution, the visiting scholar's time may be wasted. This situation benefits neither institution. As mentioned before, a company should never look at this as an opportunity for cheap labor. Yes, an intern will be an inexpensive contributor to the project, and yes, the student will be paid better than just taking classes, but the goal is for the student to learn. Sometimes university professors can be jealous of corporate resources available to the R&D organization and view a joint project as an opportunity to dip into the mythical buckets of money that don't exist. This is the wrong motivation to engage since it will damage relationships and prevent mutually beneficial results from a joint project that turns out to not be joint at all.

11.5 When to Engage, or Not to Engage

A speculative investigation that leads to problem clarification has value.

Some problems are not clearly defined, and as both Dr. Charlotte Acken and Charles Kettering have so succinctly put it: "A problem well stated is a problem half solved". As noted, the R&D organization is motivated to get to a solution that adds value. A university is motivated by the likelihood of identifying fundamental truths. It is very natural for a university to investigate a vague topic area and to refine the problem definition as part of the investigation. It is also very natural for the R&D organization to identify a critical clear problem and plan the solution. When a problem is not critical nor is it clear but would benefit the R&D organization if solved, the problem investigation phase is an ideal scenario for the R&D organization to work with academia.

When there is a schedule mismatch avoid things on the critical path.

As discussed previously, the manager of the R&D organization must work with different entities that often have different development cycle times and schedules. Within the R&D organization, this may have some regularity, but for the industry as a whole, there are a multitude of driving factors for schedules and a variety of key time frames. For example, most consumer products have a key fixed point of December for high volume delivery. On the other hand, most products connected to the government will have a key purchase date of the fiscal year starting in October. For academia, the academic calendar is key. Summer is different from the other seasons. Course work schedules (quarter or semester) drive the availability of professors and students. These schedule issues need not kill engagement unless they are ignored. Once again, the mantra is each side must see the other side's objections and constraints.

Engagement is the best in new areas.

The best overlap of opportunities for engagement between the R&D organization and a university is in new technical subject areas. As a contrast, consider a technical concept that is established and working within an R&D organization. While improvement will continue by applying the concept of learning cycles, the improvements rely upon the culture, standards, and established procedures within the R&D organization. The university is not equipped to isolate and find the type of incremental improvements that the R&D organization's learning cycles enable. On the other hand, in the case of an established and successful university research project, the university team has a large body of knowledge for published background and relevant ongoing research related to the topic that the R&D organization does not. In this case, the university meets its goal of adding to existing knowledge by identifying gaps in the current body of knowledge.

Keeping all of this in mind, the R&D organization will benefit most from the university investigating ideas that are not incremental or are too disruptive to insert into the R&D organization's continuous learning cycles. For example, the idea may have the potential to improve the performance of existing technology but will have too high a risk for any technology currently under development until it can be proven. The idea may also be an improvement in the engineering workflow or process that has too high of a risk for existing execution until it can be proven. So, to manage the risk properly, this type of research is not part of the continuous cycles of learning during development but is inserted into the research pipeline shown in Figure 8.5 at the right point.

Theoretical findings applied to real-world problems add value.

For the research produced by a university to benefit the R&D organization, some general questions and concepts must be considered. For example, the R&D organization adds value to the company's product line if the creation and delivery of technology can be done in large volume. This means that when the university discovers a new technical concept it must identify the scaling order of problem. That is, how does the cost of production scale with increasing volume. Another question that must be addressed is how applying the technical concept to implementation can be used to partition the problem. One example of partitioning the design is to use hierarchical design. The hierarchical design scales better than a flat design.

Essential to the R&D organization's value of new theoretical findings by a university is the identification of boundary conditions and suboptimal solutions. A university can afford to investigate a disruptive technology to generate new knowledge without adversely affecting the research output and the education of students. The R&D organization must be more careful to not use a disruptive technology that craters the engineering process before the new technology pays off. Therefore, the R&D organization often applies small technology changes to its processes and products until there are diminishing returns to the value-added. Then a big step of disruptive innovation can be tried to get out of a local minimum. This is the application of the theory of simulated annealing to disruptive technology.

11.6 Information Security and Publication

Information security is different for industry and academia.

The basic culture for the R&D organization must include both the openness required by academia and the information security required for the R&D organization. Within the industrial world, information is shared as needed and the default position is to protect data. Of course, for each company, there are different levels of protection for different types of information. Whereas in academia most information is readily and openly available, and only special classes of data require security. For example, a company will have documents and training sessions on internal processes that are only given to employees. Particularly sensitive is cost and pricing information, especially plans. From the university side, ideas and information are openly shared in classes and journal publications. Particularly open are tuition costs and other pricing plans.

11.7 Summary

The R&D organization's engagement with academia spans activities that include: small one time projects, education and hiring of new college graduates, continuous informal connections at conferences, major joint research projects, joint research goals within consortia (such as the SRC), through long term interaction based upon national funding organizations (such as the NSF). There is a great opportunity for both the R&D organization and a university to benefit if they engage. However, for the engagement to be beneficial, all parties must understand the objectives of the other parties. As described in previous chapters, the R&D organization's success is determined by delivering value to the customers. Academic success is determined by delivering educated students and by research that discovers truth.

11.8 Key Points

- The R&D organization's goal is to deliver value to the customer.
- The goal of academia is to educate students and discover basic principles and processes.
- Different objectives do not preclude cooperation but both parties must understand the other parties' objectives. Avoid selfish objectives.
- The R&D organization evaluates projects based on customer value and academia evaluates projects based on the potential for teaching and discovery.
- The R&D organization can benefit from joint investigations with academia on very speculative or basic research. Academic research can benefit from a connection to real-world applications.
- Engagement is the best in new areas. When there is a schedule mismatch avoid things on the critical path.
- A speculative investigation that leads to problem clarification and theoretical findings applied to real-world problems adds value.
- Information security is different for industry and academia.

11.9 Case Studies

11.9.1 *The Evolution of Academic Engagement at Intel*

In the early 1990s, the interaction between Intel and universities varied greatly from organization to organization. The interaction also varied greatly from manager to manager within various organizations. There was a central HR function for recruiting and hiring of new college graduates, primarily at the bachelor's degree level. This included campus visits for recruiting trips to individual universities and participation in university-sponsored hiring fairs. The individual student interview trips and the evaluation for hiring were in the hands of the individual managers. For research cooperation, there was a wide variety of situations, from individual managers doing one time projects with specific professors, through long term relations ships between specific departments and individual universities, through active participation in research consortiums such as SRC (Semiconductor Research Consortium) up to participation in evaluation bodies for grant funding, such as NSF.

Examples of interactions between the company and universities showed a variety of successes and failures. One mechanism used to provide research insight into new speculative ideas without revealing company propriety information into the open publishing atmosphere of the university was to have Intel employees do Ph.D. research at a university. Syracuse University was doing some research on Datapath design and Intel had an employee in the standard cell library group working on his Ph.D. at Syracuse. This provided the ideal opportunity for the long-term speculative research at Syracuse University to be injected into Intel for practical applications. The research was a success for the university in generating publications and a success for Intel by creating an internal CAD tool. Part of the mechanism for success, in this case, was the fact that the employee's Intel manager was also on his Ph.D. committee at Syracuse. Another example was an Intel employee who did his Ph.D. at Carnegie Mellon University. This was another case where the manager was on the employee's Ph.D. committee. This resulted in general interaction in the area of IC testing and fault modeling that enhanced later connections to UC Santa Cruz. Another example of speculative research was a Ph.D. student at Princeton University who was investigating Behavioral Level Synthesis with Prof. Wolf. The student was hired at Intel as an intern for several months. This was an example of speculative long-term research that was successful in a different way. The student did get his Ph.D. and there were publications, however, the success from the Intel side was a kind of negative result, in the fact that the research indicated that the Intel environment was not ripe for

Behavioral Level Synthesis at that time. An example of a longer-term interaction was with respect to testing and was between UC Santa Cruz and Intel. Over many years, Intel hired students from UC Santa Cruz as interns, an Intel manager was on several Ph.D. student research committees, and Intel hired several employees from UC Santa Cruz. UC Santa Cruz published several papers jointly with a manager at Intel and the Intel manager guest lectured at UC Santa Cruz. In each of these successes, the university and the company understood what to expect from the interaction. An example of a failure when the company hired a professor from a university to be a visiting researcher without a clear definition of the objective. The company was expecting the development of some kind of tool or methodology. The professor was expecting a relaxing, investigative effort looking at research ideas. The company was not pleased at the end of the project. An example of a failure avoided is when an Intel manager approached a professor at Stanford University looking to identify an opportunity for a joint project. After some discussion, the professor realized Intel was looking to identify a clear project on writing a CAD tool and the professor was looking for research projects. The professor summarized the mismatch with the statement, "I will not provide my students as coding monkeys". One thing all of these examples had in common was very limited funding from Intel. Each manager would fight for very small amounts of money ($20K or so) for any project. Then Intel changed its approach.

Intel's success had been very much based upon execution. When they were the industry leader, they realized they needed a more long-term strategy. Craig Barret then set aside millions of dollars just to fund university research. This funding was multiplied in its effects by a central research group that coordinated the company funding with consortia and grant funding agencies such as NSF. There was a need to keep the internal research group of the company for specific projects as well as a mechanism for bringing the university research into the company in a productive way. The long-term outlook of 3–5 years was necessary for this to be successful.

Academic research provides a large high quality and motivated pool of talent, in the form of professors and research students. This talent can be applied (or more accurately guided) to areas of interest for the company. Also, the research interaction gives a first shot at hiring talented students. This model works for large corporations, but other factors must be considered when a company decides to take this approach. One is that the company must be (or be moving to be) a technology leader. If a company is a technology follower that has the business model of doing what others have done but just better or cheaper, investment in long term university research is not productive. Another key requirement for successful university research investment by a company is identifying the leading professors in the target area. Successful research cooperation between a company and a university is very dependent upon connecting with the specific professor, not just some big-name university in general. Another essential condition for the company's involvement in university research is to leverage other funding agencies (NSF/DARPA/DOE) to amplify government funding. This requires active involvement in research review boards that the agencies have. The government can take more risks, so the leverage is important. However, it is also required that the company recognize any changes in government funding strategy. Finally, an essential part of a company's engagement with university research is active participation in relevant technical conferences. Among researchers (both corporate and academic) the personal interaction is essential for the company to stay connected with the state of the art or forthcoming disruptive technology.

A company must tap into research activities for several reasons. One is to challenge and motivate internal researchers. Another is to avoid getting disrupted by another company. Research professors also give eyes and ears for future disruptive potential.

Questions:

(1) How should an R&D manager pick a university to collaborate with?

(2) What were the common characteristics described for a successful engagement?

(3) What were the characteristics that were different for the successful engagements described?

(4) What is common to research engagement and hiring engagement?

(5) What changed to cause Intel to switch from an ad hoc approach for university engagement to a centralized and coordinated approach?

(6) What are the required characteristics for a large company to choose university engagement?

(7) What characteristics might be required for a small company to benefit from university engagement?

CHAPTER 12

Information Security

> Security is more than encryption, of course.
> — Bruce Schneier [1]

Information security protects organizational structure details, development process techniques and procedures, personnel assignments and capabilities, and technology development schedules or technical details from potential attackers. Information security is a corporate-wide issue and must be part of the R&D organization's culture. Individual privacy protects personal information. National and international laws mandate security and privacy requirements in addition to the security and privacy policies and practices for the corporation. The manager of the R&D organization has an important role to play in prioritizing security resources and that requires a rudimentary understanding of information security concepts. As a result, we devote this chapter to that topic.

The ideal goal for information security is to prevent successful attacks or to mitigate any possible losses when an attack succeeds. However, in practice, information security involves tradeoffs between the cost of protection versus the combination of the probability of the attacks and the loss due to attacks. This requires a rigorous risk assessment to determine the probability of success for each type of security attack, the loss due to each successful attack, and the cost to prevent the attacks. One needs to stay vigilant in regularly assessing the probability as attackers are always finding new and innovative means to attack.

Successful information security requires management attention, operations support, technological installations, and most importantly the correct culture. Physical and operational boundaries must be set for the protection of information, but the protection cannot depend only upon the perimeter mechanisms. Similarly, the R&D manager must include information security in the continuous strategic planning loop described in Chapter 6 to ensure evolving threats across the entire R&D ecosystem are understood and properly mitigated.

12.1 The Importance of Information Security

Information security is more important than ever.

In an article written by Joseph West [2], he states:

> "There have been a lot of changes in the world of technology over the past few years. As a result, businesses have more information than ever before on the internet. Indeed, many companies rely on the internet to carry out their daily operations. Many companies could not survive without the internet and it is important for companies to think about finding the best web hosting. Furthermore, it is also important to think about the internet when it comes to data security. This is more important than ever. There are a few reasons why data security is so important.
>
> ...
>
> One of the reasons why data security is so critical is that the threats from hackers and malware are greater than it ever has been in the past. Just like technology has improved the daily operations of businesses, hackers have changed the way they attack data as well.
>
> ...
>
> They can even hijack a company's server and release all of its sensitive information to the public.
>
> ...
>
> One of the other reasons why data security is so critical is that privacy matters.
>
> ...

Another key reason why data security is so important is that data security and integrity go hand in hand. When people talk about integrity in the world of data security, they are talking about the accuracy and reliability of the data itself.

...

Furthermore, data security is important from an accessibility perspective.

...

Finally, data security is important because it is everyone's responsibility. Whether intentional or not, it is possible that employees might end up releasing confidential information to people outside the company. In order to prevent this from happening, data security is the responsibility of everyone. Companies should educate employees on how to protect their data, how to store it, and who can see it. Companies also need to make sure that everyone knows what company information is sensitive and what isn't. In this manner, data security is important because it imparts upon everyone the responsibility of data protection."

In our book, a security attacker is an unauthorized entity (a person, group, or organization) who intentionally attempts to: gain access to information, modify information, block authorized access to services, disrupt communication, or delete information.

Information security protects organization, process, personnel and technology information.

The information to be protected includes information about the company and R&D organization, the unique or critical processes and procedures for the R&D organization, the personnel within the R&D organization, and the technology plans as well as the technology itself. For example, information about the technology could be the release dates for various features as well as the testing and evaluation procedures and results. An example of the protection of the technology itself is preventing the insertion of malicious code or hardware. As one can see, the word security can cover many items.

The point of information security planning is to set the direction, define the measurements and metrics, and describe the means to achieve it. To accomplish this, the culture and actions of the R&D organization must constantly sample the environment, challenge the assumptions and be prepared to quickly respond to events and new information. As already referenced in Chapter 6 Dwight D. Eisenhower stated,

"Plans are worthless, but planning is everything."

To paraphrase that quote for security:

"Plans cannot encompass all possible security attacks, but planning for security attacks is essential preparation for the identification of the vulnerabilities and comprehension of the state of the art of the attackers. Additionally, possible responses to an attack must be identified. Plans become stale quickly but planning equips the R&D manager with the latest knowledge to determine the most effective response."

12.2 Privacy and Information Security

Privacy is the protection of information about individuals.

Specific employee information requires the same protection as the topics of the previous section. One important issue is who has access to the information. Access control requires identifying the access requestor. Identification of the requestor can then be linked to personal information providing an avenue for privacy violations. The tradeoff between gathering personnel information for legitimate uses and protecting that information from illegitimate observation is an issue in the tradeoff between privacy and security. For example, access control, such as logging in requires the access requestor to provide something for the access granter to use to decide whether to grant access. By using a login name and password pair the access requester submits information and the access granter determines whether the information matches and allowed list. A stricter access request could include birthdate, mother's maiden name, current address, birthplace, and spouse's name. While these facts help authenticate the requestor, they also reveal private information. This is a tradeoff between increased security of access control and decreased privacy.

Another example of security utilizing a variety of levels of access control is information about the person's pay and review history. Here, the access control is not only limited to the accessor having the entry codes, but having a position that has the authority to access the information. Also, a manager at the correct level to access the information may still not be permitted access unless the employee is in their line of reports, or the manager is evaluating a transfer. In this case, access is also limited by the need to know.

Another example of limiting access based upon the need to know is the medical history of a person. The manager and fellow employees do not need this information, but the company's insurance provider and representative might. These are cases where the employee's privacy is very important, but some information must flow securely for the company to function.

As one can see from these examples, several issues exist to properly manage security while protecting privacy. Some can be generalized and some will be very specific.

Information security is necessary waste.

Protection of information security typically does not directly add value to the technology, unless the technology is explicitly an information security technology. Information security might also add value if the employees and suppliers consider this a gating criterion for them to work for a given company. However, information security is necessary for the survival of the company and the success of the R&D organization. This makes information security protection efforts a necessary waste. In other words, the customer will not pay more for information security (it is a baseline expectation) once the proper level of security is achieved. Once it is attained, adding more people to the task will not create more value. Ideally, the proper level of information security can be achieved with zero effort so resources can be spent on value-added tasks elsewhere.

The R&D manager must prioritize many different types of necessary waste including information security. Individual elements and functions for security risk assessment must be evaluated with both the likelihood of occurrence and the loss due to occurrence. Different information security attacks have a different likelihood of occurrence and different costs if the attacks are successful.

12.3 Implementing Information Security

Information security must be implemented throughout the company.

The R&D organization's security is limited by the weakest link. Therefore, the entire company must have a culture, technologies, practices, policies and procedures that include information security. Ideally, employees protect information naturally so that security attacks fail without any special effort, and should a security breach occur, it is quickly identified and responded to limit the loss. Information security encompasses much more than specific tasks during the R&D organization's typical learning cycles, however, the R&D organization must utilize those learning cycles to improve information security while also learning from the asynchronous events that typify information security failures.

One problem that occurs with successful information security protection is that past success can breed lax future behavior. Unfortunately, for information security, there is no ideal state of complete safety from attack because the attacks and attackers are constantly dreaming up and executing new threats. The analogy is an arms race where the defender achieves success, the attackers respond with something better and then the defender improves its defenses (and so on). This is how it is with security, once an attack has been defeated a new attack is devised.

The security policies and framework require a risk assessment.

Security must be formalized so each group and individual has a common basis for terms, practices, and priorities related to information security. Although the success of the security policy is dependent upon the culture, the policies and procedures provide the execution mechanism for success. Similarly, although there is not a perfect ultimate state for security, a target state must be identified. The first step is an assessment of the current state of information security within the company and within the R&D organization. This step includes identifying all of the information assets that require security. These information assets are then evaluated for levels of sensitivity. Risks are assessed and procedures for improvement are identified. The key to any information security environment is continual monitoring and assessment.

The National Institute of Standards and Technology (NIST) provides guidance on the process for the evaluation of security with an emphasis on the IT organization. The company has the responsibility for security and privacy for corporate-wide information, such as payroll, other personnel data, corporate financials, legal issues, etc. The concepts apply to R&D organizations as well. The R&D manager is responsible for the security of the technology information created and utilized within the R&D organization.

The first concept is to rate topics based upon the security triad of confidentiality, integrity, and availability. Each of these is assessed based upon the probability, cost and impact. It starts by creating a list of processes and classes of information. Then, each item is assigned a risk or a probability of occurrence and an impact or damage of occurrence for the security triplet of {confidentiality, integrity, availability} with a value of high, medium, or low. Confidentiality means protecting the information from unauthorized observation. Integrity means protecting the information against change, Availability means ensuring the information is available to authorized entities.

For example, consider the circuit schematics for some standard components. We are going to look at risk (probability of occurrence), impact (damage) and cost of prevention. For the risk factor triad, the triplet is {confidentiality = low, integrity = low, availability = low}. The risk factors are low because it is unlikely that a security breach on the standard circuit schematics will occur, although the schematics themselves needing to be shared might leak to unauthorized entities. For the impact triplet the values might be {confidentiality = low, integrity = high, availability = low}. Although the leaking of the schematics or the access to the schematics would have a low impact due to a successful security attack, an integrity attack (e.g., Inserting errors into the circuits) would have a very high impact on the R&D output. In this case, the cost of prevention is relatively minor if the protection of this content is part of the culture.

Now, consider the example of a development plan of a new highly competitive piece of intellectual property (IP) being developed in conjunction with an outside vendor. In this case, the risk triplet for security may be {confidentiality = high, integrity = medium, availability = medium}. Here, because some of the IP is stored in a part of the R&D organization that might be accessed by another vendor who is a competitor, the risk is high that a security breach might leak information. The impact triplet for this case may be {confidentiality = high, integrity = high, availability = high}. In this case, the cost will depend on who needs access to the information.

As one can see in the two examples just given and summarized in Table 12.1, a successful security breach might have serious negative consequences depending on whether the information is leaked, modified, or made inaccessible.

After the assessment of the risk and impact is performed the cost tradeoffs for protecting against the attacks are applied to

Table 12.1. Security assessment for general schematics and proprietary IP.

	Risk			Impact		
	Conf.	Integrity	Avail.	Conf.	Integrity	Avail.
Standard Schematic	medium	low	low	low	high	low
Highly Proprietary IP	high	medium	medium	high	high	high

prioritize which items need security improvements. These assessments are then performed regularly in addition to whenever a security breach has occurred. In many ways the application of the continuous improvement loops described elsewhere in this book for strategic planning, research and development are also important for security assessment.

12.4 Operations, Technology and Culture

Security involves management, operations, technology and culture.

The word security can cover many items and is typically defined based upon what is to be protected and how it will be attacked. The emphasis in this book is on information security. As we noted already, the common triad of information security is confidentiality, integrity, and availability. This is sometimes referred to as the CIA for information security. Additionally, other terms also relate to security (a small subset of examples is access control, nonrepudiation, and response).

There are four types of activities that support information security: managerial, operational, cultural, and technical. It is important that the manager of the R&D organization does not perform the technical activities nor should the technical people perform managerial functions. For example, the R&D organization manager should not be writing code and the software engineer should not be allocating total headcount for the project.

There are two general types of information that must be protected within an R&D organization. The first type of information covers the process, task requirements, organization, partners, and schedule. The manager must prioritize funding the security tasks that target improving the security of this type of information against other needs and then assign the appropriate resources. The operational part of the security task is to ensure that the required activities are executed properly within the existing R&D organization's processes to conform with the corporate and external (national and international) security standards. The technical part of the security task is the implementation and integration of the related methods and tools within the existing R&D organization's engineering flows to meet the security needs while minimizing the necessary waste. Finally, the cultural part of the security task is to educate employees on the clear value and application of the new information security tasks so that they apply the prevention, checking, and response concepts as required rather than seek workarounds and shortcuts.

The second type of information relates to the implementation of a security feature within the R&D organization's engineering technologies. Here the managerial part of the task is to prioritize the feature with the customers and assign appropriate resources. The operational part of this type of task is to properly insert the security-specific features into the R&D organization's technology. This includes the associated planning, implementation, delivery, evaluation, and improvement. The technical part of the task is to implement the software code or hardware design changes. The cultural part of the task is to encourage all employees to seek out and report weaknesses and then propose fixes to the implementation of the security features.

It is important to emphasize that the activities to create the correct culture are critical. If the culture does not properly

Table 12.2. Summary of roles and tasks.

	Managerial	Operational	Technical	Cultural
Process task requirements, organization, partners, and schedule	prioritize	create and utilize new tasks within existing processes	integrate methods and tools	educate and integrate the value of security tasks
Implementation of security features	assign resources	Implement security features into technology	write the code or design the hardware	spot and report security weaknesses and propose fixes

support information security then all of the managerial, operational and technical efforts to provide security will fail miserably. As has been stated earlier in the book, "culture eats everything for breakfast". Table 12.2 summarizes the respective roles and information security tasks. The managerial, operational, and technical roles are based upon NIST standards [3, 4, 5, 6], and we added the essential role of culture, some of which NIST includes in the other categories.

12.5 Information Security and the Law

Management must meet legal requirements for security and privacy.

An important consideration for any R&D organization is the national and international legal requirements for both information security and personal privacy that comprehend the entire R&D ecosystem shown in Figure 12.1. The figure highlights the security perimeter around the R&D organization (the heavy solid line) and the company (the lighter dotted line). Notice

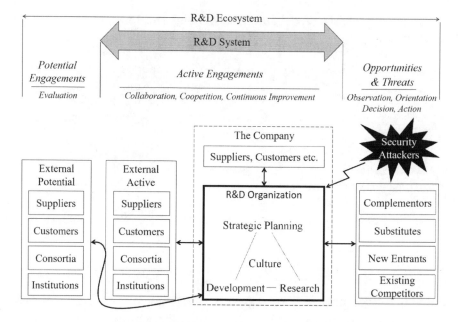

Figure 12.1.   The R&D security organization perimeter.

that much of the R&D system is outside of the security perimeter of the R&D organization and the company. Some functions that interact with other entities by their very nature are not secure. There are many information security and personal privacy issues that exist within and across the security perimeter. These include external forces such as governmental legal requirements and malicious attackers. The impact and effect of these external forces may penetrate the security perimeter. An example of a legal requirement penetrating the security perimeter is when police officers are serving an arrest warrant within the building. In that example, the police officers will be able to see plans and proprietary information.

One key legal requirement for information security is the controls on the flow of information to identified countries (and hence citizens of those countries). Just a taste of these issues will be provided here to remind managers that the complexity of this type of information security must be analyzed, planned and monitored for each specific situation. The US government lists export-controlled information and technology for which an export license is required as well as the countries and entities for which the controls apply. The information that is controlled targets very specific technology or related information and the countries are often places where normal business is otherwise going on. As a result, it very possible that the R&D organization may have operations in some of those countries.

In other cases, the restrictions are more severe and very comprehensive. Since there is no central place that coordinates all of the information on the restrictions placed on countries and other entities the R&D manager must understand the relevant sources that affect their work and stay abreast of changes. For example, as of 2019, the Office of Foreign Assets Control Regulations (OFAC) and the Export Administration Regulations (EAR) [7] are examples of two out of many government agencies that list countries that have comprehensive restrictions, such as the Crimea Region of the Ukraine, Cuba, Iran, North Korea, and Syria. As noted by the US Department of the Treasury website [7]:

> *"The Office of Foreign Assets Control (OFAC) does not maintain a specific list of countries that U.S. persons cannot do business with."*
>
> *Here's why:*
>
> *U.S. sanctions programs vary in scope. Some are broad-based and oriented geographically (i.e. Cuba, Iran). Others are "targeted" (i.e. counter-terrorism, counter-narcotics) and focus on specific individuals and entities. These programs may encompass broad prohibitions at the country level as well as targeted sanctions. Due to the diversity among sanctions, we advise visiting the "Sanctions Programs and Country Information" page for information on a specific program.*
>
> *OFAC's Specially Designated Nationals and Blocked Persons List ("SDN List") has approximately 6,300 names connected*

with sanctions targets. OFAC also maintains other sanctions lists which have different associated prohibitions."

As a result, understanding governmental international information control restrictions are especially difficult for multinational companies and companies that have a diversity of international employees. As an example, consider technology companies and the list of "controlled countries". These lists from various parts of the US government have restrictions on the type of information that can be shared with each country. Applying this to a specific example of a semiconductor company, if one has 10 employees in a controlled country, they may not be able to see the latest process technology information, technology files, std cell libraries, etc. if the proper export license does not exist. For this reason, they need to be partially isolated from the rest of the organization for certain information (even email that refers to it). If the "controlled country" nationals move to the US and join the R&D organization, they will have similar restrictions until they become a citizen. This is an example where one needs to provide fine-grain security even within the same group to address information security regulations for a specific case.

12.6 Perimeter Defense is Insufficient

A perimeter cannot be the sole or primary information security mechanism.

Traditional information security created a perimeter and relied upon the perimeter for protection and presumed everything within the perimeter was secure. Figure 12.2 is a simplified depiction of Figure 12.1 that shows this perimeter around the R&D organization and the larger one around the company. This simplified information security to a problem of access control. The simplest example of a perimeter is a building with limited access. All of the secure information is kept within the building and only the doors and windows need to be protected. For

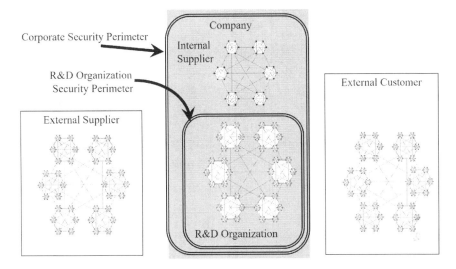

Figure 12.2. **Perimeters around the R&D organization and the company.**

electronic information, the simplest example of a perimeter is a private isolated computer network so that only computers on the isolated network have access to information.

The R&D organization has several mechanisms for information to pass back and forth across the perimeter, such as people, physical documents, and hardware storage devices. It is easy to see how policies and practices might still allow information to slip across the clear physical perimeter of the R&D organization via networks or other electronic connections. Examples of how information "slips" across the perimeter can occur include emails with attached documents, discussions during conference calls, video calls that show diagrams on the walls, employees carrying flash drives with secret information, etc. Similarly, it is straightforward to identify the risks associated with this type of security breach. Evaluating the protection effort for these cases should be done. However, this is the smallest information security vulnerability of the R&D organization.

Let us consider the case of securing intellectual property (IP) within a technology. When the technology is delivered to the customer of the R&D organization it contains some proprietary IP. If the customer is internal to the company, the IP protection is defined by the company rules. However, some extra tracking is required if the internal customer incorporates the R&D organization's technology into its technology that is subsequently sent to other customers. Similarly, when the R&D organization's technology contains IP from an internal supplier it must be tracked and appropriately controlled. Both of these cases get harder as the R&D organization technologies continue to evolve and the IP contained becomes so well integrated as to be difficult to identify or distinguish.

The next more complicated situation includes IP from organizations outside of the R&D organization's company. In the case where the IP is to be included from and external supplier into the R&D organization's technology, a legal agreement such as an NDA (Non-Disclosure Agreement) will likely be signed by both companies. The NDA will describe not only the protection of the IP itself, but may also include all related collateral documentation, such as operational test cases, application limitations, users' manuals, and design hints. The NDA also designates the access limitation for the IP, such as which internal employees and which external people have what levels of access.

An interesting scenario exists in some companies where intellectual property from different suppliers is integrated into the final technology. The incorporating company must get NDAs signed by supplier A to enable some people from Supplier B to provide support on the integrated technology (and vice versa). When you have 100 technologies integrating 100s of IPs from 5–10 different suppliers this becomes a difficult thing to manage. Legally there are approaches like the use of a master NDA that covers a complete set of IPs and is easy to update but the technical problems related to tracking them still exist as technologies proliferate. In some cases, the information and relationships that are being protected are so important that even the existence of an NDA cannot be disclosed. The R&D organization must have the policies and culture to make sure the wrong people do not accidentally get access in the future on a derivative technology that happened to be using an IP from a previous technology.

NDAs cover the legal requirements for information protection of Intellectual Property (IP), however, actual protection of the information requires information security policies, procedures, and a supportive culture to protect the information. In the simplest example, an R&D organization identifies some external IP and all of the employees are careful not to share the IP or anything about it with an inappropriate person or entity. On the other hand, an extreme case of protections prevents anyone who has access to the IP to ever interact with someone not authorized to use the IP. For example, imagine that an external supplier or development partner provides IP that is similar in function to some other IP used within the R&D organization's company that is viewed as a potential competitor. In this case, the external company may have very strict rules in who can see the IP details and require that those people not interact with anyone from parts of the company engaged in similar work for one or more years. This is often referred to as "refrigeration".

No R&D organization can have perfect information security

An R&D organization must communicate with other organizations (both inside the company and external to the company). Most of these organizations do not have the same understanding the R&D organization does regarding which information needs to be protected and at what level.

Any successful attack on information security is a failure. However, the cost of preventing or responding to a security breach should be commensurate with the damage of the breach. For example, consider a file containing a list of the engineering technologies used by the R&D organization. This is confidential information that the R&D organization does not want competitors or suppliers to see. One has a choice of extensive encryption and access control wherever the information resides or simple protection of keeping the information away from public storage. However, if a supplier or competitor does see the information, it would be a relatively small detriment to the R&D organization. Therefore, the cost to protect this information may exceed the damage should the information be compromised. In this case, extensive protection efforts would create excessive unnecessary waste. On the other hand, suppose a file contains the design details for a critical feature of technology from the R&D organization. Now a malicious party might get access to the design file and insert an error. In this case, the technology is in jeopardy, and the cost of the security breach is the cost of the technology. More extensive protection and monitoring are called for in this case.

Risk assessment is done by categories as described in the Federal Information Processing Standards (FIPS) and identifies the impact according to the traditional security triad of confidentiality, integrity, and availability. For each information category (such as employee information, medical information, design information, and market information) the risk is rated (high, moderate, and low) relative to the impact of a breach of confidentiality, integrity, and availability. For example, if an attacker gets a copy of a design file, the impact is high on confidentiality but low on integrity or availability, However, if an attacker inserts false information into the design file, the impact is high on integrity, moderate on availability, and low on confidentiality. As a result, the manager must direct the risk assessment team to identify all of the information to be secured, assess the impact of security failures, and propose appropriate fixes. The manager then considers the costs of the fixes, the impact of the security failures, and the likelihood of successful attacks to decide on the plan of action.

Technical solutions are only part of information security.

Defense against security attacks on an R&D organization requires multiple approaches to address the different types of risks and security tasks. As Bruce Schneier [8] wrote:

> *"If you think technology can solve your security problems, then you don't understand the problems and you don't understand the technology."*

As we discussed, the approaches can be managerial, operational, technical, and cultural. With this in mind, different people in the R&D organization are appropriate for different tasks. For example, it would be counterproductive to have managers perfuming technical tasks and a mistake to have technical people performing managerial tasks. The assignment of types of tasks and appropriate people were already shown in Table 12.2.

12.7 Some Information Security Technology Basics

The R&D manager needs a basic understanding of security concepts.

As mentioned, information security depends upon far more than the technical components. To provide R&D organization security, a manager needs enough of an understanding of the different components and types of attacks to assess the relevant risk and cost to allocate the resources that minimize waste while keeping the risk and cost of failure at an acceptably low level. This part of the chapter will describe several basic security concepts about which the manager needs to be aware of. This is not an exhaustive list. Such a list must be constructed using a standard (such as NIST SP 800-60) that is applied to the specific situation. Nor is it a complete description of technical details, as that should be left to the specific implementers. However, this section should be sufficient to understand reports and proposals related to the R&D organization's security plans.

The first step in any information security system is access control. Access control requires evaluating a request for access as either authorized or unauthorized. The strictness of the decision is based upon the tradeoff between the cost of loss due to mistakenly granting access to an unauthorized request vs the cost of correctly granting access to an authorized request. Figure 12.3 shows a 2×2 matrix of the access decision versus the actual authorization of the request.

If an authorized request is granted access, it is a true positive (TP is the number of true positives). When an unauthorized

	Authorized Request	Unauthorized Request
Access Granted	☺ True Positive	☠ False Positive ☠
Access denied	☹ False Negative	☺ True Negative

Figure 12.3.   Access control responses to requests.

request is granted access, it is a false positive (FP is the number of false positives). If an authorized request is denied access it is a false negative (FN is the number of false negatives). When an unauthorized request is denied it is a true negative (TN). With this in mind, the fraction of correct evaluations is given by:

$$\frac{TN + TP}{TN + FN + TP + FP}$$

Stricter authentication requirements will have lower false positives, but also higher false negatives. On the other hand, laxer authentication requirements will have higher false positives but also lower true negatives. The false-positive rate and false-negative rate can be calculated using the equations below. The trade-off between them is depicted in Figure 12.4.

$$FPR = \frac{FP}{FP + TP}$$

$$FNR = \frac{FN}{TN + FN}$$

For a very secure application, such as opening a bank vault, one accepts a high denial rate (i.e., high FNR) because the cost of letting unauthorized access is much greater than the cost of denying authorized access. However, for casual applications such as walking into a convenience store, one accepts a high false-negative rate because the cost of turning away many paying customers is much greater than letting in a few unauthorized individuals.

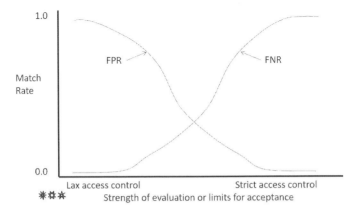

Figure 12.4. False-positive rate (FPR) and false-negative rate (FNR) tradeoff.

Traditional authentication is based on three factors:

(1) Information, such as a password or an account name
(2) Possessing a physical item, such as a credit card or a car key
(3) Physical characteristics or biometrics, such as fingerprints or voice

Given the expansion of the internet and the use of mobile devices, a fourth factor has been added:

(4) Location, such as GPS

These factors can be remembered as:

(1) What you know
(2) What you have
(3) What you are
(4) Where you are

Most identity authentication is a simple account name and password. This is a simple single-factor authentication based upon what you know. Using a bank card (what you have) and a

PIN (what you know) is an example of two-factor identification. The added factor increases security, but it also increases the effort. In the R&D organization, the access control must be set to a sufficient level of security based upon the effort or cost to implement and use the control relative to the cost or loss due to a security failure.

After access control, a key technical element of security is the protection of data. The implementation of policies, practices, and procedures to protect data depend upon some underlying technical concepts. One of these concepts is encryption. Remember that technology, and in this case, encryption is only part of the solution for information security within an R&D organization. That being said, it is still important for the manager to understand some fundamental encryption concepts to properly allocate resources for information security. There are two basic forms of encryption: private (symmetric) key encryption and public (asymmetric) key encryption.

Symmetric encryption uses the same key for encrypting and decrypting. This is sometimes called private key encryption. The current standard for symmetric encryption is the Advanced Encryption Standard (AES) [3]. In this case, the message m, called the plaintext, is input to an encrypting algorithm which applies the key to yield the encrypted message c, called the ciphertext. Then the ciphertext c is sent to the recipient via open communication. An eavesdropper can see the cyphertext but cannot read the content without the key. The recipient applies the key to the ciphertext with an inverse of the encrypting algorithm to retrieve the plaintext. The security of symmetric encryption is dependent upon protecting the key. This is called key management.

Asymmetric encryption uses a different key for encryption than for decryption. This is sometimes called public-key encryption. Common asymmetric encryption is called RSA (Rivest–Shamir–Adleman, the last names of the inventors) [9]. The encryption key is made public so that anyone can encrypt the plaintext to ciphertext. The ciphertext can only be decrypted by one possessing the private key. This simplifies the key management issue of protecting the private key.

In addition to using encryption to protect sensitive information from unauthorized reading sensitive information must be protected from malicious modifications. Secure hashing is used to identify whether information content has been modified. The idea is to create an "electronic signature" that is calculated based upon the contents of the secure document and the author or owner of the document. A common method for hashing is SHA-3 (Secure Hashing Algorithm 3).

Finally, the technical aspects of information security include types of attacks, types of encryption, and methods of authentication. As we have noted earlier, some technologists categorize security risks in terms of confidentiality, integrity, and availability (or CIA). In the security world, nicknames are often also given to the participants (Alice, Bob, Eve, and more). The manager should understand these aspects of security as well and engage the correct specialists and create a security-conscious culture.

In summary, there are many security concepts and not all security problems are the same or have the same solutions. This makes an employee culture that values information security very important.

12.8 The Importance of a Security Culture

The culture must include information security awareness.

The R&D organization's culture must fully embrace the importance of everyone comprehending information security, being aware of threats and acting to mitigate them. Efforts to train people and implement information security processes must not be perceived as the "hot topic for the day." Information security is not a one-time project, nor is it the responsibility of a specialized group. Like other processes, information security requires continuous vigilance and regular improvement. Although some security specialists are required to train and facilitate information security, the implementation of security concepts and effective protection of information must permeate the entire R&D organization's culture.

Engagements between academia and the R&D organization present special security challenges. As described in Chapter 11 , academia expressly values sharing information while the R&D organization values protecting proprietary information. In Chapter 10 we also discussed the importance of instilling a culture that enables collaboration between R&D organization and other entities in the R&D system (such as academic institutions). The major factor in ensuring successful security practices in this type of engagement is understanding the respective cultures, appreciating each partner's perspective, and modifying each participant's practices to protect what needs to protected and share what needs to be shared. Although good policies and procedures need to exist, the different cultures must understand the perspective of their collaboration partners.

Finally, as we noted in Chapter 6, a change in the competitive environment, new government regulations, suppliers, etc. can all make carefully assembled set of assumptions crumble. Like other aspects of the R&D organization, information security requires a strategy that comprehends the forces that act on it. Therefore, security topics should be included in the continuous strategic planning loop and utilize the learning cycles and continuous improvement processes embedded in the engineering workflows to reduce waste.

12.9 Summary

Information security means the protection of organizational, process, personnel, and technology information while ensuring privacy means the protection of information about individuals. The point of information security planning is to set the direction, define the measurements and metrics, and describe the means to achieve it. Although no company can maintain perfect information security, continuous monitoring and planning for attacks will help the R&D organization respond quickly and effectively. This also means establishing a culture where all employees are aware of the importance of information security so they can block threats at the source. With this in mind, the manager of the R&D organization must have a good working knowledge of information security and privacy technology and practices to properly allocate the resources needed and to establish the right balance of access control. This also requires setting the right culture, conducting regular risk assessments and establishing the right legal and technical framework for managing IP entering and exiting the company. Similarly, the R&D manager must include information security in the continuous strategic planning loop described in Chapter 6 to ensure evolving threats across the entire R&D ecosystem are understood and properly mitigated. With respect to information security, eternal vigilance is required more than ever since, "only the paranoid survive".

12.10 Key Points

- Information security protects organization, process, personnel and technology information. Privacy protects information about individuals. Information security must be implemented throughout the company.
- Security involves management, operations, technology and culture.
- The security policies and framework require a risk assessment.
- No R&D organization can have perfect information security and a security perimeter cannot be the sole or primary security mechanism.
- Technical solutions are only part of information security. The organization's culture must include information security awareness.
- The R&D manager needs a basic understanding of security concepts.

12.11 Case Studies

12.11.1 *Insider Threat*

Information security includes policies and procedures to prevent attackers from gaining access to any proprietary information anywhere, including any internal networks. The technical support for controlling access to the system includes firewalls, user authentication (with passwords or multiple factor authentication), physical protection of servers and password files, and special network address assignments. Of course, there are also different levels of access. System administrators must have access to much more of the computer network than the typical user.

Many years ago, an employee of ours was working on the weekend and had brought a friend to visit while they worked. That Monday, someone from the corporate IT group called the person's manager and said some suspicious activity was observed on one of the department's computers. An investigation was started and the manager met with HR, a company lawyer, and the IT security person to discuss the options. That afternoon the manager met the employee, asked some questions and put them on paid leave during the investigation. The next day the employee called the manager and said they confronted their friend and found that they had accessed the system while at the company, and knowing the standard locations for Unix password files, they downloaded the files, took them home and later cracked the encryption with some standard hacking methods. They then accessed the system. As a result of this, the employee was terminated. Unbeknownst to the former employee, the friend was a well-known Silicon Valley hacker who believed that all code and computer information, once the people are paid to create it, should be free and open to all.

This example is a variation of the classical insider threat. Usually, the insider threat is an employee or person with authorized access to the information who misappropriates the data. In this case, it was a naïve employee allowing a trusted person to access the system.

Questions:

(1) Since the employee came clean and confessed, should they have been terminated?
(2) What policy could have prevented this?
(3) Making a system secure does not preclude the need for constant vigilance. What technical monitoring is needed to detect security breaching?
(4) Some places have installed cameras all over the workplace. When is extreme monitoring appropriate and when is it overboard and a violation of privacy?
(5) What key component of corporate culture is needed to avoid this particular situation?
(6) What technical mechanisms can be used to prevent insider threats to information security?
(7) What policies, training, and management practices can be used to prevent insider threats to information security?
(8) What can corporate culture do to prevent insider threats to information security?

12.11.2 *Wide Mix of Data Security*

Organizations have a wide variety of data and information requiring a wide variety of security measures. The national nuclear weapons laboratories create and have access to very sensitive information. They have double fences topped with barbed wire around all of the buildings. The access gates have armed guards that check each individual's badge upon both entry and exit from the secure area. There are clean desk policies that mean that at the end of the workday no material can be laying on the desk or any flat surface to prevent accidentally leaving classified documents unattended overnight. Guards check the buildings at night and report any violations. Employees that have classified documents must have a safe in their office to lock up the documents at night. All of the employees are required to have top-secret clearances and new hires are investigated by the FBI and until their clearance comes in, they are not allowed to work in the secure areas without and escort. Even if you had a top-secret clearance, you were not allowed access to classified documents unless you had a clear "need to know".

Even in this environment, some work is not classified, so there are some buildings outside of the secure area where routine activities can occur, such as the parking lot or a cafeteria. The employees are strictly instructed that no classified discussions are allowed in the uncleared areas. Also, the integrated circuit design methods and tools are not classified while the design details related to design to a classified program are classified. This creates a mix of completely open information (such as a logic simulator and its user manual) and completely secret information (such as the parameters for a specific VLSI circuit and its application or uses).

In the late '70s and early '80s, most integrated circuit design and the associated Computer Aided Design (CAD) tools were developed in house. These were versions of generic tools such as the SPICE circuit simulator from UC Berkeley, or they were versions of tools also available from suppliers such as a logic simulator, place and route tool, or a testability analysis program. These in-house tools and algorithms were developed with significant interaction in the open research community via conferences and joint university research efforts.

The mix of open and secret information presented a problem supporting highly classified projects that required custom integrated circuits. Those projects needed the development of new features in the "in-house" CAD tools to do their work at the same time that similar design tools and algorithms were being developed and designed by different researchers in academia. To keep up to date with progress happening in there, the internal CAD tools developers needed to attend conferences and interact with researchers working on similar IC design concepts. During this time, a professor from Stanford University visited the laboratory, but because he was a foreign national, he had to have an armed guard follow him everywhere — even to the restroom. Needless to say, this was a bit embarrassing to the hosting CAD organization and not conducive to setting a culture of collaboration. To overcome this problem in the future, the labs then set up some conference rooms external to the secure area for such future events.

It was in such a special meeting place that the lab's logic simulation developers also began to meet with external users of the logic simulator. This was before the advent of commercial CAD companies, so the lab's logic simulator (SALOGS) had been released to outside companies, such as Tektronix, which modified it and used it as their internal logic simulator. The external interaction with academia and companies (like Tektronix) was essential to the continuous improvement of the internal CAD tools. The interaction also benefited the external world as demonstrated by the testability analysis tools called SCOAP (Sandia Controllability Observability Analysis Program). The basic algorithms invented in the Sandia Labs CAD group that are still taught in modern VLSI design books and used today in modern test generators.

One method for advancing the internal tools involved the National Lab to send some of its developers to universities (for the CAD group it was Stanford) to study for their PhDs. This was done to ensure that state-of-the-art academic research could be brought back into a secure environment without the threat of leaks. The key to enabling the mixture of secure information and open public information was a culture that was embraced by the employees of the lab. Of course, policies were specifying the proper handling of classified information. And there were still practices to protect sensitive information such as searching briefcases. And there was technical monitoring of phone and data transmissions. But the culture reinforced the individual sense of responsibility.

Questions:

(1) What does the employee need to do if they are not sure whether some information is classified or not?
(2) When is it safe to tell a secret?
(3) Compare the policies and procedures to protect national security secrets to corporate intellectual property secrets.
(4) Compare the damage due to national security leaks versus the damage due to corporate intellectual property leaks.
(5) How do you as an individual know what part of your work is secret or proprietary versus that which is open and needs to be discussed with outsiders?
(6) National Labs have armed guards — how should a company protect is intellectual property?

References

[1] Schneier, B. (2016). The Value of Encryption, Schneier on Security, April 2016. Available from: https://www.schneier.com/essays/archives/2016/04/the_value_of_encrypt.html [June 2020].
[2] West, J. (2020). Why Data Security is More Important Than Ever, iTECH POST, Feb 20, 2020. Available from: http://www.itechpost.com/articles/102013/20200220/why-data-security-more-important.htm [June 2020].
[3] NIST (2001). Federal Information Processing Standards Publication 197. United States National Institute of Standards and Technology (NIST). November 26, 2001. Archived (PDF) from the original on March 12, 2017. Retrieved October 2, 2012.
[4] NIST (2010). Special Publication 800-37 Revision 1, Guide for Applying the Risk Management Framework to Federal Information Systems: A Security Life Cycle Approach. Gaithersburg, MD: National Institute of Standards and Technology.
[5] NIST (2013). Special Publication 800-53 Revision 4, Security and Privacy Controls for Federal Information Systems and Organizations. National Institute of Standards and Technology. Available from: http://dx.doi.org/10.6028/NIST.SP.800-53r4 [June 2020].
[6] NIST (2017). Security and Privacy Controls for Information Systems and Organizations, Security Draft NIST Special Publication 800-53

Revision 5. August 2017. U.S. Department of Commerce. National Institute of Standards and Technology.

[7] US Department of the Treasury. Resource Center. Available from: https://www.treasury.gov/resource-center/sanctions/Programs/Pages/faq_10_page.aspx).

[8] Schneier, B. (2015). Secrets and Lies: Digital Security in a Networked World. (Wiley).

[9] Rivest, R.L. and Shamir, A. and Adleman.L. (1978). A method for obtaining digital signatures and public-key cryptosystems. Communications of the ACM 21, 2, February 1978, pp. 120–126. Available from: http://dx.doi.org/10.1145/359340.359342 [June 2020].

Other Reading

FIPS (2006). Minimum Security Requirements for Federal Information and Information Systems. Federal Information Processing Standards Publication, FIPS PUB 200. Computer Security Division, Information Technology Laboratory, National Institute of Standards and Technology (NIST). Gaithersburg, MD.

FISMA (2014). Federal Information Security Modernization Act of 2014. Available from: https://www.congress.gov/bill/113th-congress/senate-bill/2521 [June 2020].

ISO/IEC (2005). Information technology — Security techniques — Information security management systems — Requirements. International Organization for Standardization (ISO) and International Electrotechnical Commission (IEC). ISO/IEC 27001 First edition 2005-10-15.

Kohnke, A., Sigler, K., and Shoemaker, D. (2017). Implementing cybersecurity: A guide to the National Institute of Standards and Technology Risk Management Framework. CRC Press, an imprint of Taylor & Francis Group.

U.S. DOE. (2018). Family Educational Rights and Privacy Act (FERPA). United States Department of Education. December 2018. Available from: https://www2.ed.gov/policy/gen/guid/fpco/ferpa/index.html [June 2020].

U.S. DOE. (2018). Protecting Student Privacy, By Audience: Researchers. United States Department of Education, December 2018. Available from: https://studentprivacy.ed.gov/audience/researchers [June 2020].

CHAPTER 13

Leadership

> Optimism is about creating possibilities and pragmatism is about knowing probabilities, an effective leader needs both.
> — **Siva Yerramilli**

Throughout this book, we have highlighted the key responsibilities of the R&D manager when it comes to the foundation and core functions of the R&D organization and system. We dedicate this chapter to reinforce the critical leadership behaviors that must be developed and applied to successfully implement what we have discussed so far.

Put succinctly, leadership is a combination of dreaming and doing. It spans the fuzzy front end of articulating a big and bold vision, translating that into a strategy that creates differentiable value and then motivating, funding and orchestrating all of the entities in the R&D system to work harmoniously together to deliver results with the greatest velocity. To be an effective leader the R&D manager must balance the traits of curiosity, learning and having an open mind with a problem-solving mindset and a laser focus on results. Most of all, an effective leader knows that their success is based on the trust people have in them to do the right thing, the right way, for the right reason and they work hard to maintain that trust.

The terms "leadership" and "management" are described differently in the literature depending on what source is used. Andy Grove, Edgar Schein and Peter Drucker believe that management is an umbrella term for doing whatever it takes (including providing leadership) to deliver results within the bounds of good business practices and ethics. Meanwhile, John Kotter and others separate leaders and managers. Their view is that leadership is about change and management is about complexity. They believe companies are often overmanaged and under-led. We believe, the R&D manager needs to be both a strong manager and a strong leader. The act of leadership is required to motivate (the why) and set a clear direction (the what) and is a prerequisite to achieving the best results. On the other hand, the act of management is required to turn the direction into clear roles, goals, and enable the team to achieve the results (the how). Therefore, when we refer to either the leader or the manager we are referring to the same person. It is someone whose acts of leadership and management are performed with equal diligence.

13.1　Articulating Why, What and Enabling How

Leaders articulate what needs to be done and why and then enable it.

The primary job of a leader is to clearly articulate what needs to be accomplished and why and then become an active enabler to accomplish the goals. This is referred to as the "Why", "What" and "How". A model of that is depicted in Figure 13.1. In that figure, leadership skills are needed to establish the "why" and "what" and to establish a strong agreement between "what" and "how".

In Figure 13.1, "Why" captures the vision and reasoning behind it. This is an example of possibility thinking that starts

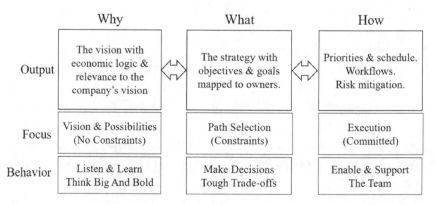

	Why	What	How
Output	The vision with economic logic & relevance to the company's vision	The strategy with objectives & goals mapped to owners.	Priorities & schedule. Workflows. Risk mitigation.
Focus	Vision & Possibilities (No Constraints)	Path Selection (Constraints)	Execution (Committed)
Behavior	Listen & Learn Think Big And Bold	Make Decisions Tough Trade-offs	Enable & Support The Team

Figure 13.1.　R&D manager behavior for why, what and enabling how.

without any constraints. It requires the R&D manager to listen, learn, think big and act boldly. "What" comprehends constraints and translates the vision into a feasible strategy with objectives and owners. It requires the R&D manager to make decisions based on tough trade-offs. "How" is where the execution of the strategy occurs and requires the funding and management of workflows and processes, to meet a given schedule. It also clearly defines the roles (the who) and when the primary tasks will be completed (the when).

We can also map the R&D manager behaviors onto the OODA loop for continuous strategic planning that we introduced in Chapter 6 as shown in Figure 13.2.

The rest of this chapter dives into the Why, What and How in greater detail and also addresses three topics that apply to all of those (balancing velocity and efficiency, learning, and people):

Section 13.2: Articulating Why
Section 13.3: Articulating What
Section 13.4: Enabling the How

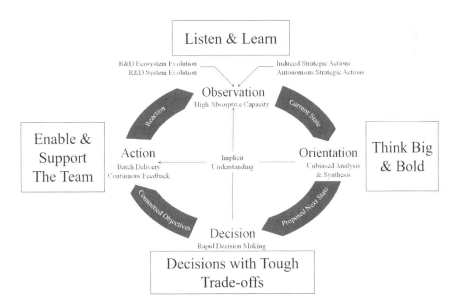

Figure 13.2. R&D manager behavior and the OODA loop.

The three topics in Section 13.5 apply to everything above.

Section 13.5.1: Balancing Velocity and Efficiency
Section 13.5.2: The Leader and Learning
Section 13.5.3: The Leader and People

13.2 Articulating the Why

Leaders are bold and think of possibilities when setting the vision.

A vision is a clear and vivid description of the aspiration of an enterprise and sets the context for the strategy of the R&D organization. As depicted in Figure 13.1, discussions in this phase are unconstrained and this is why it is sometimes referred to as the fuzzy front end. It may not be clear yet what it will take to succeed and that is okay. This phase is about dreams and unconstrained ambition; it is about thinking big and being bold. To do so requires the R&D manager to develop personal skills that enable them to anticipate, learn and adapt.

Leaders develop the ability to anticipate, learn and adapt to change.

As referenced in Chapter 5, "He who can handle the quickest rate of change survives". This applies to the R&D organization and also the manager of the R&D organization themselves. If the R&D manager develops the personal skills to embrace change faster than others, they will learn and adapt faster than their peers. When done properly, this can be professionally and personally rewarding. When this is not done then changes in the eco-system will surprise the R&D manager and will prove to be frustrating and detrimental.

A role model for personal learning is Bill Gates who reads scores of books and does other learning activities every year. There are also many examples of companies that benefited from continuous learning and adaptation. We have discussed some of them in other parts of this book. Corporate examples include Netflix who anticipated the need for a new business model and fundamentally changed its video rental business to take advantage of streaming over the internet while competitors like Blockbuster and others did not (and went out of business). Netflix continued to learn and has evolved further to become a premier content producer. They have evolved to develop core competencies to reduce their critical dependence on content producers while staying relevant to their core customer base. In many ways, the Netflix and Blockbuster examples show that *irrelevance is a much bigger problem than incompetence*. In other words, it does not matter how good you execute if your products and/or services are irrelevant. This also holds for the R&D manager as well who must develop three skills needed in that journey:

(1) Anticipation
(2) Learning
(3) Adaptation

1. Anticipation
People who identify an important change in the eco-system early can be the greatest beneficiaries. The key, of course, is to identify the correct change at the earliest possible time. This gives the R&D manager time to adapt their organization before the change arrives like a freight train. The earlier the R&D manager makes the observation, the more time they will have to drive necessary changes in their organization.

The first key trait the R&D manager must develop to enable anticipation is inexhaustible curiosity. Curiosity opens doors to new possibilities that may result in change. The second trait is direct observation. This allows the R&D manager to see the source of the change for themselves. The third trait is having an open mind. The combination of curiosity, direct observation and an open mind is key to developing the skill of anticipation. Intellectually understanding them is one thing, practicing them requires a tremendous commitment to developing the skills and traits required. We explain each of those three parts in detail.

Curiosity: This desire opens the gateway to learning. Learning something well is really hard work. Charles Duhigg, the author of *The Power of Habit* [1], argues that it takes a tremendous amount of energy to learn something new, far more than executing a pre-existing mental script or habit. Like in physics, more force is needed to get an object at rest moving, than is needed to keep it moving once it is in motion. True learning also requires a long-term commitment. The individual must realize that learning causes extreme mental fatigue and plan for it. It is best to learn something new when one is full of energy and mentally fresh and active (e.g., in the morning, after a good night's sleep). It is also important to keep at least a day allocated for learning every week.

With so many online resources, access to learning is easy; the hard part is scheduling the time. The plethora of easily accessible sources of information also presents a problem. Today, the ability to generate data often exceeds the ability to analyze it. With so much information available at the fingertips, it is hard to decide what is most relevant. Also, some information is just plain wrong or contradictory. Sifting through that can be time-consuming. As described in the introduction to *The Master Algorithm* [2], technology such as Machine Learning (ML) can sometimes help sift through enormous data to find patterns. If that is not applicable, then referring to trusted sources such as peer-reviewed publications to separate the wheat from the chaff is a good practice.

Direct Observation: One learns far more through personal direct observation than viewing charts and graphs and even learning from the observations of others. This is why people still pay money to watch sports in-person in a large stadium when they could also watch the same event on their television for a lot less. This remains true even though the TV watching experience is quite rich these days. When one watches sports live, one watches exactly what one wants to watch as opposed to what the TV camera operators and the editor of the TV program want you to watch. Watching the tennis star Raphael Nadal at Roland Garros in person means the fan can focus on how he moves his feet for an entire game if desired. In the next game, the fan can focus on how he moves his hands. The point of the analogy is simple: direct observation allows the R&D manager to focus on what is important and ignore distractions. It is irreplaceable when it comes to learning.

Having an Open Mind: One can't learn fully until they observe events with an open mind. In the words of Leo Tolstoy [3]

> *"The most difficult subjects can be explained to the most slow-witted man if he has not formed any idea of them already; but the simplest thing cannot be made clear to the most intelligent man if he is firmly persuaded that he knows already, without a shadow of doubt what is laid before him."*

One of the authors of this book once gave a lecture about having an open mind at a seminar with hundreds of people. He asked the audience if they have run into *close-minded people.* Almost all the people raised their hands. Then, He asked if they think they are *"open-minded."* Almost all of them raised their hands – maybe a couple fewer than for the first question. He asked the audience how it could be that none of the close-minded people they knew were in such a large audience? He then defined the state of having an open mind as being:

(i) *Stateless:* This means one brings zero knowledge or data to the observation. It is akin to having a fresh experience like smelling a rose as if one is smelling it for the first time, every time.

(ii) *Egoless:* This means checking one's ego at the door before one enters to make observations. This is easy to understand

and hard to do. This is akin to being a fly on the wall with no pride, no power, and no designations.

(iii) *Filter-less:* This means looking through a camera without any filters. It is akin to being unaware of the level of education, designations, accomplishments, gender, race, job type, country, etc. of the source of the information.

After articulating this definition, our co-author asked the audience if they were truly *open-minded*– none of them raised their hands. This is because having a truly open mind a very hard task. The R&D manager does no need to be open-minded all the time – it may not be prudent to do so in some cases. However, the R&D manager who possesses that trait can apply that mindset when appropriate to accelerate their learning through direct observation.

A simple example that highlights the power of direct observation and an open mind is the case of a paint company that was aiming to increase its differentiation to charge a premium for its product. The staff proposed improving the adhesion of the paint, offering more colors, or improving the integrity of the colors. The CEO did not accept any of these suggestions and, instead, asked them to go observe people using their products. The staff observed painters in action and found that they would paint the 1st coat, wait several hours for the paint to dry, and then do a 2nd coat. Based on what they observed, they came back to the CEO with a different proposal. They proposed to make the paint dry faster so painters would not need to wait so long between applying coats. The proposal was accepted and it became a very profitable differentiation for them.

2. Learning

Like many things in life, if one strives to maintain their skills, one must develop and tune them regularly. Skills required for learning that enable curiosity, direct observation and an open mind atrophy over time as a person's professional repertoire and pride grow. The remedy is to keep those learning skills in shape by exercising them. This can be done by seeing through the eyes of a perpetual novice. Developing those skills means taking a goal to learn something completely new. For example, learning a new language, a new sport, or a completely new subject. This helps keep one humble by rejuvenating the atrophied learning skills we just discussed. In this way, one will continuously practice learning.

Doing this as a group is also very beneficial. A group tends to learn faster than an individual due to diversity of thought, observations, and debates. As a practical example, when someone makes a recommendation or proposal in a meeting it is good to ask them to articulate the views of others who disagree with the proposal in the same manner and tenor those who disagree would describe it. This forces the person to fully understand the views of those who disagree. This is a good litmus test for exercising observation with an open mind.

3. Adaptation

Once one recognizes and accepts that a change is imminent, the adaptation process begins. This is relatively easy if everyone else also sees and accepts the change. However, it is rarely so straight forward and change management is an occupational hazard. Two excellent resources on this topic are, *Leading Change* by John P. Kotter [4] and *Managing Transitions* by Bill Bridges [5].

Bill Bridges highlights in his book that change management is a journey and may take a very long time depending on the scope of the change. The main strength of the transition model described in his book is the focus on transition and not change. The difference between these is subtle but important. Change is something that happens to people, even if they don't agree with it. Transition, on the other hand, is internal: it's what happens in people's minds as they go through change. Change can happen very quickly, while transition usually occurs more slowly. Specifically, the model highlights three stages of transition that people go through when they experience change. These are:

(1) Ending, losing, and letting go
(2) The neutral zone
(3) The new beginning

Bridges notes that people will go through each stage at their own pace. For example, those who are comfortable with the change will likely move ahead to stage three quickly, while others will linger at stages one or two. As a result, R&D managers leading the change need to understand what is described as the marathon effect. They may be in stage 3 while others will be in stage 1 or 2. Comprehending the marathon effect and implementing a complete transformation framework is critical to making the change initiative successful.

R&D managers who regularly practice the key leadership skills of anticipation, learning and adaptation turn them into habits that become part of their routine. They chase opportunities and learn with excitement. They observe seemingly unconnected events or inconsequential results in their eco-system and see trends. They identify required changes early and transform their organizations. When developed, the skills become a critical competitive advantage.

Using an analogy from sports again we can look at Wayne Gretzky's famous quote that we described more fully in Chapter 6 [6],

"I skate to where the puck is going to be, not to where it has been".

The anticipation needed to accomplish that goal was not cultivated through repeated drills but via the combination of observation, free-flowing, unstructured play and unfettered passion. In various interviews, Gretzky recalls the hours he spent alone working on the game he loved as well as the times spent on the backyard rink, playing shinny with his brothers and neighbors for nothing but pride. He also recalls the nights spent watching games on the television and tracing the path of the puck as it moved around the hockey rink to see patterns. He fell in love with the game through play and curiosity and this fueled his desire to learn more.

13.3 Articulating the What

Leaders turn vision into a strategy

Setting the strategy starts with the vision. Strategic planning takes that bold ambition as the guiding light to drive its due diligence and explore actions that may be required to succeed in the context of the existing eco-system. At some point, the strategic planning process proposes the required actions needed to accomplish the vision. This is when the strategy starts to become real and the leadership of the R&D manager becomes very important. Since the strategy only becomes real when the required resources are assigned, the R&D manager needs to provide the leadership to make trade-offs based on the feasibility, risk and available resources to fund the strategy. After this critical juncture, the role of an R&D manager becomes providing the support required to drive the projects to completion in the most effective manner possible. If the eco-system changes and the related strategy itself changes then the R&D manager needs to re-evaluate whether the originally conceived projects are still needed. This may mean that projects that no longer align with the revised strategy are to be re-directed or canceled.

Leadership in setting the strategy means making sure the strategic planning process gets the attention it needs. The R&D manager needs to ensure that the entire spectrum of options is presented and enough time is allocated to fully comprehend and analyze the options. These discussions won't be popular with everyone, as they take time away from the pressing and urgent execution challenges. The result is that the leader must often legislate the attendance and allot adequate time so that strategic discussions get quality mindshare of the leadership team.

It is also critical for the R&D manager to empower senior technical leaders to take ownership of topics and enable them to carve out enough of their bandwidth to dive deep into highly technical topics. This often means having them make proposals to the senior management staff for feedback and refinement. In organizations we have led, the senior technical experts of the R&D organization were given the responsibility to investigate and educate themselves on critical strategic topics and then propose a strategy to the R&D manager and their direct staff.

Involving senior technical experts and empowering them to make proposals help shape a technology-based view of the environment. This prevented middle managers from dominating the strategic discussion and pushing narrow strategies based on parochial interests. In that model, it was critical to fully involve the R&D manager's senior staff in the strategic discussions to enable their feedback before making the final decision (with the GM having veto power). The senior staff acted as a sanity check on the feasibility of a strategic proposal. They also needed to be personally invested in the process and the final decision to ensure quick action on the execution that would be needed later. Similarly, the R&D manager must ensure there is good participation from other managers who must implement the strategy. Their buy-in is critically important to establish a strong connection between "what" and "how".

Leaders invest in communication.

Leadership requires communication. Communication is the fabric that ties the entire organization together and ensures effectiveness and efficiency. The health of the organization thrives on relevant and timely communication. Employees tend to fill in blanks when official communication channels don't provide a complete picture or adequate details — in the corporate world they refer to them as rumors. Rumors are an organization's way of filling the voids when communication is lacking, or not timely. Therefore, it is critical to have a skilled person dedicated to communication who can help the R&D manager. If the organization size is small, the communication specialist's job might be part-time, but it must exist. The communication expert should have full visibility of the entire organization's systems, processes, and operations; otherwise, their effectiveness will be diminished.

The R&D manager should spend at least a quarter of their time on communication. The favored communication type varies from person to person. As a result, to ensure critical messages reach every employee in the organization, the R&D manager should communicate through multiple redundant channels, such as audio, video, email, e-newsletters and face to face meetings. The type of communication we discuss in this section is mostly large group communication, which we might call broadcast or one-way communication. That can be divided into three categories:

(1) Communication of business results
(2) Communication of execution status and plans
(3) Communication for employee well-being

Communication of Business Results: It is the responsibility of the R&D manager to ensure that the strategy, goals & roles of the organization are crystal clear to every employee in the organization. It is also important for every employee to understand how they connect to the corporate goals. If employees understand their roles and the connection to the corporate mission and goals then they will work with greater autonomy and improve faster. Clarity on this also increases employees' enrollment and commitment to the corporate and R&D organization goals. Therefore, this communication mechanism should address updates on business goals, results, customers, key decisions and their rationale. This kind of information is communicated mostly in a broadcast style communication (live or recorded). Often, recognition of major group or individual accomplishments is mentioned too. The preparation for this kind of meeting is of paramount importance. The leader should spend significant time preparing for it.

Communication for execution status and plans: This category is mostly about driving execution efficiently. This communication can be done at various levels in the organization for speed and efficiency. The goal of this type of communication is to ensure every employee has the necessary information in a timely fashion to do their job efficiently. This includes relevant project indicators that show progress and issues. Key decisions should also be communicated to all relevant employees promptly. This could be done in a daily stand-up meeting similar to what is done in a Scrum, with individual contributors or in weekly execution meeting for all managers. In this case, the communication frequency should match the rate of change in the execution. In a factory or 2-week Scrum, daily stand-up reviews may be needed. In the development of a microprocessor, the correct frequency may be weekly since the information turns are longer.

Communication for employee well-being: Communication in this category relates to employee recognition, development, and satisfaction. Communication regarding recognition refers to formally acknowledging and rewarding accomplishments and role models for the organization. This should be done at all levels of the organization. The level at which the accomplishment is rewarded depends on the size and scope of the impact. Communication regarding employee development refers to the dissemination of information for all employees so they can take full advantage of the available classes, seminars, conferences, and other events to advance one's aspirations. Communication regarding employee satisfaction also needs to be done regularly. The results give a picture of the health of the R&D organization and their feedback is critical to make the organization better. Feedback on customer satisfaction can be shared as well. The Net Promoter Score (NPS) survey from Bain Consulting Group [7] is one way to measure this. One element of leadership is the ability to share this type of information in a transparent way to generate an open and frank discussion.

It would be remiss of us if we didn't mention two other types of communications that are critical to providing the leadership needed in the R&D organization.

Informal small group communication: This type of communication is interactive and less structured. These types of conversations help leaders pulse the organization, learn challenges first hand, and gather a variety of opinions to help synthesize information from elsewhere. It also gives the manager of the R&D organization direct feedback on how well their messages are being understood by everyone in the organization. This type of communication is very critical during the fuzzy front-end stage of development. These types of conversations take more time but have the biggest impact when strong alignment is needed. The best case is when they can be done by the senior leaders of the organization and not just the manager of the whole organization since it forces them to understand the messages. It also reinforces to the rest of the R&D organization that the leadership is aligned and listening.

One on One communication: This type of communication is required when there are key individuals that need to align to a decision or are affected by a decision. This is also a very critical channel for individuals to give feedback or provide new ideas to the manager since many people tend not to express their ideas/opinions candidly in a large group setting. Part of the traits of a good leader is the willingness to identify key people with knowledge and expertise and meet with them one on one to ensure alignment and get feedback.

13.4 Enabling the How

Leaders have a strong orientation to results.

Actual results are what matter most. They contribute to velocity and are the visible indication of an organization's performance that can be compared to other similar organizations. In this context, leadership means ensuring that the R&D organization delivers against the committed goals predictably and reliably. Earlier in the book, we described Intel's corporate values. They recently changed but for several decades, results orientation was arguably the strongest value of them of all. A leader who demonstrates a results orientation is not satisfied until they get to the desired results and the full set of learnings have been internalized. When in doubt, a leader who demonstrates results orientation steers the decision-making and resource allocation towards achieving results. It starts with clear goals that enroll the employees. After that, true leaders become subservient to the needs of the engineers doing the execution and use their position to make decisions that remove obstacles.

Leaders set clear objectives.

The objectives of the corporation should be clear. The R&D manager must derive similar objectives for the R&D organization. As we noted in Chapter 6 the goals may be induced top-down from the corporate objectives or derived bottom-up through autonomous actions. When they are induced top-down from a corporate objective or strategy the R&D manager needs to ensure there is a strong agreement between the "what" of the corporate objective and "how" the R&D organization will implement it. Ideally, every employee will be clear on their objectives and how they directly connect to the objectives of the corporation. However, practically speaking, not every objective in the R&D organization should map to a corporate objective since some degree of autonomous strategic action or disruptive innovation should be tolerated. As we discussed in Chapter 8 a key leadership trait of the R&D manager is the ability to set the right balance between the selection of induced and autonomous actions. The R&D manager should allocate about 5–10% of the resources to encourage autonomous innovation even if they cause dissonance with the current corporate direction. As also mentioned in Chapter 8, the R&D manager should use the research pipeline and pathfinding processes to frequently review those projects.

As we discussed in Chapter 6, the objective of goal setting is to drive execution with simple, clear and memorable objectives that will enable the strategy. As we noted in that chapter, a good test of any objective is whether it is SMART [8]:

Specific: Clear and simple to understand.
Measurable: Meaningful progress can be tracked.
Achievable: The risk is acceptable and resources exist.
Relevant: It is connected to the strategy and goals.
Timely: The schedule will maximize economic benefits.

Leaders optimize revolutionary approaches over multiple cycles.

Project planning assumes that experts in the organization know what it takes to complete a project and feasibility is not a question. Project planners use the theory of constraints to identify and eliminate all bottlenecks to the extent possible. Project managers build some buffers into their plans based on historical judgment to address potential surprises that might crop up during execution. By the time project planning is completed, the project manager is generally convinced that all known risks have been mitigated or comprehended in the plans. They also may have secured excess resources or a schedule buffer to deal with the surprises based on their engineering judgment.

However, if the project is a greenfield project, and this is the first time it is being done, then the inherent risk is higher. Although project managers deal with known risks quite well, these projects have many unknown unknowns. Since most of the planning accuracy depends on learnings that were incorporated from past projects, these types of projects have much higher uncertainty. In these cases, the likelihood of hitting the schedule with the expected results and with exactly the planned resources is unlikely. Planning is still very useful, but the plans often are not. They will change to deal with the realities on the ground and inevitable surprises.

For such revolutionary projects, a lot is learned by doing the execution itself so it is important not to over-optimize the execution plan at the beginning. In reality, most revolutionary projects follow the rule of 3. The 1st development cycle tends to become a proof of concept of the value with incomplete implementation. The 2nd cycle is complete but with some suboptimal implementation. The 3rd cycle is a complete implementation with a much more optimal implementation. Several new technologies in history have taken at least three cycles of development to hit their stride. A visible example is Microsoft's transition to the Windows operating system that took three generations before it adequately met the needs of the consumer.

Leaders connect what and how and make informed decisions.

One of the most important leadership traits is the ability to drive timely engineering decisions. The difficulty in making decisions largely stems from incomplete data, and the multitude of dissenting views. To overcome incomplete data, the R&D manager often needs to demonstrate leadership and apply their engineering judgment. As a result, the R&D manager not only needs to understand all views but must also adequately comprehend the data and engineering problem to make a judgment themselves. The commitment of the team is far greater when they believe the leader fully comprehended all the relevant data and experts' views and the R&D manager can explain the rationale for the decision and the path forward. A few additional principles that will help improve the quality of decisions are:

Learn before deciding (aka data-driven decisions): Obtain a thorough understanding of the data before deciding on the actions.

Wait until the last responsible moment to decide: As discussed in Chapter 7 concurrent engineering allows decisions to be made at the last responsible moment without jeopardizing the delivery milestones. This maximizes knowledge and minimizes waste due to rework. As noted in Chapter 7, this requires high confidence in the execution time and effort once a decision is made. Therefore, waiting for the last responsible moment (and not any longer) requires a thorough understanding of the execution capacity of the organization and suppliers.

Strongly connect the "what" and "how": The R&D manager needs to insist on achieving the objectives (the what) but must be willing to listen to the implementation challenges (the how) to fully understand the trade-offs involved. Sometimes, the realities on the ground lead to sub-optimal results. When this happens the R&D manager must provide the leadership to engage with the development team to understand the trade-offs and decide on the best course of action. These conversations are very pragmatic and non-judgmental to find the best possible answer. Leadership often requires stretching the team to achieve more than what they think is possible, but if the leader stretches them too far, it may break their confidence in the project and the R&D manager.

Leaders provide focus and role models a problem-solving mindset.

Execution is about sticking to the plan or getting back to the plan in the face of surprises. Execution critically relies on SMART goals as described earlier. Also, execution depends on the R&D manager providing focus, and clear ownership while enabling employees to report surprises without delay, use a problem-solving mindset, and harvest learnings through retrospection.

Focus: Once execution begins the R&D manager should fully commit their organization and provide their disciplined and undivided attention until completion. This is required for predictable execution and velocity.

Ownership: The R&D manager must assign clear ownership of tasks. The employees then take ownership until completion. This reinforces accountability. Apple corporation supposedly calls this a Directly Responsible Individual (DRI).

Report surprises without delay: The R&D manager must enable everyone to recognize problems or surprises early and immediately report them. The R&D manager must do the same for their management. Bad news should be communicated without delay (e.g., a phone call, not email).

A problem-solving mindset: When a problem is first encountered, the R&D manager should not obsess about who was at fault. The focus should be on how to best resolve the problem. When this happens, employees will be more willing to report problems or surprises in the future.

Retrospection: Once the project is complete or a problem has been resolved, the R&D manager should conduct a retrospective review to identify improvements. This is a critical step in continuous improvement discussed in Chapter 5.

Leaders set stretch goals to motivate and reward employees.

A "stretch goal" or "bet" refers to a set of objectives that involve most people in the organization and provide clear incremental improvement beyond what is needed to keep the business running. To be effective, the stretch goals should also be achievable and have a meaningful reward for the employees. This is a very good way to make broad organizational goals highly visible and meaningful to all employees. Since the objective is to both improve and reward everyone, having unachievable or vague targets defeat the purpose. Also, the impact of achieving the stretch goal is far more valuable to the organization than the cost of the award so the R&D manager should not be timid in determining the reward.

Leaders take informed risks and mitigate other risks.

Taking informed risks is a requirement of R&D leadership. Quite often, the R&D manager doesn't have a choice and must take a risk order to push the frontiers of technology. As a result, the R&D manager must become comfortable with being uncomfortable. When the R&D manager starts to feel too comfortable it is an indication that improvement has stalled and atrophy may be occurring. In the words of Beyoncé [9],

> *"I get nervous when I don't get nervous. If I'm nervous I know I'm going to have a good show."*

As an example, the commitment the semiconductor industry took to roughly double the number of transistors that can be manufactured on a chip every two years required risk-taking on many levels. Although the engineers working on advancing silicon process technology had clear goals, they didn't know how to accomplish some of those goals at the time of definition. This led to rigorous and disciplined experimentation with clear metrics and targets to mitigate the risks. As described in the case study in Chapter 5 titled, "Manufacturing, CAD and Design Learning Cycles" this also created a virtuous cycle of continuous improvement across semiconductor manufacturing, integrated circuit design and computer aided design. This would not have happened without the willingness to take informed risks. With that in mind, leadership requires informed risk-taking and this gets enabled in four ways:

(1) Understanding the value
(2) Using the value to drive research & development
(3) Eliminating unnecessary risk
(4) Emboldening necessary risk-taking

Understanding the value: The concept of value for the engineering R&D organization was discussed in Chapter 4. Analyzing opportunities for creating incremental value derived from direct customer engagement is described in that chapter as well as Chapter 7. Analyzing opportunities for creating bigger and more consequential value based on the evolving forces in the eco-system was described in Chapter 6. This included opportunities inherent in creating new markets. As we discussed in Chapter 8 and earlier in this chapter, the R&D manager should provide 5–10 % of their resources to foster and monitor autonomous or disruptive innovation. However, whether it is induced or autonomous the underlying principle is that the R&D manager has a clear understanding of the existing or potential value of a proposed activity and is willing to pursue the effort in the face of uncertainty and other risks to take the next step and learn more.

Using value to drive research and development: As we discussed in Chapter 8 induced research topics are tied closely to the strategy and objectives of the company and therefore have a clear value proposition. On the other hand, autonomous research topics are not tied directly to the current strategy of the company and the value is less obvious to senior management. Investment in autonomous or disruptive innovation often comes in the form of a dedicated research effort, joint research with another institution, or engagement with a start-up through the venture capital arm of the company. The bottom line, however, is that decisions to invest in this type of research are still determined by the R&D manager's perception of the potential value of the topic. The difference is in the greater risk or uncertainty that the R&D manager is willing to accept to learn if the value is realizable. A way of managing this was explained in detail in Chapter 8 when we discussed the research pipeline and pathfinding.

Eliminating unnecessary risk: The continuous loop of strategic planning described in Chapter 6, concurrent engineering described in Chapter 7 as well as the research pipeline and pathfinding processes described in Chapter 8 all provide opportunities to re-assess the value of ongoing efforts and their associated risks. As we described in each case, the R&D manager must provide the leadership and oversight to ensure that the value, risk and underlying assumptions are continuously challenged so unnecessary risk can be eliminated.

Emboldening necessary risk-taking: As we noted, risk-taking is a requirement of R&D. A value-based innovation and risk reduction culture does not preclude taking risks. A transparent process like the ones described in previous chapters that are based on clear expectations, a consistent set of criteria and different levels of risk tolerance will encourage engineers to invest their time on the right topics. However, for the behaviors to take root and thrive, the entire leadership team must uphold the same value-driven and risk reduction principles across the board. Decision-makers can't play favorites or make exceptions. As a result, the R&D manager and the rest of the leadership team should be transparent in explaining decisions to reinforce the proper culture and existing processes. Once this takes root, it aligns the work-force while emboldening them to take informed risks.

13.5 Common Topics for Why, What and How

13.5.1 *Velocity and Efficiency*

Leaders focus on velocity when innovating to create new value.

As we have noted throughout the book, the rate of deployed value (velocity) is critical for the R&D organization to thrive. As we define it, velocity is a function of time and not the effort or cost. In previous chapters, we detail how to create new value and the journey often consists of a series of learning cycles that involve experimentation. If done correctly, the experiments enable the R&D organization to be first to market and gain a leadership position. This can significantly increase value. However, the urgency and need for experiments require more effort than a typical development effort where the implementation is already known. Therefore, as we discussed in Chapter 4, enabling high velocity, in this case, will come at the cost of efficiency.

The bottom line is that the R&D manager should focus on velocity first when new value is being created and shift the focus to efficiency once it is deployed. Demanding greater velocity and greater efficiency simultaneously during the creation of new value can be very confusing for the rest of the R&D organization since the metrics and targets will often conflict with one another. To enable execution with clarity requires a clear mandate between velocity and efficiency when they conflict.

Leaders emphasize efficiency when executing a known plan.

Although efficiency should not be a major consideration when creating new value it is critical for execution once the value has been initially deployed and subsequent development plans are clear. The ideal state for efficiency is to do everything perfectly, every time and with the lowest possible effort. Therefore, achieving efficiency during development often includes a quest for perfection.

A sports analogy comes from tennis great Roger Federer. A tennis match is not a place for experimentation. This happens during training. During the match, a tennis player must exert a tremendous amount of effort for each point. However, they also must conserve as much energy as possible to last the entire match. Any wasted motion accumulates to the ultimate detriment of the athlete. Roger Federer does that better than most, if not all. He has learned to take the ball as early as possible, take the least possible number of steps, eliminate unnecessary movement and still win the point. He is a role model for being deliberate about every movement and eliminating wasted energy. By seeking perfection in positioning and motion he has become very efficient in the use of his energy.

Another analogy highlights that seeking perfection may delay the result or cause it to never happen. Free solo mountain climbing refers to climbing cliffs without any ropes and anyone else. It is something that requires perfection. Alex Honnold completed a free solo on one of the most difficult climbs, El Caption, on June 3, 2012. He completed it in 3 hours and 56 minutes. He was perfect in every step from beginning to end. However, Alex planned diligently to find the perfect weather conditions and had to cancel many attempts due to weather. As a result, he did not do it in the least amount of time or at the lowest cost.

As highlighted by the last example, there will be some tension between maximizing velocity and efficiency during execution. For example, achieving perfection may mean delaying the introduction of new technologies until they are complete and have no defects. On the other hand, increasing velocity may mean delivering an incomplete solution that is not perfect but still delivers clear value.

With this in mind, excellence during execution is about keeping an eye towards improving efficiency (and hence perfection) while also steering resources towards efforts that maximize value.

13.5.2 *Leadership and Learning*

We touched on the importance of learning as part of the leader's role in determining and articulating the "why" in Section 1.2. In this section, we look at the importance of learning for the leader in more depth.

Leaders have a learning mindset.

A prerequisite for continuous learning is a growth mindset. This means believing that there is always a better answer. The R&D manager must continuously be in search of that. In this mindset, there is no room for arrogance or ego. However, learning requires tremendous commitment, endurance, and energy. Also, the belief that there is a better answer often feels uncomfortable with people since it devalues their existing expertise. As Charlie Munger [10] states, to overcome the natural resistance to continuous learning:

> *"You need to have a passionate interest in why things are happening. That cast of mind, kept over long periods, gradually improves your ability to focus on reality. If you don't have the cast of mind, you're destined for failure even if you have a high I.Q"* —

With this advice in mind, it is critical for the R&D manager to develop the passion to learn about what is happening to get a clear picture of reality and help drive the many decisions they need to make.

Leaders use first-order thinking to zoom in on key attributes of a problem.

As we discussed in Chapter 9 the variety of problems the R&D manager faces is large and requires treating different types of problems in a unique fashion. Even with this help, the problem space may seem very large. One tool that is often used to simplify understanding is the 2×2 matrix used in several places in this book. The 2×2 matrix helps wring out unnecessary details and keeps the minimum information required to visualize the problem. The 2×2 formulation forces the R&D manager to identify the two most important and 1st order attributes of the problem. An example of a 2×2 is shown in Figure 13.3. It is a representation of how Steve Jobs may have viewed the business segments at Apple when he rejoined the company and how he drastically reduced the number of products [11].

> *"… What he found out was that Apple had been producing multiple versions of the same product to satisfy requests from*

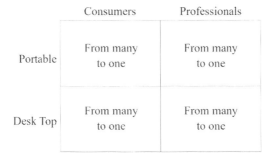

	Consumers	Professionals
Portable	From many to one	From many to one
Desk Top	From many to one	From many to one

Figure 13.3. Thinking model to help with the prioritization.

retailers. For instance, the company was selling a dozen varied versions of the Macintosh computer. Unable to explain why so many products were necessary, Jobs asked his team of top managers, "Which ones do I tell my friends to buy?" When he didn't get a simple answer, Jobs got to work reducing the number of Apple products by 70 percent. Among the casualties was the Newton digital personal assistant"

Leaders work to accelerate their learning.

In Chapter 8 we discussed how the Wright brothers used technology such as the wind tunnel to economically increase the rate of their learning. Similarly, the R&D manager needs to seek out ways to budget their time to accelerate their learning.

Every human spends their time and vitality on three things:

(1) Joy
(2) Fun
(3) Waste

Thinking back to our definitions of value and waste, activities that provide joy and fun can be categorized as adding value to the quality of life for an individual. Similarly, activities are waste provides no value to the individual. How the R&D manager expends their energy and time on these three things depends on one's ambition and discipline.

Activities that provide joy: This results in the satisfaction of diligently doing your job to the best of your abilities, achieving goals, developing new skills, or helping others do the same.

Activities that are fun: These activities usually re-energize an individual's physical and mental vitality through such things as relaxation, exercise, recuperation, time with family and friends.

Activities that are waste: These are the activities that, when one is healthy and capable, aren't directed towards the self-defined activities that create joy or fun.

Most people derive joy after a hard day's work if they accomplish what they set out to do. Most people also derive fun when they spend their time doing their favorite activities, spending time with their loved ones, or helping others. On the other hand, activities that are waste are activities that do not result in fun or joy. In the ideal case, joy and fun apply to everything one does and there is no waste. An example of this is someone who loves their work. To others, they may seem to be obsessed with their work but that is because to them it fits the definition of both joy and fun.

Leaders balance multiple perspectives and levels of abstraction.

A mature R&D manager can fully comprehend the full spectrum of possibilities in any given vector. In comprehending a subject, the R&D manager must be able to zoom out to extract the key attributes of the problem (as described earlier) and zoom in to comprehend the details that affect such things as the complexity, feasibility and risks. Good leaders also embrace seemingly opposing ends of the spectrum. For example, they combine optimism and paranoia as well as thinking long-term and short-term. Maturity means striking the right balance to achieve the desired results. To comprehend a subject fully, the R&D manager may need to zoom-in and zoom-out multiple times as part of the learning process. By doing this, the R&D manager keeps the big picture in mind when making decisions on the details and vice versa. This can be very useful when a complex technology that is being engineered also has to be very easy to use. Striking the right balance between complex implementation and ease of use often requires iterative cycles of zooming in and zooming out.

13.5.3 *Leading People*

Leaders keep people proud, satisfied, and committed.

People are essential to the creation and deployment of value. We remember Andy Grove (the former CEO of Intel) state in the early 90s that the intellectual capital of the work-force at Intel exceeded the capital investments (factories, buildings, etc.).

The ideal employee is intellectually & emotionally committed to the strategy and values of the R&D organization and will provide the extra effort needed for meeting all of the objectives. On the other hand, from the employee's standpoint, the ideal employer gives them opportunities to work in a safe environment, on activities that give them the joy we discussed earlier and properly reward them for their effort. Not surprisingly, there is sometimes tension between these expectations. Maximizing value steers one towards the employer's mindset while making the R&D organization attractive to prospective employees steers it towards the employee's mindset.

The manager of the R&D organization provides the leadership needed to strike the correct balance. To thrive, the R&D organization needs to both achieve aggressive objectives and attract and retain highly skilled employees who will be committed and passionate participants.

Leaders attract and retain the best talent.

The R&D manager must provide the leadership needed to make the recruitment of talented engineers a key priority in the organization. This means investing in building a robust pipeline of candidates from many sources that can be sampled when needed to pick the best people available. Many corporations continuously cycle between freezing and opening hiring based on financial results. This presents a difficult challenge for the R&D manager since this is not conducive to creating a dependable pipeline of candidates. The R&D manager must resist the urge to shut down their own recruiting and wait for the hiring window to open again. Doing that will lead to making it very difficult to find and recruit the best available candidates before the window closes again.

Despite what happens at the corporate level, the R&D manager needs to provide the leadership to maintain a deliberate and continuous recruiting process. Ideally, hiring can be synchronized with the availability of the best talent. Even if it is not possible to do any hiring due to a corporate hiring freeze it is still valuable to keep the pipeline full so the best candidates available can be pursued with the hiring window opens again.

It is of paramount importance that incoming employees are compatible with the R&D organization's values and the manager of the R&D organization needs to set an example for everyone else in the people they choose. Admiral Hyman G. Rickover was famous for the culture of safety that he instilled in the US Nuclear Navy program. The Admiral, himself, interviewed every candidate for every job in the program to make sure they would fit in with the culture he had created. Rickover understood that misalignment on the basic values of the program he ran could be fatal and, at a minimum, was very difficult to rectify.

Another key aspect of Admiral Rickover's emphasis on culture in the nuclear navy was the intensive training everyone underwent when they joined. Dedicating a few weeks to train new people on the values, methods, systems and practices will ensure they get started in the right direction. The R&D manager and other leaders in the organization should be a key part of that education.

Leaders know that retaining good employees is better than having to hire.

Existing employees are trained, proven and productive. To get a new employee to that state requires time and effort. As a result, if the R&D manager senses that somebody is unhappy and may consider leaving, they must make it their top priority to resolve the matter in a satisfactory way for the employee and the organization. Of course, doing things such that it never reaches this point is also important and we discussed that in Chapter 9.

A key ingredient for retaining people is maintaining their level of emotional commitment. Psychologist Jonathan Haidt [12] introduced a useful analogy of an elephant and its rider (shown in Figure 13.4) for thinking about how humans handle change. Haidt argues that people have an emotional side (the

Figure 13.4.   Elephant and the rider.

Elephant), and an analytical, rational side (its rider). In Haidt's analogy the rider is rational and can, therefore, see a path ahead while underneath him, the elephant is irrational and driven by emotion and instinct. Given the magnitude and criticality of emotional commitment, the R&D manager needs to provide the leadership in the organization to gain the emotional commitment of employees (the elephant) as well as getting agreement on the strategy and direction (the rider).

Leaders create a diverse and inclusive workforce.

In Chapter 3 we captured feedback from several people on the importance of diversity and inclusion. The role of the R&D manager in enabling a diverse work-force is critically important to the success of the R&D organization and requires leadership.

As we discussed in Chapter 3, leadership means taking the time to comprehend the worldview of the diverse work-force and understanding what they perceive is fair and acceptable. This will help understand how to get the emotional commitment from all employees. Diversity is critical to the success of the R&D organization since the R&D organization must utilize all accessible intellectual capital to come up with the best decisions. An example of why this is important is given by the paraphrasing of an ancient Indian parable and depicted in Figure 13.5.

> It is a story of a group of individuals with their eyes closed who have never come across an elephant before and who learn and conceptualize what the elephant is like by touching it. Each person feels a different part of the elephant's body, but only one part, such as the side or the tusk. They then describe the elephant based on their limited experience and their descriptions of the elephant are different from each other. The moral of the parable is that humans tend to claim absolute truth based on their limited, subjective experience.

Keeping the parable in mind, the leader will learn more by encouraging and collecting all viewpoints. In a diverse group,

Figure 13.5. Individuals with their eyes closed and an elephant.

the observations are richer, deeper and wider and one makes higher quality decisions. For this to work, the team must have mutual trust. If they do not have trust, they might see their differences in an unproductive manner.

Leaders create a psychologically safe environment for conversations.

We already discussed the importance of synthesizing intelligence from many sources as part of strategic planning, development and engineering research. Even though there may be official processes for that, it is important that the manager of the R&D organization hold informal but regular conversations with the key experts and thought leaders in their organization. To make those conversations effective, they need to create a psychologically safe environment. Psychological safety means employees feel they can contribute ideas and challenge the current thinking without fear of being ridiculed, discounted, or otherwise punished. A leader who creates this type of environment and has also mastered looking at problems through an open mind will benefit greatly. Conversations they have with employees will help employees understand the importance of their work and the employee will gain insights that they can share with their peers.

While holding these conversations, a leader should ask many questions but must resist the temptation to give direction. Micro-understanding is when the R&D manager asks detailed questions to learn This is okay. Micro-management is when the R&D manager tells people deep in the organization how to do their job when they know the work far less than the employee. This is not okay.

Leaders understand that every employee is different.

The personality assessment tool called the Myers–Briggs Type Indicator, MBTI [13] is a popular means to understand the different personalities at play within a group of people. It can be used to help the R&D manager understand each employee's natural preference. When a manager understands how an employee thinks and processes information, they can be more effective in bringing out the best in them. The MBTI has four dimensions

(1) Extrovert or Introvert
(2) Intuitive vs Sensing
(3) Thinking vs Feeling
(4) Judging vs perceiving

As a very general overview, extroverts are ok with sharing their ideas without thinking deeply about them, whereas, introverts would like advance notice and think through ideas before expressing them. Intuitive personalities are comfortable with concepts and ideas in abstraction, whereas, sensing personalities are pragmatic and want detailed data. Thinking personalities worry about what is the right thing to do, whereas, feeling personalities tend to think through their heart and are sensitive to people's feelings. Judging personalities are about planning and efficiency, whereas, perceiving personalities are about flexibility and are willing to change plans easily. Conducting an MBTI assessment or something similar with the

R&D manager's leadership team is a good way for the team to get to know each other and respect each other's differences. It is also a very good way for the R&D manager to better understand how each member of their leadership team absorbs and processes information.

Leaders develop their employees.

The R&D manager must take a personal interest in developing their direct employees and enabling other managers in the organization to do the same. To do that, the employees need to be clear on their career objectives. The R&D manager can facilitate that process but an employee needs to think it through and make decisions for themselves.

Most employees are very clear about what they don't like. But are rarely clear on what they would love to do in the long-term. It is a difficult question since people tend to change their interests over time. It is the R&D manager's responsibility to facilitate that journey with their direct reports and other leaders in the organization. This can be done through periodic conversations that augment the regular performance reviews that are typically held. The conversations will allow the R&D manager to act fast when opportunities come up in the R&D organization or in the rest of the company that provides an opportunity for an employee to grow in a direction they have already outlined together.

Leaders build trust.

Employees perform best when they trust their management. The trust comes when an R&D manager demonstrates the commitment to do things right. This means,

> *Do the right thing, the right way, for the right reason*

Doing the right thing is about pursuing the best results for the company. Doing it in the right way means doing the work with transparency and according to the R&D organization's values. Doing it for the right reason means that the intent is pure and not biased by parochial or self-serving interests. When the R&D manager role models all three rights in the organization people will tend to trust them and give their best effort. Following the three rights also provides evidence that the R&D manager is committed to a psychologically safe environment.

Leaders sometimes have to change their personnel.

Very rarely, an R&D manager gets to select their entire team from the ground up. Often, they inherit a leadership team. As a result, any bold vision or change initiative is usually met with a decent amount of resistance. The R&D manager should spend adequate time to explain the vision and to enroll his leadership staff in their vision for the organization. In that process, an R&D manager may also have to deal with an array of conflicts with their direct reports and amongst peers. It is critical to spend the required time to sort through those differences upfront rather than dribbling those nagging differences and misalignments forward. We learned from experience that it is better to spend the time to sort out those differences as soon as possible to gain alignment and hence momentum towards the new direction. To achieve that, the R&D manager may have to let go of some of the leaders who do not agree with the direction and vision. It is important to have a leadership team that shares a passion for the mission.

This can be done by working with those affected to move to a good place in another organization where their passion better matches the strategy and mission of the organization. A constructive disagreement on the new vision or direction should not be punished but that does not mean the R&D manager should accommodate an unhappy employee.

Leaders strive to be functional in a dysfunctional environment.

Politics is the use of power and networking to achieve results that benefit the organization or individuals within it. Politics exists wherever human beings exist. Politics used in the wrong way often involves an unstated agenda that serves the interests of a few instead of the interests of the R&D organization and company. This creates dysfunction in a group for the benefit of a few. Politics used in the right way involves aligning and influence others to achieve a good outcome for the R&D organization and company.

When the R&D manager is faced with a bad political environment, they must still strive to be highly functional and do the right thing, the right way for the right reason. They must be skilled enough to recognize what is happening and resolve it productively to best serve the R&D organization and company. They must not let political dysfunction paralyze or demotivate them. Instead, they need to remain a leadership role model for others. Bad politics thrives when good leaders stay quiet.

13.6 Summary

Leadership is a combination of dreaming and doing. It spans the fuzzy front end of articulating a big and bold vision, translating that into a strategy that creates differentiable value and then motivating, funding and orchestrating all of the entities in the R&D system to work harmoniously together to deliver results with the greatest velocity.

To be an effective leader the R&D manager must balance the traits of curiosity, learning and having an open mind with a problem-solving mindset and a laser focus on results. Most of all, an effective leader knows that their success is based on the trust people have in them to do the right thing, the right way, for the right reason and they work hard to maintain that trust.

We believe, the R&D manager needs to be both a strong manager and a strong leader. The act of leadership is required to motivate (the why) and set a clear direction (the what) and is a prerequisite to achieving the best results. On the other hand, the act of management is required to turn the direction into clear roles, goals, and otherwise enable the team to achieve the results (the how). When we refer to either the leader or the manager we are referring to the same person. It is someone whose acts of leadership and management are performed with equal diligence.

13.7 Key Points

- Leaders articulate what needs to be done and why and then enable it.

When articulating "why", leaders must be

- Bold and think of possibilities when setting the vision.
- Develop the ability to anticipate, learn and adapt to change.

When articulating "what", leaders:

- Turn vision into strategy.
- Invest in communication.

When enabling "how", leaders:

- Have a strong orientation to results and set clear objectives.
- Optimize revolutionary approaches over multiple cycles.
- Connect what and how and make informed decisions.
- Provide focus to inculcate a problem-solving mindset.
- Set stretch goals to motivate and reward employees.
- Take informed risks and mitigate other risks.

When balancing between velocity and efficiency, leaders:

- Focus on velocity when innovating to create new value.
- Emphasize efficiency when executing a known feasible plan.

To improve themselves, leaders:

- Have a learning mindset.
- Use first-order thinking to zoom in on key attributes.

- Work to accelerate their learning.
- Balance multiple perspectives and levels of abstraction.

To harvest the full potential of their organization, leaders:

- Attract, retain, and develop the best talent.
- Create a diverse and inclusive workforce.
- Create a psychologically safe environment for conversations.
- Understand that every employee is different.
- Strive to be functional in a dysfunctional environment.
- Build trust.

13.8 Case Studies

13.8.1 *Challenges in Aligning Multiple Teams*

Mark was the R&D manager of an engineering group in a very large technology company named XYZ Corporation. The company has been very successful in producing outstanding silicon products and achieved >80% market segment share in their category. As XYZ grew in size, its engineering productivity methodology was lagging behind the best of the breed. Mark was hired by the Executive Vice President of product development and was chartered to develop brand new tools and methodology to double engineering productivity. Mark had carte blanche authority to do whatever was required to regain its leadership in design methodology. Mark started a new geographic site and hired his employees from outside the company. Mark's team consisted of engineers from a variety of silicon development companies. They didn't understand the culture or methods of XYZ corporation and they were set on proving their methodologies could double productivity.

Due to slowing business conditions, however, the company soon enforced a hiring freeze and Mark couldn't hire any more engineers from outside the company. Mark's boss was still willing to give him the team he needed so Mark got a few hundred engineers from elsewhere in the company. They happened to be located at a different geographic location, Peach Garden, with a few hours of time difference. Mark was located at a different site called Rose Garden. Mark was very disappointed since he was promised that he could build his entire team at Rose Garden; that was one of his conditions for joining the company.

Now, Mark's team consisted of his existing new hires located at Rose Garden and an inherited team from Peach Garden. He didn't care for the skills of the employees in Peach Garden. Mark had handpicked his hires and they were highly skilled in developing new products and development methods. The Peach Garden employees are skilled at doing product revisions or proliferations with an extreme focus on cost and schedule. They didn't possess the skills to create new methods or new products. Nevertheless, Mark was still chartered to double productivity and create a brand-new leadership silicon product from scratch to prove the methodology.

The Problem

Mark was a trailblazer and was hired to change the product development methodology and improve engineering productivity. The product was "just" a means to prove his methodology and productivity gains. To him, the new methodology was the main goal, not the product. However, for the team located at Peach Garden, the production schedule was golden, and the cost was of paramount importance. The peach Garden team had taken dozens of products to market and were highly skilled in production methods that were tuned to the company's existing policies, procedures, and financial expectations. In effect, Mark's team was composed of two very different groups with different goals and success metrics.

To make things even more complicated, Mark was given a partner who was asked to co-manage the project with him. The co-manager, George, was from Peach Garden. He didn't have much experience in developing new products but was skilled at creating new methodologies. The two parts of the teams argued about the work partition. Although the Peach Garden team is less skilled, they wanted to have a fair partitioning of interesting and challenging engineering work across both teams. The Rose Garden team, however, didn't trust them enough to give important work to them. After a lot of deliberations, the Rose Garden team relented and partitioned the development ownership based on available skills in the teams.

Once the work partition got sorted out, the teams started debating on design methodologies. The Rose Garden team was trying to drive new methodologies to improve productivity — that was their mission. The Peach Garden team already had a working methodology for dozens of products they took to market. They were skilled at their methodology and didn't have any interest in changing to a new and unproven methodology. Furthermore, the Rose Garden team wasn't very schedule or cost-focused. For Peach Garden, the cost was everything — it was critical for them to hit the cost targets to achieve the gross margin expectations the company mandates. It didn't help to have two different managers since their goals, values and culture were very different.

After about a year into the execution of the project, William, the manager of Mark and George, was very unhappy that the team wasn't working well together. William kept tabs on the project by making multiple visits to both sites to gain insight into the problems. William concluded that Mark wasn't doing enough to align the teams. For example, Mark never traveled to Peach Garden. Furthermore, the feedback from teams suggested that George was thoroughly committed to the project and was striving to bring the teams together to establish the team mission (one team, one goal and one success). So, William made the hard call to give full control of the project to George. Mark stayed for a couple of quarters but ultimately decided to leave the company.

Actions Taken

George realized that the teams are not aligning due to some fundamental reasons:

- They didn't have the same goals. One wanted to implement and prove a new methodology, the other wanted to engineer a product with high gross margins.
- The Rose Garden team felt very threatened as they were a new site for the company. They were worried that the company would stop the project and lay them off.
- The teams didn't trust each other. They were assembled as the result of a hiring freeze. It wasn't their choice to work together.
- The Rose Garden team didn't trust the leader from the Peach Garden site. George didn't hire them and he was a stranger to them. He was not their leader.
- The Rose Garden team didn't understand the culture of the company. It is hard for them to make connections and align with other engineering groups. They felt like they were on an island.

George understood the problems and tackled them head-on. He decided to travel to Rose Garden regularly. George spent even weeks at Peach Garden and odd weeks at Rose Garden — no exceptions. When he stayed in Rose Garden, in the evenings he would buy the 1st pitcher of beer and invite the team members. George's 1 × 1s were frank and honest. His stated mission was "one team, one goal, and one success".

After a few months, Rose Garden team members recognized that George was competent, genuine and committed and this generated mutual trust. George also started a series of classes to assimilate all team members to embrace the mantra of "one team, one goal, one success". The classes created a common vocabulary, common goals, and common set of values. George held monthly leadership Face-to-Face (F2F) meetings to resolve project-level issues with key leaders from both sides.

During the F2F meetings, George took the initiative to invite all key leaders to his house for team building. During these events, people from both sites got to know the leaders from the other site. They realized that they are all very competent, committed, and similar in many ways. This started building pockets of trust between the sites resulting in a "one team" mentality. George insisted that they make decisions together at the F2F meeting and then abide by them with the sole purpose of making the project successful.

George also convinced the leaders from Rose Garden that nobody at the company will embrace their new methodologies until they are proven on a successful product and meet the expectations of the corporation. George also convinced the team in Peach Garden that this was a great opportunity for them to take ownership of a major new product development effort. Once George gained Rose Garden's trust, he gave assurances from William (his boss) and himself that Rose Garden will always have a bright future with a solid charter in the company to alleviate their concerns about the vulnerability of their isolated location.

Results

By investing the time to build trust across the combined organizations, George was able to mobilize his team around the motto, "one team, one goal, one success" as their rallying cry.

Questions:

(1) What was the problem Mark faced?
(2) What was the problem George faced?
(3) Whose argument would you support, Rose Garden, Peach Garden, or neither?
(4) Do you agree with George on the actions he took to align the teams?
(5) What would you do if you were in George's position?
(6) What would you do if you were in Mark's position?
(7) What would you do if you were in William's position?

References

[1] Duhigg, C. (2012). The Power of Habit: Why We Do What We Do in Life and Business. (Random House).

[2] Domingos, P. (2015). *The Master Algorithm by Pedro Domingo* (Basic Books).

[3] Tolstoy L. *The kingdom of God is within you* Available from: https://kwize.com/quote/11767) [June 2020].

[4] Kotter, J.P. (2012). *Leading Change.* (Harvard Business Review Press).

[5] Bridges, W. (1991). Managing Transitions, Making The Most of Change (Hachette Books).

[6] Polsky, G. (2018). In Search of Greatness, Nov. 2018 (Gravitas Ventures, Director Gabe Polsky).

[7] Reichheld, F.F. (2003). One Number you Need to Grow, HBR, December 2003. Available from: https://hbr.org/2003/12/the-one-number-you-need-to-grow [June 2020].

[8] Doran, G.T. (1981). There's a S.M.A.R.T. way to write management's goals and objectives (*Management Review*).

[9] Knowles, B. (Beyonce). Available from: https://www.brainyquote.com/quotes/beyonce_knowles_405120 [June 2020].

[10] Munger C. Available from: https://www.azquotes.com/quote/598838 [June 2020].

[11] Fell J. (2011). How Steve Jobs Saved Apple, Entrepreneur Magazine, October 27, 2011. Available from: https://www.entrepreneur.com/article/220604) [June 2020].

[12] Heath, C. and Heath D. (2010). Switch: How to Change Things When Change Is Hard, (Random House of Canada).

[13] Kroeger O. and Thuesen J.M. (1993). Type Talk At Work (Dell Publishing).

CHAPTER 14

Closing Thoughts

> Live as if you were to die tomorrow. Learn as if you were to live forever.
>
> — **Mahatma Gandhi [1]**

Holistic engineering management is a broad topic. In this book, we discussed the foundational concepts of culture, diversity, inclusion, value, waste, velocity and learning cycles that are key attributes of the R&D organization. We then discussed the core R&D functions of strategic planning, development, engineering research and how they can be organized and work with the rest of the R&D system. We ended with discussions on the special topics of academic engagement, information security and leadership.

In this chapter, we provide a summary of the important directives for the manager of the R&D organization from each chapter. Since people are the key to the success of a holistic engineering R&D organization, we also provide some additional thoughts on how the manager should look after themselves and their employees to ensure everyone remains healthy, happy and full of energy. Finally, we touch on recent events in the world and their implications on the continuing evolution of engineering work.

14.1 Key Directives for the R&D Manager

14.1.1 *Foundational Concepts*

Chapter 2, Culture: Ensuring that the actual culture of an organization aligns with the strategy and values needed to win is the most important thing the R&D manager can do to guarantee success. This requires that the manager and their leadership team role model the values every day, in all situations, every time. Any separation between what they say and what they do will be viewed as evidence that the values do not need to be taken seriously.

Chapter 3, Diversity and Inclusion: Whereas diversity is measured by the variety of the workforce, inclusion is measured by the perceptions of the individuals within the workforce. Therefore, the content for this chapter summarized the conversations with several people on their experiences and how to be more inclusive. A key message is that the R&D manager must reinforce the right behavior in others and be a role model by ensuring that everyone they work with feels included and truly appreciated.

Chapter 4, Value, Waste and Velocity: A clear understanding of value enables everyone to analyze their workflows and reduce the time spent on wasteful tasks. Less time spent on wasteful tasks means more time can be spent on the tasks that provide value and increase velocity. Agreeing on what is value, unnecessary waste and necessary waste in different workflows or processes is not always straightforward and requires conversations. It all starts with a clear message from the R&D manager on what value the organization provides.

Chapter 5, Learning Cycles: The completion of each task generates information on the effectiveness of the engineering decisions and the operation of the workflows and processes. When this knowledge is used to improve subsequent engineering decisions or incrementally improve the workflows and processes, then a learning cycle has been completed. As a result, learning cycles are a critical part of continuous improvement. The R&D manager must create the culture and training needed so that everyone can become a scientist and continuously conduct experiments on how to improve their own engineering decisions, workflows and processes.

14.1.2 *Core Engineering Functions and their Organization*

Chapter 6, Strategic Planning: Strategic planning comprehends the competitive and cooperative forces in the entire R&D ecosystem and determines the best mix of differentiable value for the R&D organization and the means to achieve it. The R&D manager must actively manage a continuously repeating strategic planning loop that observes the R&D ecosystem, synthesizes new information, makes strategic decisions, translates those into specific objectives and enables action.

Chapter 7, Development: Development focuses on the implementation of the engineering work and the required coordination with other entities in the R&D system. It also includes the continuous improvement of the whole value stream to increase velocity as measured by the customer. To succeed, the R&D manager must establish standards and a development cadence. They must also use a framework for making the right trade-offs that will keep the business running, maximize value, increase velocity and reduce risk across the whole R&D system.

Chapter 8, Engineering Research: Engineering research increases the power of engineering by creating new knowledge and more powerful modeling, analysis and optimization methods to solve design problems. The R&D manager must create the culture and processes where induced and disruptive research thrives and value is realized without unnecessary delay. This means ensuring research that is not dependent on others can proceed unencumbered and more complex research, that is dependent on others, follows a process that engages the correct people at the right time.

Chapter 9, The R&D Organization: The R&D organization includes the employees and other assets in the direct chain of command of the manager. It also includes the governance, capabilities, structure, networks and reward systems needed to execute the strategy with high velocity. The R&D manager must master working in a matrixed environment and needs to create the right balance of power using networks and the formal reporting hierarchy while instantiating processes that speed decision-making and knowledge sharing. The R&D manager must also take responsibility for inculcating the right culture.

Chapter 10, The R&D System: The R&D system is the specific collection of entities, assembled and directed by the manager of the R&D organization to work together and deliver the engineering product or service. This includes the R&D organization along with its suppliers, customers, institutes, consortia, and others. The manager of the R&D organization must decide when and how to engage. This requires investing in the culture, infrastructure and processes needed to enable collaboration and the discipline to focus on areas where the objectives are clear and the return on investment is high.

14.1.3 *Special Topics*

Chapter 11, Engagement with Academia: There is a great opportunity for both the R&D organization and a university to benefit from such things as informal conversations, joint projects and student internships. Whereas academic success is determined by delivering educated students and the discovery of truth, the R&D organization's success is determined by delivering value to the customer. Because of this, the R&D manager must select the right projects to engage with academia and ensure everyone understands the objectives of the other parties.

Chapter 12, Information Security: Information security protects such things as the details of the technology, organization, process and development schedules from potential attackers. The manager of the R&D organization must have a basic understanding of security concepts to continuously assess the probability and potential impact of attacks and correctly allocate resources. Most of all, they must make the information security mindset part of the culture of the R&D organization.

Chapter 13, Leadership: Leadership is a combination of dreaming and doing. It spans the fuzzy front end of articulating a big and bold vision, translating that into a strategy that creates differentiable value and then motivating, funding and orchestrating all of the entities in the R&D system to work harmoniously together to deliver results with the greatest velocity. To be an effective leader the R&D manager must balance the traits of curiosity, learning and an open mind with a problem-solving mindset and a laser focus on results. Most of all, an effective leader knows that their success is based on the trust people have in them to do the right thing, the right way, for the right reason and they work hard to maintain that trust.

14.2 Life is More Than Work

The successful R&D manager realizes that success is dependent upon the people. The first person the R&D manager must consider is themself! Not surprisingly, this is the one person that can be most directly influenced by the manager. In addition to understanding and implementing all of the concepts described in the book, getting the peak performance out of the R&D manager is dependent upon maximizing their mental and physical health. The R&D manager is very good at planning and prioritizing the workflows for others, and they must also apply those skills to their own lives. Physical and mental exercise, eating well, family life, and hobbies increase stamina and broaden perspectives. This makes the R&D manager better able to provide the effort needed to direct a complex R&D system. Not to mention, add value to their life in general.

In addition to recognizing and embracing the need for managing their well-being, the R&D manager must also encourage and enable the employees to do the same. Of course, the manager cannot (nor should they) force employees to take care of themselves outside the workplace. However, the manager does have a responsibility to encourage and enable healthy individual behavior by being flexible when employees are conscientiously trying to do this for themselves. As we have noted several times in this book, the successful engineering R&D manager, understands that a high-velocity organization depends on healthy, happy and energized employees.

14.2.1 *Investing in Personal Health*

Throughout this book, we have mentioned that it is critical to have a mindset that enables continuous learning and adaptation. This can be exhausting and is difficult to sustain unless the R&D manager is mentally and physically fit. Doing so requires discipline and healthy habits. If the tyranny of the urgent dominates one's life, then looking after one's health will take a backseat. This results in a negative spiral of work and health-related problems. To avoid that, the R&D manager should build physical fitness into their routine. As we already discussed, learning is hard and can cause extreme mental fatigue. Physical fitness is a prerequisite to enable the stamina and mental freshness to think clearly and learn effectively.

R&D managers should devise their regimen for staying healthy. This means a healthy diet, restful sleep and regular and adequate exercise. The best way to support that goal is to commit oneself to an activity they thoroughly enjoy. For example, the doctor of one of the coauthors of this book told them that walking 3 miles per day was a good proxy for sustaining good health. This is about 20 miles per week. The co-author translated that into 1,000 miles/year (assuming 2 weeks of vacation). They committed themselves to this goal about 15 years ago and track progress every week to ensure they hit the 1,000-mile annual goal. Similarly, another co-author has a daily exercise goal that includes either a 50-minute walk, swimming 1,800 yards or a 40-minute stationary bike ride. Both can attest to the fact that regular exercise like that has increased their ability to stay focused during long days at work. The first step is to take a specific goal and then to track progress much like the R&D manager would do with a project they managed at work. Tracking progress creates motivation and is a catalyst to help achieve the goal.

Sleep is also important. It is about maintaining a rhythm and sticking to it. One co-author uses a Fitbit tracker to monitor their sleep. This helps identify and avoid the circumstances that led to a bad night's sleep. Another co-author uses an application on their phone that is designed to help users fall asleep by sharing stories read soothingly. In general, the R&D manager must plan the day's activities to allow 8 hours of sleep to feel fresh every day.

A good healthy diet is a topic by itself. There are plenty of books written about this subject that can be used as a resource. At a minimum, the R&D manager should commit to healthy eating habits. Replacing the jar of candy in the office with something healthier is a good start. The R&D manager should also commit to the right number of physical check-ups every year. Even if the R&D manager feels fine, this is a good idea since it will identify and address health issues proactively.

Finally, to refresh the learning skills it is important to become a life-long learner and explore your curiosity on subjects that may not be related to your work but excite and interest you. For a healthy balance, we believe in a model called 3H (Head, Heart, and Hands). The word *Head* stands for the intellectual stimulation of the mind which potentially comes from work and hobbies, Heart is about the generosity of giving to others, Hands is about physical well-being. We believe that the R&D manager who maintains a healthy long-term balance of three H's will be mentally and physically fit enough to meet the challenges of the job.

14.2.2 *Enabling Employees Balance*

This is a critically important but tricky subject as the onus of finding this balance is left to the discretion of an employee. As every employee goes through the stages of life, the balance may vary significantly. When one of the co-authors started working for Intel, they were so enamored by the nascent technologies and limitless possibilities at work that they were working 14 hours/day and every weekend. They had no balance (it was all work), but they were having a time of their life working, learning and adapting to the new world. They were living alone and had no other responsibilities or commitments at that time. It was extremely rewarding to learn, contribute and make a substantial impact on the company's products. They wanted to learn rapidly and had aspirations to grow in scope and responsibility. It was their choice, and nobody demanded them to work those hours.

Much later in their life, they became senior leaders of hundreds of employees. They also had to become an expert at juggling their family, work and health. They had to learn to trust, empower and enable their direct reports to work independently to accomplish the objectives. They also had to learn to become extremely disciplined, focused and efficient with their time so that they could accomplish their work duties in a 60-hour work-week. To do this they had a personal assistant diligently schedule important family milestones and occasions and block that time on their calendar. They had an agreement with their manager to support their balance and give them flexibility when needed. By doing so, they were able to attend important personal life events without adversely affecting their work. For more on this, we highly recommend a book written by one of our role models, Pat Gelsinger (who is currently the CEO of VMWare and wrote the foreword for this book), it is titled *The Juggling Act : Bringing Balance to Your Faith, Family, and Work* [2].

Ultimately, the R&D manager should encourage every employee to comprehend this balancing act early in their career. The R&D manager can work with the employee to help them to make a choice that best fits their needs and aspirations at that time in their career. It is critical that an R&D manager genuinely believes and supports the choice that is made. For an employee, it is a balancing act between work-related aspirations and family commitments which vary widely over time.

We strongly believe one can thrive at work and meet all family commitments with the right planning and discipline. However, it requires flexibility from the employer and employee and the R&D manager can be instrumental in enabling the employee to make the choices and allowing them to work effectively when conflicts arise. With today's mobile devices, high bandwidth connectivity, video conferencing applications and flexible work policies, it is easier than ever to pursue both work aspirations and family/personal commitments.

We also recommend that every employee should have a trusted mentor who can also help them evaluate and choose the appropriate path that fits their needs. A great resource for this is Andy Grove's article "How to be a Mentor — If you turn teaching into a routine, 'you screw it up'"[3]. In this case, the R&D manager educates and enables employees while the mentor helps the employees make decisions by providing valuable counsel and perspectives on an on-going basis. Great companies have many good senior role models to choose from. Corporate programs for finding a good mentor rarely work so the onus of finding a mentor and developing a trusting and productive relationship falls squarely on the shoulders of the employee. Sometimes, a mentor from outside the company is suitable.

14.2.3 *The COVID-19 Disruptive Event*

Occasionally, a major worldwide catastrophe occurs that changes everyone's life. The two world wars were examples. Recently, the COVID-19 pandemic has also changed human interaction. Disruptions of this magnitude cannot be predicted and mitigated in advance. However, in the words of Joseph Heller [4]:

> *"Just because you're paranoid doesn't mean they aren't after you."*

This is also true for the COVID-19 pandemic. Just because many people have been incorrectly predicting some kind of disaster was just around the corner, it did not prevent one from happening.

One place where the change due to the pandemic is visible is in the employees' work environment. The assumption up until now for most R&D organizations has been that the workforce was physically at a company location. Many companies had explicit policies discouraging employees from working from home except for personal emergencies. However, during the COVID-19 pandemic, working from home became the dominant work environment and this is a trend that will likely continue. A change like that tests the organization's culture. A culture that emphasizes enabling autonomy and individual responsibility will continue to succeed even when employees work remotely. However, a culture based on strict oversight and micromanagement will run into difficulties.

For example, a strict micromanagement culture cannot adjust for the fact that labs were not designed for social distancing. However, a culture that enables autonomy and is aligned with the R&D organization's strategy and values will find a way to get the job done in the absence of direct oversight.

The new work model defined by the pandemic also presents an information security problem since it increases the probability of confidentiality breaches and security attacks. For example, there are existing technologies that can fake the identity of a person in a video conference (including their video) by stitching the identity of a legitimate attendee to an intruder. There also is a greater possibility of IP leakage due to the ability of people to take pictures of their screens at the other end of a virtual meeting. All of this points to the need to change the approach used for information security and IP protection.

During an event like the pandemic, the R&D manager must also modify policies, capital expenditures, communication channels and priorities to enable the team to still deliver value.

In the past, most equipment needed to stay on the company premises. Now, items that enable a person to perform their tasks may need to be taken home. Capital equipment still needs to be accounted for and COVID-19 has forced a rethinking of how to allocate and position resources. Ultimately, the manager still must decide what is necessary and sufficient for a task to be completed vs what is just nice to have, but the context has changed.

In terms of the employees' productivity, working from home can also present problems. Some people have a home office and a relatively straightforward transition from working at the office to working from home. However, most people do not have a natural location to work at home. The internet might be available, but it is a family resource, not an individual resource. The computer might be right next to the home entertainment center. The chairs and tables used may not be properly set up for ergonomic usage. Also, employees may have members of a family at their home who may expect attention.

The new work environment also strains the ability to be inclusive since it is difficult to replace the intimacy of informal face-to-face, one-on-one meetings in a room or even those that happen by chance in a hallway. Informal symbols of acceptance and inclusion are also absent, such as the old fashioned handshake. Also, simple things once taken for granted have changed. Communication may be intermittent. Meetings can be disrupted when the local connection to the internet fails and the rest of the meeting continues without even noticing a key employee is gone.

As we stated, the right culture can help individual employees still do the right thing when an event like COVID-19 occurs. However, the engineering manager must also show leadership when making decisions based upon information or the lack of it. For example, during the pandemic, different cities had a different stay at home rules. These were outside the manager's control and changed quickly. In that situation, it did not help to complain. When things like that happen, the R&D manager needs to work within the environment to keep the team producing value. The manager must mix a focus on execution and results with the reality that the employees are in a very difficult position.

In summary, COVID-19 has demonstrated that a major worldwide event like a pandemic will challenge even the most successful engineering R&D manager who understands all of the concepts discussed in this book. Such events may make it more difficult to build the relationships that form the basis for holistic engineering management. However, the bottom line is still that the culture combined with the manager's leadership will ultimately provide the essential ingredients for success when the R&D organization is faced with disruptive events that are beyond their control.

References

[1] Goalcast (2017). Top 20 Most Inspiring Mahatma Gandhi Quotes, March 20, 2017. Available from: https://www.goalcast.com/2017/03/20/top-20-inspiring-mahatma-gandhi-quotes/ [accessed on July 2020].

[2] Gelsinger, P. (2003). *The Juggling Act: Bringing Balance to Your Family, Faith & Work*. (Colorado: David C Cook Publishers).

[3] Grove, A. S. (2011). How to be a Mentor — If you turn teaching into a routine, you screw it up. *Businessweek*, September 22, 2011. Available from: https://www.bloomberg.com/news/articles/2011-09-22/andy-grove-how-to-be-a-mentor.

[4] Heller, J. (1961). *Catch-22*. (New York: Simon & Schuster).

Glossary

A3 Problem Solving Process: A systematic process for problems solving named after the one-page summary of the description of the problem, current state, ideal state, countermeasures and action plan.

Absorptive capacity: An entity's ability to identify, assimilate, transform, and use external knowledge, research and practice. It is a measure of the ability to learn. — The Oxford Review.

Academic Research: Creates new knowledge that provides a basis for new understanding, new models, and new procedures and relates the new knowledge to the existing body of knowledge.

Access control: Access control is a mechanism that determines whether to allow or deny access to the requested data, equipment, or facility.

Academia: A term used to describe the community professors, students and schools that conduct research and educate. In this book it refers to research Universities.

AES: Advanced Encryption Standard.

Agile Manifesto: Twelve principles describing agile software development.

Agile: A set of practices based on the values and principles expressed in the Manifesto for Agile Software Development and the 12 Principles behind it.

Amdahl's Law: a formula which gives the theoretical speedup in latency of the execution of a task at fixed workload that can be expected of a system whose resources are improved.

Always Shippable Code: Software that is continuously integrated and validated such that at any point in time the latest contour can be shipped to a customer for use.

Asymmetric Encryption: An encryption algorithm that has a different key for encryption than decryption, thus the encryption key is made public so anyone can send a secret but only the possessor of the second key (private) can read the secret. It is also known as public key encryption. An example is RSA.

Authentication factor: An authentication factor is data or a measurement used to validate an identity claim.

Autonomous Strategic Action: Actions that originate bottom-up within the R&D organization and are not aligned with the current strategy.

Backlog: The ordered list of everything that is still needs to be completed to complete a project.

Backlog Grooming: The continuous effort to update the list of requirements in the backlog.

Burn Down Chart: The number of tasks in the backlog left vs. time.

Breach: A successful attack on a secure system.

CAD: Computer Aided Design.

Ciphertext: Ciphertext is an encrypted message.

Collaboration: The act of working with another person or entity to create something.

Company: A commercial business composed of people and assets that conduct the business.

Competitor: An entity that is engaged in commercial competition with the R&D system or its company.

COMSEC: Communication Security.

Complementor: An entity that provides a product or service that increases the value of the product or service of another entity. A commonly referred to example is Microsoft and Intel.

Concurrent Engineering: A process of refining and narrowing down multiple engineering options in parallel to converge on the final answer when uncertainty of market success is lowest.

Consortia: A group of companies and other entities that pool resources to work together on specific topics of mutual interest such as standards and pre-competitive research. Examples include the Semiconductor Research Corporation (SRC) and the World Wide Web Consortia (W3C). Consortia can be the vehicle for coopetition where entities that are usually competitors work together on topics that benefit each other or the broader industry.

Continuous Integration: Changes in a unit are immediately added to and tested in the system.

Coopetition: Cooperation between competing companies.

Culture: The beliefs of a group of employees such that when faced with ambiguous or incomplete information they instinctively know what to do.

Current State: The description of how everything related to the topic actually works today.

Customer: An individual, group or company that receives and uses the product or technology. In this book the customer usually refers to the entity that receives the engineering product or service delivered by of the R&D organization.

Cyber Physical System: A Cyber Physical System is a physical system with an embedded computer control. It contains various physical devices being controlled by system software and measurement devices.

DARPA: Defense Advanced Research Projects Agency.

DDOS: Distributed Denial of Service.

Decryption: Decryption is the algorithmic modification of a ciphertext using a key to recover the plaintext content of the message.

Denial of Service attack: A Denial of Service (DoS) attack is flooding a server with repeated service requests, thereby denying its services to other requestors.

DES: Data Encryption Standard.

Development: All of the engineering activities required to deliver the specifications needed to build the technology and meet the requirements of the customer. It also includes continuous improvement of the development processes to increase velocity across the entire value stream.

Disciplined Collaboration: Collaboration that is based on measurable objectives.

Distributed Denial of Service attack: A Distributed Denial of Service (DDoS) attack uses multiple clients on the Internet to repeatedly send excessive numbers of requests to a server. It prevents the recipient from providing services to other requestors. Frequently this causes a website to completely crash.

Diversity: Diversity is measured as the variety of characteristics (e.g. attitude, culture, ethnicity, gender, geography etc.) of a work group. See also Inclusion.

DOS: Denial of Service.

Eavesdropping: A security attack where the Attacker (Eve) intercepts data traveling along the network between the sender (Alice) and the recipient (Bob), but does not alter the data. This is also called a passive attack.

Ecosystem: See R&D eco-system.

EDA: Electronic Design Automation.

Efficiency: A measure of how well an input produces for a specific output. Usually expressed as a ratio Output/Input. The units of measure for the inputs and outputs will differ depending on the field and industry.

Electronic signature: An electronic signature is a secure hash of a message with its author's identification. This electronic hash is often used as a replacement of a physical signature on a paper document.

Encryption: Encryption is the algorithmic modification of a message using a key to convert plaintext to ciphertext. This makes it difficult for an attacker to read the content of the message except with a key.

End of Life (EOL): The point when a technology or product is no longer worked on or supported. Customers us the technology at their own risk.

Engineering: "The systematic application of scientific knowledge in developing and applying technology". — AAAS.

Engineering Research: Creates new knowledge and more powerful analytical, modeling and optimization methods that increase the power of engineering and enables solving design problems that give the company a competitive advantage.

Entity: This is a common term for a company, group, or individual.

Error Checking hash: Error checking hash is the calculation of a short fixed sized number based on the content of a message. It is used to check whether the integrity of a data or message has been compromised.

Execution: The act of completing a task.

External Company: A company that is separate from the R&D system's parent company.

External Customer: A customer that is not in the same company as the R&D organization. Also see: customer.

External Supplier: A supplier that is in a different company than the R&D organization. Also see: supplier.

External Complementor: A complementor that is not in the same company as the R&D system.

Extreme Programming: An agile software development framework that aims to produce higher quality software, and higher quality of life for the development team.

Extrinsic Motivation: Behavior that is driven by direct compensation, rewards or punishment regardless of whether or not it is personally satisfying.

Fab: Fab refers to a fabrication plant for manufacturing Silicon Chips.

False Match Rate: False Match Rate (FMR) is the rate at which authentication evaluation allows an imposter access because measured data incorrectly matched expected data.

False Non-Match Rate: False Non-Match Rate (FNMR) is the rate at which the authentication evaluation prevents an authorized user access because the measured data didn't match the expected data.

Fault Tolerance: Fault tolerance is the property of a computer system to keep functioning properly in presence of a hardware or software fault.

Flow: The series of connected tasks that are usually based on a specific methodology.

Flow Testing: The testing of complete flows using representative test cases.

Gate Keeper: A barrier that is opened when a specific criterion is met (e.g. no defects discovered).

Gemba: A Japanese term meaning "the real place". Go to Gemba is lean phrase that refers to going to where the work is being done to directly observe and understand the current state.

Gratuitous Development: The act of creating a new feature, technology or product that does not provide any real value in the eyes of the customer. These usually happen when the developer's perception of value is based on their own opinions and not those of the customer.

Group: Formal organizational unit with direct reports. Groups exist at all levels of the company hierarchy.

Hash Checksum: Hashing for error checking or authentication produces a checksum. This can be a simple count or parity of the data stream or be calculated using a Linear Feedback Shift Register (LFSR).

Hashing: Hashing is the calculation of a short, fixed size number based upon the content of a message. Hashing can be used for error checking, authentication, and integrity checking.

Holistic: Relating to or concerned with wholes or with complete systems rather than with the analysis of, treatment of, or dissection into parts. — Merriam Webster Dictionary.

Holistic Engineering Management: The authors perspective on how the engineering manager should direct an R&D system that includes suppliers, customers and others to continuously increase velocity. At its heart is a common culture of learning, adaptability and value creation and overseeing it all is a leader that includes and fully utilizes the contributions of everyone.

Ideal State: What the complete fulfillment of the vision looks like. It may not be totally achievable but it is used to guide decisions to incrementally move closer and closer to it.

Identity Authentication: Identity authentication is determination that a claimant is who it claims to be. Identity authentication uses a process or algorithm for access control to evaluate whether to grant or deny access.

Inclusion: Inclusion is measured by the perception of the individual that their contributions are valued and that they are valued and their comments are heard. See also Diversity.

Individual: A person who reports into a group and may participate in working groups. Individuals exist at all levels of the company hierarchy.

Induced Strategic Action: Actions that are influenced by the current strategy and expectations for how everything will evolve going forward.

Information Security: Information security is the protection of information confidentiality, integrity and accessibility.

Information Turn: The time taken from specifying the experiment, conducting the experiment, analyzing the results and defining the next experiment based on those results.

Internal Customer: A customer that is in the same company as the R&D organization. See: customer.

Internal Supplier: A supplier that is in the same company as the R&D organization. See: supplier.

Intrinsic Motivation: Behavior that is driven by a sense of personal satisfaction and not extrinsic items such as compensation, rewards or punishment.

Institution: A society or organization founded for a purpose. In this book it refers to societies or organizations formed that further the objectives of engineering and technology. They may be public or private. Some are a source of research (e.g. universities and colleges) and others help with the development of engineering community and set standards (e.g., the Institute for Electrical and Electronic Engineers (IEEE)).

Intellectual Property (IP): Property (such as an idea, invention, or process) that derives from the work of the mind or intellect. — Merriam Webster Dictionary.

Kaizen: A philosophy of continuous improvement of workflows.

Kaizen Event: A continuous improvement effort that is usually face to face and done in a week.

Kanban: A pull system that helps visualize work in progress and deliver results just in time.

Key Distribution: The method or algorithm to solve the problem of getting the keys in a secure system that uses encryption to only the desired participants.

Key Generation: The method or algorithm to generate the keys in a secure system that uses encryption.

Key Management: The problem and solutions involving key generation, key distribution, key protection and key updating.

Knowledge Brief: One-page summary of the problem, current state, ideal state, proposed next state and required actions. The K-Brief is a concise means to communicate and track progress.

LAMDA: Look-Ask-Model-Discuss-Act. An example of how to complete a learning cycle. Initially described by Allen C. Ward.

Leadership: Defining and communicating the "what" and "why" of a vision in a way that aligns employees. Driving capabilities that enable the vision and removing obstacles that get in the way. See also management.

Lean: A product development philosophy based on the principles observed at Toyota.

Lean Principles: Principles that guide the lean product development philosophy.

Lean Rules: Basic rules for how employees construct and improve the value stream.

Learning Cycles: A loop that starts with action and ends with absorbing feedback on the results.

Little's Law: Asserts that the time average number of customers in a queueing system is equal to the rate at which customers enter the system × the average waiting time of a customer.

Management: Selects which tasks will be done and then defines and directs "how", "when" they will get accomplished. See also leadership.

Management Waste: Waste that results from the actions of management.

Manufacturing Waste: Waste that is related to manufacturing.

Manager: Someone who leads a group with direct reports that combines the management and leadership roles mentioned above. Also see: R&D manager.

Mission Critical Tasks: Tasks that must be completed to ensure the day to day operation of the company is not put at risk.

Meltdown: Meltdown is a side channel attack that takes advantage of some speculative execution implementation details.

Multi-factor authentication: Multi-factor authentication uses more than one factor to determine authorization. Commonly used factors are: information, biometrics, physical device, and location. Example factors are: password, fingerprints, car key, and GPS coordinates.

Necessary Waste: Time spent on tasks that do not create value but are still required.

New Entrants: New competitors that enter an existing industry.

NIST: National Institute of Standards and Technology.

NSA: National Security Agency.

OODA (Observe, Orient, Decide, Act): A decision model developed by USAF Colonel John Boyd.

Overproduction: Producing more of a product that is needed or wanted.

Pair Programming: A technique in which one programmer writes code while the other reviews each line of code as it is typed in.

Pathfinding: A jointly owned process between those doing research, the consumers and those who will need to implement and deploy the result. The goal of pathfinding is to enable the shortest path to adoption for research with realizable value. It uses short steps to build consensus based on data.

Plaintext: Plaintext is data to be encrypted to yield ciphertext.

Plan Do Study Act (PDSA): A process improvement model defined by Dr. W.E. Deming.

Plan Do Check Act (PDCA): A popular variation of the PDSA problem solving model.

Privacy: Privacy is the protection of personal information.

Product: The usable form of technology that is provided to a customer.

Productivity: A metric used to measure worker or group efficiency. It is the volume of production vs. the input (often time or labor costs). Productivity = Production/Input. The units of measure for the inputs and outputs will differ depending on the field and industry.

R&D: Research and Development.

R&D Ecosystem: The entire network of entities that are affected by each other even if they do not work together directly.

R&D System: A subset of the R&D eco-system. A specific collection of entities, assembled and directed by the R&D organization to work together and deliver the engineering product or service.

R&D Manager: The name used to represent the manager of the R&D organization. The R&D manager manages the R&D organization and assembles and directs the R&D system. Sometimes referred to as the "manager".

Regression Testing: Testing to discover defects after changes are made to the unit.

Release Candidate: A module that is ready to be integrated with the rest of system for release.

Release Contour: The list of all software utilities and versions released with the system.

Research (Engineering): See Engineering Research.

Research (Academia): See Academic Research.

Research and development: The tightly interdependent processes of strategic planning, research and development that result in the design of new or improved technologies.

Retrospective: A review of completed work to learn and apply improvements for the next cycle.

RSA: Rivest–Shamir–Adleman.

SCRUM (aka Scrum): A development practice based on iterative development cycles and self-managed teams.

Scrum Master: Facilitates the day to day work performed in a Scrum.

Secure Hash Function: A function that generates a relatively small unique digest from a very large set of data used to verify that the large set of data is authentic and unaltered. An example is SHA3.

Security Attackers: Entities attempting to read, modify, or otherwise interfere with information with the intent to do harm.

SHA3: Secure Hash Algorithm 3.

Side channel attack: An attempt to create a security breach by indirectly attacking the secured information, such as guessing a secret by measuring power supply current.

Single factor authentication: Single factor authentication uses only one factor to determine authorization. Example factors are: password, fingerprints, car key, or GPS coordinates.

Six Sigma: A statistical measurement of quality or likelihood of an even 6 standard deviations away from the mean.

Software Development Waste: Waste that is related to software development.

Speaker identification: Speaker identification (or voice recognition) uses audio signals to identify a speaker. One use of speaker identification is for authentication.

Spectre: A side channel attack that takes advantage of some speculative execution and indexed addressing implementation details.

Spoofing: Pretending to be someone else in order to gain access, circumvent authentication, or trick a user into revealing secrets.

Sprint: Usually described in terms of a Scrum. It is time boxed development that includes planning, execution and delivery of working software followed by a team retrospective that then feeds into the next sprint. Sprints are typically 1–2 weeks in duration and the final complete software product may take several sprints to complete.

Standard Deviation: A measure of the spread of values for data, referred to as sigma.

Standards: Something that has been established by an organization or institution as a model to which others should conform.

Strategy: A direction that addresses the forces that have the potential to materially affect the company's destiny.

Strategic Action: consequential actions related to the strategy that cannot be easily undone.

Substitutes: Technologies that does not immediately compete with a product but makes it irrelevant. It becomes a substitute for the prior technology.

Supplier: An entity that delivers a product or service to a consumer. In this book it usually refers to entities that currently or potentially may deliver to the R&D organization. An Internal supplier is from within the same company, an external supplier is from another company.

Symmetric Encryption: An encryption algorithm that has the same key for encryption and decryption, thus the key is kept private and must be privately communicated between the sender and receiver. It is also known as private key encryption. An example is AES.

System testing: Testing the complete, integrated system in a way the user will utilize it.

Tactics: Actions whose outcome do not significantly affect subsequent freedom to act.

Technical Debt: The accumulated work that is required after a new feature is released to test, fix and update it on subsequent releases.

Technology: A capability given by the practical application of knowledge. — Merriam Webster Dictionary.

Technology Development: Provides the knowledge needed to create and build the technology.

Test Driven Development (TDD): Requirements are defined using specific customer test cases and data. Development completes when all tests pass.

The Company: The company that the R&D system is part of. In the case of a small company where the R&D system is the company then those two terms are equivalent.

Threat Model: Threat modeling is an approach for analyzing the security of an application. It is a structured approach to identify, quantify, and address the security risks associated with an application.

Time Boxed: Time is allocated for a task. Work stops when the time expires to allow retrospection and improvement for the next cycle. Uncompleted tasks will move to the next cycle.

Total Quality Management (TQM): A set of management practices to help companies increase their quality and productivity based on Dr. W.E. Deming's 14 key points for management.

Toyota Production System: Toyota's product development philosophy and practices.

True Match Rate: True Match Rate (TMR) is the rate at which the authentication evaluation allows an authorized user access because the measured data correctly matched expected data.

True Non-Match Rate: True Non-Match Rate (TNMR) is the rate at which the authentication evaluation prevents an imposter access because measured data didn't match the expected data.

Trusted Compute Boundary: Trusted Computed Boundary (TCB), or Trust Boundary, refers to a region that consists of devices and systems on a platform that are considered safe. Sources of data outside of TCB are untrusted and thus checked for security violations, such as network ports. In general, TCB on a system should small to minimize the possibility of security attacks.

Trusted computing: Trusted Computing refers to a situation, where users trust the manufacturer of hardware or software in a remote computer, and are willing to put their sensitive data in a secure container hosted on that computer.

Unit Testing: Testing individual units of hardware or software with associated data. Passing unit testing usually gates inclusion into the system for complete system testing and release.

Unnecessary Waste: Time spent that adds no value and should not exist in an ideal world.

Value: Contributes to the company's vision in a way that the customer will pay.

Values: The principles that define an entity's desired culture and employee behavior.

Velocity: A measure of the rate of deployed value vs. time (dV/dt).

Verification: Verification is the process of establishing the truth, accuracy, or validity of something.

Waste: Time spent on tasks that do not create value.

WIP: Work in Progress.

Zombie: A computer system that has been successfully infected by an attacker to launch robot attacks such as DOS.

Index

A3 process, 132
academia, 4, 13, 239, 286, 288,
  320, 323, 339, 357–364, 366,
  367, 369, 370, 398, 402, 454
academic community, 66
academic research, 360, 359,
  364, 370, 373, 403
access control, 14, 378, 379, 383,
  388
actual culture, 30, 37, 38, 40, 43,
  46, 50, 267, 289, 451
Advanced Encryption Standard
  (AES), 396
After-Action Review, 134
Agile, 7
agile development, 4, 21, 117, 142
Agile 'Software Development'
  Manifesto, 118
Apple Way, 141
applications support, 199, 210,
  344
applied research, 359
asymmetric, 396
Asymmetric Encryption, 396
attackers, 12, 16, 375, 378, 381,
  386, 400
authentication, 394, 395, 400
autonomous strategic action,
  153, 168, 179, 225, 228, 229,
  294, 422

bad tension, 44, 269
bias, 74, 75, 76, 100, 245
blue ocean, 202, 231

ciphertext, 396
collaboration premium, 316,
  319, 323, 326
color celebrate, 75
communication, 38–40, 48, 65,
  78–80, 100, 122, 136, 169, 248,
  276, 304, 313, 331, 377, 396,
  418, 419, 444
Company Vision, 83–89, 223,
  308
competitor, 12, 16, 25, 56, 105,
  149, 151, 155, 164–167, 177,
  184, 208, 225, 226, 231, 234,
  302, 303, 307, 318, 363, 382,
  391, 411
complementor, 165, 166, 231
concurrent engineering, 9, 123,
  124, 182, 201, 204, 205, 210,
  211, 425, 428
concurrent learning cycles, 121
conditioned oblivion, 76
consortia, 3, 12, 13, 15, 155, 183,
  184, 224, 225, 234, 239, 276,
  277, 288, 301, 302, 306, 307,
  323, 327, 353, 357, 369, 372,
  454

continuous improvement, 8, 11–13, 18, 34, 35, 49, 51, 77, 100, 101, 117, 118, 125–128, 130, 131, 135, 142, 143, 181, 182, 193, 194, 199, 201, 202, 209, 262, 271, 276, 277, 280, 286, 288–290, 292, 294, 296, 301, 302, 305, 314, 320, 325–327, 383, 398, 403, 426, 427, 452, 453

coopetition, 1, 12, 13, 15, 165, 301–303, 327

COVID-19, 459–462

culture, 3–8, 10, 11, 13, 14, 21, 22, 29, 31, 32, 34, 36–45, 47–50, 53, 54, 58, 61, 65, 69, 72, 73, 80, 83, 96, 102, 105, 107, 118, 127, 128, 143, 154, 160, 169, 193, 209, 222, 228, 231, 232, 261–263, 265, 267, 269–271, 276, 278, 279, 289, 290, 296, 301, 304, 307, 314, 315, 319, 322, 324, 326–328, 335, 338, 344, 352, 361, 367, 369, 375, 378, 380–385, 390, 391, 397–399, 401, 403, 429, 436, 445, 447, 448, 452–455, 460–462

customer, 2, 3, 5, 6, 11–13, 21–23, 32, 33, 35, 36, 39–43, 46, 52, 54, 55, 58–60, 65, 67, 83–93, 95–97, 100, 103, 104, 107–109, 112, 114, 115, 124, 125, 130, 132, 136, 139–141, 145, 146, 149, 151, 155, 160, 162–167, 176, 177, 179, 181–184, 191–199, 201–205, 207–211, 218, 221–223, 226, 229, 231, 233, 238, 239, 241, 242, 250, 252, 256, 268–270, 274–277, 283, 285, 286–288, 301–310, 318, 326, 327, 330, 340, 349, 353, 357, 358, 361–363, 369, 370, 380, 384, 390, 394, 411, 419, 421, 428, 453, 454

Cynefin framework, 291

decryption, 396

design automation environment, 190

Design Thinking, 130, 139, 130, 141, 143

development, 1–3, 8, 10, 11, 15, 19, 21–23, 31, 44, 48, 83, 84, 87, 90, 92, 93, 96, 97, 107, 109, 112, 117–121, 123, 124, 128, 132, 133, 136, 137, 139, 142, 143, 146, 150, 155, 166, 167, 181–188, 190, 193, 195–197, 199–211, 217, 218, 221, 224, 227, 231, 235, 244, 248–254, 256–259, 261, 263, 267, 272, 273, 277, 286–288, 294, 298, 304, 305, 311, 313, 315, 317, 322, 324–326, 336, 337, 339, 340, 342, 343, 346–348, 350, 351, 358, 366, 368, 372, 375, 382, 383, 391, 402, 420, 421, 424, 425, 430, 439, 445–447, 449

   development flow, 19, 93, 109, 132, 191, 192, 194, 196, 208, 209

   development velocity, 181, 191–193, 196, 209, 210

development environment, 187
development function, 9
diet, 456, 457
discrimination, 73
disruptive innovation, 1, 225,
   230–233
disruptive research, 10
diversity, 3–6, 10, 41, 52, 65–71,
   73, 74, 76, 77, 79–81, 181, 261,
   315, 387, 388, 415, 438, 451

eavesdropper, 396
ecosystem, 1, 8, 12, 15, 149,
   151–154, 156, 158–162,
   167–169, 176–179, 184, 266,
   278, 301–303, 318, 320, 321,
   323, 328, 348, 376, 385, 399,
   453
educate, 358, 365, 370, 377, 384,
   385, 418
efficiency, 45, 83, 84, 94, 95,
   103–108, 114, 129, 212, 261,
   270, 271, 276, 286, 310, 409,
   410, 418, 420, 429, 430, 431,
   440, 444
egoless, 413
electronic signature, 397
encryption, 375, 391, 396, 397,
   400
engagement, 4, 13, 303, 316,
   321, 323, 352, 357, 359–361,
   364, 365, 367, 369, 370, 374
engineering, 2, 9
extrinsic motivation, 278, 279,
   290, 296

false-negative, 394, 395
false positive, 394, 395

filter-less, 414
framework for organizational
   transformation, 45
fundamental research, 359
fuzzy front end, 407, 410, 421,
   443, 455

Genchi Genbutsu, 35
Generative Innovation, 199, 210
global collaboration, 313
good tension, 178, 269, 295
governance, 13, 38, 261, 264,
   270, 286–289, 295, 296, 315,
   318–320, 322, 324, 326, 453
group, 57

hashing, 397
Head, Heart, and Hands, 457
high velocity, 454
holistic, 4, 5,14, 50, 451
human resource management,
   262, 266, 281, 295

identity authentication, 395
inclusion, 4–6, 10, 32, 54, 41,
   65–78, 80, 81, 261, 438, 451,
   461
inclusive, 51, 52, 65, 66, 68, 69,
   72, 77, 80, 81, 140, 172, 438,
   445, 452, 461
inclusive culture, 6, 65, 68, 72,
   81
induced strategic actions, 152,
   153, 222, 225, 292, 319, 321,
   322
institution, 3, 12, 13, 15, 155,
   183, 224, 276, 277, 306, 307,
   365, 398

intellectual property, 14, 360,
    382, 389–391, 404
Intel Values, 32
interdependent research, 221
internal customer, 6, 199, 390
intrinsic motivation, 272,
    278–280, 290, 296
intrinsic network, 10, 16, 17, 19,
    125, 126,143, 262, 264 270, 271,
    278, 286, 290, 292, 294, 318

Kaizen, 35, 312
key management, 397
Keep the Business Running
    (KTBR), 202, 294

LAMDA, 130, 139, 143, 242, 243
lean thinking, 4, 7, 117, 142, 208
learning cycle, 4, 7, 18, 95, 105,
    117, 118, 120–126, 128, 130,
    135, 136, 139, 142–144, 146,
    156, 172, 181, 182, 190, 195,
    199, 209, 210, 222, 224, 242,
    250, 251, 261, 263, 271, 276,
    282, 283, 289, 339, 358, 367,
    380, 398, 427, 429, 451, 452
lines of communication, 19, 24

matrix, 10, 35, 44, 261, 268–270,
    272–275, 286, 295, 296
mental health, 455
mentor, 459
metrics, 95, 99, 104, 106, 175,
    266, 267, 273, 280, 281, 294,
    378, 398, 427, 430, 446
Michael Porter's five forces, 162

necessary waste, 95, 276, 379,
    380, 384

new entrant, 12, 16, 162, 163,
    177, 302, 318

objectives and key results, 157,
    169, 172–176
offensive, 67, 81
OODA loop, 137, 138, 143,
    155–159, 161, 177, 179, 320,
    409
open mind, 407, 411, 413–415,
    439, 443, 455
organizational health, 277, 283

patent, 1
pathfinding, 10, 205, 222, 227,
    228, 231, 236–238, 241, 242,
    244, 246–248, 250, 251, 254,
    255, 258, 259, 272, 277, 286,
    288, 323–326, 423, 428, 429
paying customer, 3, 8, 84, 85, 87,
    90–94, 103, 182, 191, 250, 357,
    358, 394
physical health, 455
plaintext, 396
Plan-Do-Check-Act, 130, 131
Plan-Do-Study-Act, 130, 131
planning, 125
practical, 2, 72, 358, 359, 365
privacy, 13, 314
productivity, 44, 58, 83, 84, 91,
    94, 104, 105, 108, 118, 137,
    145, 212, 275, 280, 310, 330,
    335, 339, 347, 445–447, 461
property, 360

quality, 23, 33, 44, 51, 52, 54–56,
    59–62, 83, 87, 95, 96, 99,
    104–108, 112, 118, 130–132,
    198, 208, 229, 249, 259, 280,

286, 310, 311, 331, 418, 424, 433, 439
quality assurance, 208

racism, 75
RAPID model, 173, 272
rapid prototyping, 242, 251
R&D ecosystem, 12
R&D functions, 261, 263, 451
R&D processes, 278
R&D system, 2–6, 8–15, 17, 21, 26, 65, 66, 81, 83, 84, 90, 94, 95, 99, 105–108, 117, 124–126, 129, 142, 221, 224, 225, 228, 232, 234, 236, 238–240, 242, 256, 264, 275–277, 289, 296, 301–303, 305–308, 312, 315, 316, 318, 319, 322, 323, 325–328, 330, 338, 344, 345, 352, 353, 386, 398, 407, 443
reporting hierarchy, 16, 17, 19, 21, 271, 454
research, 3, 4, 9, 10, 13, 15, 87, 121, 123, 125, 184, 205, 261, 272, 277, 286–288, 357, 359, 360–362, 363, 364, 367–373, 383, 402, 423, 428, 429, 453
    academic research, 234
    applied research, 221, 249, 251, 252
    disruptive research, 9, 222, 231, 232, 250, 453
    distinct research, 235
    engineering research, 9, 10, 84, 118, 128, 142, 150, 155, 183, 202, 221–224, 226–228, 231, 235, 237, 244, 249, 250, 261, 263, 322, 439, 451, 453

interdependent research, 235, 242, 247, 254
    pure research, 221
    sustaining research, 238, 250
researcher, 360
research maturity model, 240
research pipeline, 9, 10, 222, 227, 234, 236–238, 240, 242, 244, 246, 250, 277, 286, 288, 304, 322, 423, 428, 429
results orientation, 46, 47, 52, 54, 422
retrospective, 120, 124, 126, 136, 193, 195, 426
reward systems, 10, 38, 249, 261, 262, 266, 273, 295, 313, 453
risk taking, 46, 47, 51, 98, 99, 279, 427, 429
ROI framework, 201, 205, 211
role model, 65, 70, 81, 295, 410, 430, 443, 452
Rivest, Shamir, & Adleman (RSA) algorithm, 396
rubber band model, 168, 169, 266, 267

scrum, 118, 119, 130, 135, 143, 193, 195, 196, 294, 295, 420
security attackers, 303
single-factor authentication, 395
sleep, 456, 457
SMART, 174, 423, 425
sprint, 119, 120, 126, 135–137, 193–195, 217, 295
standard, 76
standardization, 128, 130, 143, 164, 188, 190, 191, 210, 211, 218, 345

standards, 7, 9, 13, 15, 41, 43,
51, 76, 118, 130, 142, 181,
185–190, 206, 210, 271, 273,
275, 277, 286, 288, 292, 294,
303, 305, 306, 313, 328, 335,
343, 353, 381, 384, 385, 392,
453
star model, 263, 264
stateless, 413
strategic action, 8, 139, 149, 150,
152, 169, 176, 179, 226, 229,
266, 267, 279, 292, 294
strategic capabilities, 264, 265
strategic planning, 3, 4, 7, 8, 10,
31, 84, 118, 121, 128, 130, 137,
142, 143, 149–158, 160, 162,
163, 169, 177–179, 183, 184,
201, 205, 207, 225, 237, 239,
261, 263, 266, 275, 277, 288,
304, 320–323, 352, 376, 383,
398, 399, 409, 417, 428, 439,
451, 453
strategy, 5, 8, 10, 11, 14, 29–31,
35–37, 44, 48–50, 56, 57, 87,
96, 99, 100, 120, 121, 138,
149–154, 157, 159–161, 167–
170, 174–176, 178, 179, 225,
229, 231, 238, 261–267, 269,
270, 273, 274, 276, 278–281,
286–290, 294–296, 298, 299,
307, 316, 317, 328, 330, 332,
341, 350, 351, 372, 373, 398,
407, 409, 410, 417–419, 422,
423, 428, 435, 438, 442–444,
451, 454, 455, 460
substitute, 12, 16, 163, 164, 166,
177, 303, 318

supplier, 15, 21, 22, 25, 48, 62,
90, 123, 151, 163–167, 184,
192, 194–198, 206, 207, 239,
277, 286, 288, 303–306, 308,
309, 311, 312, 324, 325, 343,
344, 352, 353, 390, 391
supplier management, 286, 288,
310
supplier-partnering hierarchy,
311, 312
suppliers, 1, 3, 11–13, 15, 21,
22, 35, 48, 52, 55, 58, 65, 102,
124, 125, 146, 149, 151, 155,
160, 163–166, 177, 183, 193,
197, 198, 202, 205–207, 231,
276, 277, 301–303, 306, 307,
310–312, 318, 320, 324, 325,
327, 330, 339, 342, 344–346,
352, 353, 355, 379, 390, 391,
398, 402, 425, 454
sustaining innovation, 202, 225,
227, 233
symmetric, 396
symmetric encryption, 396

tactics, 149, 152, 167
theoretical, 359, 368, 370
tokenism, 71, 80
Toyota Production System, 97,
127, 310
true negative, 394
true positives, 393
trust, 14, 31, 33, 35, 51, 52, 54,
69, 72, 181, 193, 205, 210, 310,
340, 354, 407, 439, 441–443,
445, 447–449, 455, 458
two-factor identification, 396

university, 13, 71, 224, 313, 335, 357, 359–374, 402, 403, 454
unnecessary, 392

value, 2–10, 13, 14, 16, 18, 29, 31, 41, 43, 46, 48, 53, 65, 66, 70, 73, 81, 83–86, 90, 91, 99–109, 111, 112, 114, 117, 121, 142, 149–151, 153, 163, 166, 167, 169, 170, 176–179, 182, 188, 191, 199–203, 206–208, 210, 224, 237, 238, 245, 246, 261, 263–265, 270, 276, 279, 281, 283, 286, 287, 289, 294, 295, 299, 306, 308, 309, 315, 316, 319, 320, 323, 327, 340, 352, 354, 357, 358, 360–364, 366, 368–370, 379–382, 384, 385, 397, 398, 407, 422, 424, 427–431, 433, 435, 436, 443, 444, 447, 448, 451, 452, 455, 460, 462
   deployed value, 3, 6, 83, 95, 103, 104, 106–108, 308
   differentiable value, 306, 453, 455
   existing value, 6, 86–88, 92, 159, 236, 239, 241, 244, 246, 248, 256, 257
   potential value, 86, 89, 92, 223, 228, 236, 238, 239, 241, 247, 249
   raw value, 247, 253–256
   realizable value, 10, 236, 320, 323
   unaligned value, 86, 87, 89
value net model, 165, 166
   Apple Values, 33

Enron Values, 38
Intel Values, 32, 46, 53, 54
stated values, 5, 29–31, 37, 38, 40, 43, 44, 47–51, 289
Toyota Way, 34, 35
Wells Fargo, 40–43
value stream map, 18, 132, 194
velocity, 3, 4, 6, 8, 9, 13, 14, 24, 26, 29, 31, 48, 50, 83, 94, 95, 100, 103–108, 167, 181, 188, 194, 195, 201, 210, 222, 250, 261, 263, 270, 271, 286, 289, 290, 294, 301, 306, 308, 309, 325–328, 407, 409, 410, 422, 426, 429, 430, 431, 443, 444, 451–453, 455, 456

waste, 4, 6, 19, 26, 72, 83, 86, 87, 90, 93–95, 97, 99, 102, 103, 105, 107, 108, 112, 115, 157, 171, 188, 193, 194, 204, 210, 242, 245, 255, 261, 263, 274, 276, 289, 290, 309, 362, 379, 392, 393, 398, 425, 433, 434, 451, 452
   necessary waste, 83, 96, 101, 105, 107–109, 111, 112, 309
   unnecessary waste, 83, 95, 96, 99, 101, 108, 111, 112, 309
   waste elimination, 6, 101–103, 108
   wastes in a manufacturing system, 97
   wastes of management, 97–99
   wastes of software development, 97, 98

waterfall, 21, 118, 195
workflow, 7, 8, 18, 22–24, 26,
   117, 118, 120–130, 142, 143,
   172, 182, 186–190, 193,
   197–199, 209, 263, 275, 289,
290, 292–294, 303–305, 309,
   310, 312, 319, 320, 325, 326,
   328, 398, 409, 452, 455
working from home, 460, 461

About the Authors

Robert J Aslett is a former Intel VP and the founder of 3e8 Consulting LLC, who has over 30 years of experience in technical and management positions. While at Intel, he led organizations responsible for strategic planning, research, and development of design technologies for microprocessors and other product development. The organizations ranged in size from 40 to over 1,000 engineers and worked closely with suppliers, partners, and customers to deliver a complete set of electronic design automation solutions for a broad range of customers. In 2018, Robert left Intel and founded 3e8 Consulting LLC with the goal to help organizations deliver more value. Through 3e8 Consulting, Robert provides consulting on technical, leadership, and management topics. Robert is also active in helping the non-profit community. During 2018 and 2019 Robert was an Encore Fellow at Friends of the Children of Portland who empower youth who are facing the greatest obstacles through relationships with professional mentors. In 2020, Robert became an Investor Partner with Social Venture Partners of Portland Oregon who are a leader in venture philanthropy and using a business approach to create social impact. Robert received his Bachelor of Science degree in physics and astronomy from the University of Victoria, Canada in 1984.

John M Acken received his Bachelor of Science and Master of Science degrees in Electrical Engineering from Oklahoma State University in 1976 and 1978 respectively. He received his Doctor of Philosophy degree in electrical engineering from Stanford University in 1988. He is currently a research professor at Portland State University in Portland Oregon. From 1999 through 2015, he was at Oklahoma State University where he was advisor for several Master of Science theses on testing and information security. He held several jobs throughout the 80s and 90s including at Sandia National Labs in Albuquerque, New Mexico, Intel Corporation in Santa Clara, California, Schlumberger Palo Alto Research, Valid Logic systems, and Crosscheck Technology. He was in the US Army from 1970 through 1973 in the Army Security Agency. He has over 50 publications. His current research interests include digital testing, memristor circuits, information security for cloud and edge computing, and speaker recognition. Dr Acken has served on several conference committees including: IEEE ICCAD (Finance Chair for ICCAD90, Audio-Visual Chair forICCAD87, Treasurer for ICCAD86, ICCAD85, and ICCAD84,and Registration Chair for ICCAD83, Session Chair for ICCAD93), IEEE ITC (Session Chair for ITC90, ITC91 and ITC92), Session Chair for CCCT 2008, Publicity Chair for IEEE IDDQ Testing Workshop 1996, Registration chair for BAST90, and the ACM/SIGDA Workshop on Logic Level Modelling for ASICs (Session Chair in 1995, treasure in 1991, 1993, and 1994, and Program chair in 1989). Dr. Acken also served as a TWG member for Design and Test Road map of SIA (Semiconductor Industry Association) in 1994.

Siva K Yerramilli is Corporate VP of Strategic Programs in the Silicon Engineering Group at Synopsys. In this role, he is responsible for driving cross-business unit initiatives including Design Technology Co-Optimization (DTCO), which is aimed at achieving better results (performance, power, and area) in the face of ever-increasing complexity by utilizing Synopsys' capabilities from silicon to software. Siva is also responsible for the photonics IC development business, which is well poised to power the high-speed I/O roadmap. Siva joined Synopsys in 2019. Prior to Synopsys, Siva held several technical and management positions across various aspects of technology enablement and silicon product development at Intel. He has more than 30 years of experience in the technology industry. Siva received his Bachelor's degree in electronics and telecommunications from Jawaharlal Nehru

Technological University, India. He also studied computer science at the Indian Statistical Institute in Kolkata, India. Siva received his Master's in computer engineering at Syracuse University and later completed the Advanced Management Program for executives at Harvard University.